Lecture Notes in Computer Science 16225

Lianwen Jin · Richard Zanibbi · Veronique Eglin
Editors

Document Analysis and Recognition – ICDAR 2025 Workshops

Wuhan, China, September 20–21, 2025
Proceedings, Part I

Springer

Editors
Lianwen Jin
South China University of Technology
Guangzhou, China

Richard Zanibbi
Rochester Institute of Technology
Rochester, NY, USA

Veronique Eglin
National Institute of Applied Sciences
of Lyon
Villeurbanne, France

ISSN 0302-9743 ISSN 1611-3349 (electronic)
Lecture Notes in Computer Science
ISBN 978-3-032-09367-7 ISBN 978-3-032-09368-4 (eBook)
https://doi.org/10.1007/978-3-032-09368-4

This Springer imprint is published by the registered company Springer Nature Switzerland AG
The registered company address is: Gewerbestrasse 11, 6330 Cham, Switzerland

Preface

The ICDAR 2025 Workshops were held as satellite events of the 18th International Conference on Document Analysis and Recognition (ICDAR 2025) on September 20–21, 2025, in Wuhan China. These workshops are vital platforms for fostering collaboration, exchanging ideas, and advancing the frontiers of document analysis and recognition research. Comprising seven specialized workshops, the event brought together researchers, practitioners, and industry experts from around the world. Among the seven workshops, five solicited paper submissions, while two (AMDU 2025 and IWCDF 2025) focused exclusively on invited presentations and discussions.

A total of 74 submissions were received for the five workshops: WML 2025, GREC 2025, M3RD 2025, VTG-TIP 2025, and DALL 2025. These workshops covered a broad spectrum of topics at the forefront of document analysis and recognition, including machine learning for document understanding, graphics recognition, multi-modal mathematical reasoning, visual text generation, text image preprocessing, and the analysis of low-resource languages. All paper submissions to these workshops underwent rigorous peer review, using either a single-blind or double-blind review process. Each paper was reviewed by approximately three reviewers. Ultimately, 46 workshop papers were accepted for publication in the ICDAR 2025 workshop proceedings. The accepted papers present novel methodologies, benchmark datasets, and real-world applications, reflecting the community's continued interest in developing models and systems used to construct efficient, robust, and interpretable document intelligence systems.

The ICDAR 2025 Workshops successfully facilitated discourse and collaboration. The diversity of topics and methodologies presented reflects the evolving landscape of document intelligence. The workshops also identified promising directions for future investigation, including the integration of large language models within document analysis models, the enhancement of cross-modal reasoning capabilities, and the development of more inclusive technologies for underrepresented languages and scripts.

We extend our sincere gratitude to the ICDAR 2025 general chairs and the organizing committee for their invaluable support and guidance in hosting these workshops. We also deeply appreciate the dedication and hard work of all workshop organizers, program committee members, and reviewers, whose expertise and commitment ensured the scientific rigor and success of this year's program.

September 2025

Lianwen Jin
Richard Zanibbi
Véronique Eglin

Organization

General Chairs

Lianwen Jin	South China University of Technology, China
Richard Zanibbi	Rochester Institute of Technology, USA
Véronique Eglin	INSA Lyon, LIRIS, France

Fifth ICDAR International Workshop on Machine Learning (WML 2025)

Workshop Chairs

Umapada Pal	Indian Statistical Institute, Kolkata, India
Venu Govindaraju	University at Buffalo, SUNY, USA
Xiao-jun Wu	Jiangnan University, China

Workshop Program Chairs

Michael Blumenstein	University of Technology Sydney, Australia
Eric Granger	École de Technologie Supérieure, Canada
Shivakumara Palaiahnakote	University of Salford, UK
Mickael Coustaty	University of La Rochelle, France

Program Committee

Alireza Alaei	Southern Cross University, Australia
Arthur Hemmer	University of La Rochelle, France
Aythami Morales	Universidad Autónoma de Madrid, Spain
Brian Kenji Iwana	Kyushu University, Japan
Chiranjoy Chattopadhyay	Flame University, India
Daichi Haraguchi	CyberAgent, Japan
Denis Coquenet	IRISA, France
Donato Impedovo	University of Bari, Italy
Ishita De Ghosh	Barrackpore Rastraguru Surendranath College, India
Kaushik Roy	West Bengal State University, India

Killian Barrere	INSA Lyon, France
Liangcai Gao	Peking University, China
Mickael Coustaty	University of La Rochelle, France
Muhammad Hammad Saleem	University of Salford, UK
Muhammad Muzzamil Luqman	University of La Rochelle, France
Nicolas Sidere	University of La Rochelle, France
Palash Ghosal	Sikkim Manipal University, India
Pedro Garcia Freitas	University of Brasília, Brazil
Raghavendra Ramachandra	Norwegian University of Science and Technology, Norway
Ruiqing Wu	University of Electronic Science and Technology of China, China
Saumik Bhattacharya	Indian Institute of Technology Kharagpur, India
Shivakumara Palaiahnakote	University of Salford, UK
Siladittya Manna	Indian Statistical Institute, Kolkata, India
Srikanta Pal	Maynooth University, Ireland
Sukalpa Chanda	Østfold University College, Norway
Suresh Sundaram	Indian Institute of Technology Guwahati, India
Umapada Pal	Indian Statistical Institute, Kolkata, India
Zi-Rui Wang	Chongqing University of Posts and Telecommunications, China

ICDAR 2025 Workshop on Multi-modal Mathematical Reasoning in Documents (M3RD 2025)

Workshop Chairs

Qiufeng Wang	Xi'an Jiaotong-Liverpool University, China
Yang Liu	Sun Yat-sen University, China
Kun Zhang	Carnegie Mellon University, USA
Cheng-Lin Liu	Institute of Automation, Chinese Academy of Sciences, China

Program Committee

Kaizhu Huang	Duke Kunshan University, China
Yu Zhou	Nankai University, China
Fei Yin	Institute of Automation, Chinese Academy of Sciences, China
Jian Xu	Institute of Automation, Chinese Academy of Sciences, China

Xiaobo Jin	Xi'an Jiaotong-Liverpool University, China
Tonghua Su	Harbin Institute of Technology, China
Meng Fang	University of Liverpool, UK
Qingxing Cao	Sun Yat-sen University, China
Zhicheng Yang	Hong Kong University of Science and Technology (Guangzhou), China
Wenda Li	University of Edinburgh, UK
Jindong Wang	William & Mary, USA

Additional Reviewers

Dahan Wang	Xiamen University of Technology, China
Kecheng Liang	Sun Yat-sen University, China
Xiangdong Su	Inner Mongolia University, China

16th IAPR International Workshop on Graphics Recognition (GREC 2025)

Workshop Chairs

K. C. Santosh	University of South Dakota, USA
Xizhao Wang	Shenzhen University, China

Program Committee Chairs

Lianwen Jin	South China University of Technology, China
Jean-Christophe Burie	La Rochelle University, France

Program Committee

Eric Anquetil	IRISA – Rennes, France
Bertrand Couasnon	RISA/INSA Rennes, France
Mickael Coustaty	La Rochelle University, France
Alicia Fornes	Universitat Autónoma de Barcelona – CVC, Spain
Alexander Gribov	Environmental Systems Research Institute, USA
Romain Raveaux	University of Tours, France
Motoi Iwata	Osaka Metropolitan University, Japan
Bart Lamiroy	Université de Reims Champagne-Ardenne, France
Josep Llados	Universitat Autónoma de Barcelona – CVC, Spain

Sekhar Mandal	Indian Institute of Engineering, Science and Technology, India
Wataru Ohyama	Tokyo Denki University, Japan
Alexander Pacha	TU Wien, Austria
Oriol Ramos Terrades	Universitat Autónoma de Barcelona – CVC, Spain
Christophe Rigaud	La Rochelle University, France
Richard Zanibbi	Rochester Institute of Technology, USA
Akram Bennour	Larbi Tebessi University, Algeria
Mohammad Javed	IIIT Allahabad, India
Nibaran Das	Jadavpur University, India
Debanga Raj Neog	IIT Guwahati, India

ICDAR 2025 Workshop on Visual Text Generation and Text Image Processing (VT-TIP 2025)

Workshop Chairs

Yu Zhou	Nankai University, China
Gangyan Zeng	Nanjing University of Science and Technology, China
Hongtao Xie	University of Science and Technology of China, China
Xu-Cheng Yin	University of Science and Technology Beijing, China

Program Committee

Chengyang Fang	Jiangxi University of Finance and Economics, China
Chun Yang	University of Science and Technology Beijing, China
Gangyan Zeng	Nanjing University of Science and Technology, China
Huawen Shen	Institute of Information Engineering, Chinese Academy of Sciences, China
Jin Wei	Lenovo Research, China
Gengluo Li	Institute of Information Engineering, Chinese Academy of Sciences, China
Min Lu	Inner Mongolia University of Technology, China
Shancheng Fang	University of Science and Technology of China, China

Weichao Zeng	Institute of Information Engineering, Chinese Academy of Sciences, China
Xugong Qin	Nanjing University of Science and Technology, China
Yan Shu	Institute of Information Engineering, Chinese Academy of Sciences, China
Yan Zhang	Institute of Information Engineering, Chinese Academy of Sciences, China

ICDAR 2025 Workshop on Documents Analysis of Low-Resource Languages (DALL 2025)

Workshop Chairs

Yong Tso	Xizang University, China
Brian Kenji Iwana	Kyushu University, Japan
Harold Mouchere	Nantes University, France

Program Committee

Nyima Trashi	Xizang University, China
Brian Kenji Iwana	Kyushu University, Japan
Harold Mouchere	Nantes University, France
Jian Cheng	University of Electronic Science and Technology of China, China
Anna Zhu	Wuhan University of Technology, China
Yongbin Yu	University of Electronic Science and Technology of China, China
Yong Tso	Xizang University, China
Rinchen Dongrub	Xizang University, China
Dorje Tashi	Xizang University, China

Additional Reviewers

Chunxiao Fan	Hefei University of Technology, China
Hai Min	Hefei University of Technology, China
Bo Zhou	Hefei University of Technology, China

ICDAR 2025 Workshop on Advancing Multimodal Document Understanding: Challenges and Opportunities (AMDU 2025)

Workshop Chairs

Xipeng Qiu	Fudan University, China
Yuliang Liu	Huazhong University of Science and Technology of China, China
Anton van den Hengel	University of Adelaide, Australia
Shuang Chen	Fudan University, China

ICDAR 2025 Workshop on Computational Document Forensics (IWCDF 2025)

Workshop Chairs

Nicolas Sidere	La Rochelle Université, France
Victor Sanchez	University of Warwick, UK

Contents

ICDAR 2025 Workshop on Multi-modal Mathematical Reasoning in Documents (M3RD 2025)

The Fifth ICDAR International Workshop on Machine Learning (WML 2025)

PBa-LLM: Privacy- and Bias-Aware NLP Using Named-Entity Recognition (NER)

Gonzalo Mancera(✉), Aythami Morales, Julian Fierrez, Ruben Tolosana, Alejandro Peña, Miguel Lopez-Duran, Francisco Jurado, and Alvaro Ortigosa

School of Engineering, Universidad Autonoma de Madrid, 28049 Madrid, Spain
{gonzalo.mancera,aythami.morales,julian.fierrez,ruben.tolosana, alejandro.penna,miguel.lopezd,francisco.jurado,alvaro.ortigosa}@uam.es

Abstract. The use of Natural Language Processing (NLP) in high-stakes AI-based applications has increased significantly in recent years, especially since the emergence of Large Language Models (LLMs). However, despite their strong performance, LLMs introduce important legal/ethical concerns, particularly regarding privacy, data protection, and transparency. Due to these concerns, this work explores the use of Named-Entity Recognition (NER) to facilitate the privacy-preserving training (or adaptation) of LLMs. We propose a framework that uses NER technologies to anonymize sensitive information in text data, such as personal identities or geographic locations. An evaluation of the proposed privacy-preserving learning framework was conducted to measure its impact on user privacy and system performance in a particular high-stakes and sensitive setup: AI-based resume scoring for recruitment processes. The study involved two language models (BERT and RoBERTa) and six anonymization algorithms (based on Presidio, FLAIR, BERT, and different versions of GPT) applied to a database of 24,000 candidate profiles. The findings indicate that the proposed privacy preservation techniques effectively maintain system performance while playing a critical role in safeguarding candidate confidentiality, thus promoting trust in the experimented scenario. On top of the proposed privacy-preserving approach, we also experiment applying an existing approach that reduces the gender bias in LLMs, thus finally obtaining our proposed Privacy- and Bias-aware LLMs (PBa-LLMs). Note that the proposed PBa-LLMs have been evaluated in a particular setup (resume scoring), but are generally applicable to any other LLM-based AI application.

Keywords: Privacy · Bias · Fairness · Responsible AI · NLP · Large Language Models · LLM · Named Entity Recognition · NER

1 Introduction

During the past decade, automated decision-making systems have increasingly permeated sensitive domains such as education [6,9], healthcare [28,55], finance

L. Jin et al. (Eds.): ICDAR 2025 Workshops, LNCS 16225, pp. 3–20, 2026.
https://doi.org/10.1007/978-3-032-09368-4_1

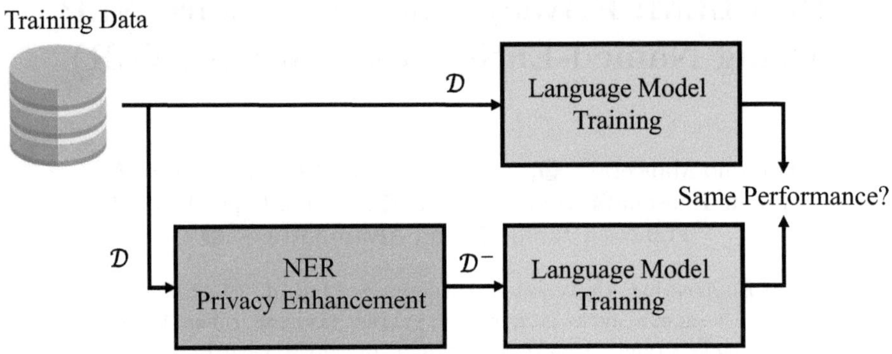

Fig. 1. Graphical abstract of the framework evaluated in this work.

[54], and recruitment [4,45,46], driven by rapid advances in AI technologies. As these systems assume tasks traditionally requiring human judgment, they raise critical questions about personal data protection [2,13,21,64]. In particular, the field of NLP has seen unprecedented growth, powered by the Transformer model [62], which relies on self-attention mechanisms and has enabled the development of powerful LLMs such as GPT [1], LLaMA [60,61] and Mistral [24]. These models are now embedded in numerous applications, including those that process sensitive information, such as automated resume screening systems [18].

Although offering operational advantages, these technologies also face challenges related to fairness [12,42,53] and privacy [7,29,63]. Text data often contains personally identifiable information (PII) [51] - names, addresses, and demographic details that, if mishandled, could expose individuals or introduce bias into decision-making processes. Recent regulatory initiatives, such as the AI Act[1] approved by the European Parliament in June 2024 [15] and the White House Memorandum on AI leadership [59], seek to address these risks by promoting transparency, accountability, and fairness in AI systems.

The intersection of privacy and fairness in multiple AI applications based on human data (e.g. biometrics and NLP [41]) has fueled interest in privacy enhancement technologies (PET) for multimodal human-centric AI including NLP modules [26]. Techniques such as differential privacy [23], de-identification [19], crypto-biometrics [22], data minimization [47], and federated learning [20, 31] are being explored to protect sensitive information while maintaining system performance. Nevertheless, achieving a balance between privacy preservation and model accuracy remains a significant challenge, particularly for complex tasks like natural language understanding.

This paper addresses this challenge by proposing a generalizable approach to anonymize input text while preserving system performance, in a similar way related works remove unnecessary sensitive data while maintaining good performance in a main AI task [36], in this case particularly applied to NLP tasks

[1] https://artificialintelligenceact.eu/.

based on LLMs (e.g., document layout analysis [43] or document topic classification [44]). Our method uses NER to detect and remove sensitive entities, minimizing privacy risks. The main idea is depicted in Fig. 1, where we show two branches: with and without removing the sensitive entities found using Named-Entity Recognition. In addition to that privacy enhancement via NER, we also apply and experiment the bias reduction methods recently proposed in [42] to derive our final Privacy- and Bias-aware models (PBa-LLMs).

Although recruitment is our primary case study, the framework is designed to extend to other domains, offering a versatile solution for privacy-preserving NLP applications.

The remainder of the paper is organized as follows. Section 2 reviews the related literature on privacy-preserving technologies in NLP. Section 3 details our proposed framework and its integration of NER-based anonymization. Section 4 describes the experimental setup. Section 5 presents the results and discusses their implications. Finally, Sect. 6 summarizes our contributions.

2 Related Works

Our work falls under the development of privacy-enhancing technologies to handle sensitive data, particularly for NLP applications. Techniques such as *NER* have been widely used to anonymize personal information while maintaining the utility of NLP data for various tasks. Recent advances in *NLP*, including the integration of pre-trained models such as *BERT*- and *GPT*-based models, have further improved entity recognition and anonymization capabilities [55].

2.1 Privacy in NLP Systems

Data privacy has become an increasingly critical issue in our data-driven society. With exponential growth in the collection and processing of sensitive personal information, the risks of breaches and misuse are more pronounced than ever. Personal data such as names, ages, locations, and organizational affiliations raise significant concerns about acquisition, storage, and use. Mishandling personal data can inadvertently reveal more than intended, leading to discrimination [53], identity theft, and other harm. Thus, managing sensitive information, particularly personally identifiable details, requires rigorous measures to ensure confidentiality and integrity.

Traditional privacy protection methods often fail to address the complexities of contemporary data use. Techniques such as pseudonymization, access control, and basic encryption are common. Pseudonymization replaces identifiable data with pseudonyms, but privacy can still be compromised if mappings or contextual clues are accessible [56]. Access control restricts sensitive data to authorized individuals but may fail if poorly implemented [49]. Basic encryption is crucial, but its effectiveness is based on strong algorithms and proper key management [39]. Moreover, machine learning models are vulnerable to adversarial attacks that can expose training data [11,34].

As machine learning and artificial intelligence (AI) technologies continue to advance, the reliance on large datasets presents a double-edged sword. While driving progress, it increases the risk of exposure to sensitive information. Therefore, developing robust privacy-preserving techniques to safeguard sensitive data has become more critical than ever.

2.2 Named-Entity Recognition

Named-Entity Recognition (NER) is a fundamental task within the broader field of *NLP*, aimed at detecting and categorizing entities in textual data. The primary goal of NER is to identify specific entities within a body of text and classify them into predefined categories [30].

The NER process plays a pivotal role in converting unstructured data into structured and meaningful information by automatically extracting relevant entities from large datasets. This is particularly valuable in applications such as Machine Translation [5], Information Retrieval [65], Question Answering [35], or Sentiment Analysis [14,25].

Identifying and categorizing entities significantly improves the performance of these applications, making it easier to process and analyze textual data effectively. Beyond these applications, NER serves as a cornerstone in the broader analysis of textual data, enhancing the ability to extract, structure [37], and interpret information meaningfully [16]. By providing a clear framework for understanding entities within text, NER facilitates further analysis, such as trend detection [48], relationship mapping [38], document layout analysis [43], topic classification [44], and even predictive analytics [17].

As the volume of unstructured data continues to grow exponentially in the current digital age, the importance of NER in processing and obtaining insights from these data becomes increasingly critical. Consequently, continued development and refinement of NER technologies remain essential to harness the full potential of big data and drive informed decision making in various sectors.

3 Privacy-Preserving Learning Approach

In this section, we outline our contribution, which focuses on improving privacy in various tasks. By implementing privacy-preserving techniques, we aim to protect sensitive information across multiple domains. Our approach ensures that personal data remains protected while maintaining the functionality of the tasks, reducing potential exposure to privacy risks without sacrificing performance or utility. This work contributes to ongoing efforts to balance data privacy and task effectiveness, which is critical in an era of increasing concerns about data privacy and individual rights.

3.1 Problem Formulation

Consider a language model M, defined by its parameters $\mathbf{w}$, trained with a dataset $\mathcal{D}$ to optimize a performance metric P. Each data sample i in $\mathcal{D}$ consists of a raw text $\mathbf{x}_i$ and optionally its respective label $\mathbf{y}_i$. Text data can be

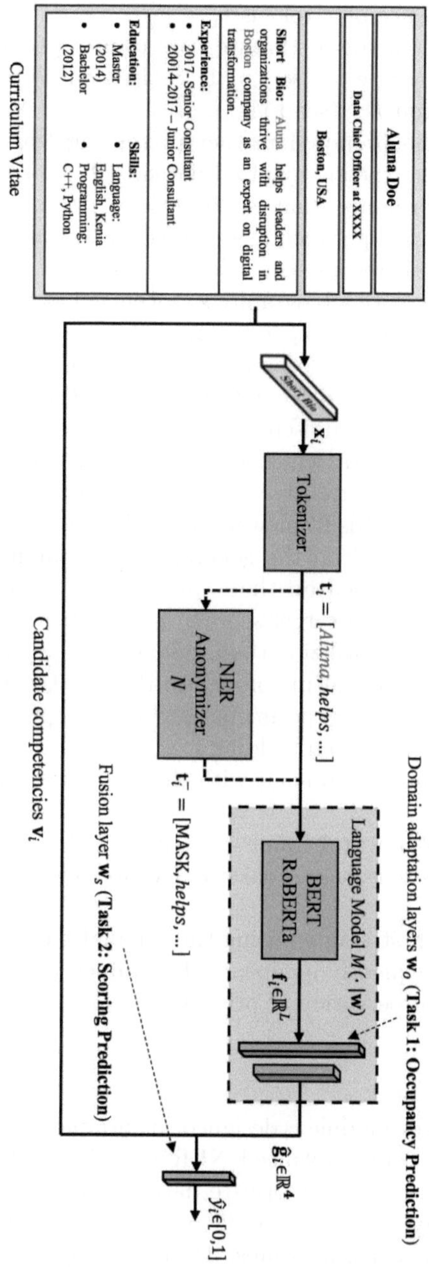

Fig. 2. Learning framework for the automatic recruitment tool. The system takes a multimodal input resume, consisting of a candidate competencies vector $\mathbf{v}$ and a text biography. An anonymization module N removes two types of identities: locations (LOC) and persons (PER). A pre-trained Transformer processes the anonymized biographies to extract a feature vector $\mathbf{f}$. This feature vector is fused with the candidate competencies using the fusion module $\mathbf{w}_s$ to generate the final score $\hat{y}$.

represented as a sequence of τ tokens $\mathbf{t}_i = [t_i^{(1)}, \ldots, t_i^{(\tau)}]$. Assume that there is sensitive information z_i within $\mathbf{t}_i$ that we want to protect. We use the anonymization function N to generate $\mathbf{t}_i^-$, so $z_i \notin \mathbf{t}_i^-$. Based on N, we transform $\mathcal{D}$ into a new anonymized version $\mathcal{D}^-$ (see Fig. 1).

The purpose of the proposed privacy-preserving learning framework is to train a model M^- with the anonymized dataset $\mathcal{D}^-$ to ensure the removal of z while preserving performance, that is, $P(M) = P(M^-)$.

Specifically, in our experiments we consider a dataset $\mathcal{D}$ of multimodal resumes, where the i-th resume consists of a vector of attributes or candidate competencies $\mathbf{v}_i$, along with a biography $\mathbf{x}_i$ tokenized as $\mathbf{t}_i$. In this dataset, sensitive information z_i is only present in biographies (e.g., names, locations, companies). Text data is subsequently processed through the anonymization module N, responsible for protecting sensitive data based on *NER* models. Concretely, this module identifies the entities present in $\mathbf{t}$, and produces an anonymized sequence $\mathbf{t}^-$ by removing sensitive data, which will typically be personal information of the candidate, embedded inadvertently in the short biography (e.g., name, surname, locations, companies). In this framework, a machine learning-based resume scoring tool is developed based on the general learning architecture illustrated in Fig. 2. This system is trained on the dataset $\mathcal{D}$ to accurately estimate a new score $\hat{y}_i$ that approximates a target score y_i (e.g., supervised labels produced by human resources experts) utilizing the input resume $\{\mathbf{v}_i, \mathbf{t}_i\}$. It is important to note that the scoring mechanism comprises: i) a pre-trained Transformer (e.g., BERT and RoBERTa in our experiments), which accepts a tokenized biography as input and generates a text embedding $\mathbf{f}_i \in \mathbb{R}^L$; and ii) a fusion module, which further processes $\mathbf{f}_i$ and integrates it with the candidate attributes $\mathbf{v}_i$ to predict the score $\hat{y}_i$.

Based on the experimental frameworks proposed in [41] we incorporate the analysis of two different learning tasks: *i*) Occupancy prediction; *ii*) Score prediction.

The models were first trained using the original $\mathbf{t}$ to establish a non-private baseline and are subsequently independently trained on $\mathbf{t}^-$ to assess differences in performance and preservation of privacy.

3.2 NER Models

The NER-based privacy module is designed to automatically anonymize sensitive information in texts using advanced NER techniques. Its primary goal is to identify and replace personal or private data, such as the names, locations, or other entities of individuals that could compromise privacy. This module can be categorized into two main groups: 1) specifically designed for NER tasks, which utilize tailored algorithms and models to accurately identify entities within the text, ensuring high precision and recall for effective detection of sensitive information, and 2) LLMs, which are primarily developed to handle a diverse set of tasks in NLP. Here, we experimentally compare both categories: the use of specialized NER models versus general LLMs for anonymization. The following approaches are utilized:

- **Presidio**[2] is an open-source tool by *Microsoft* for detecting and anonymizing sensitive information. Its architecture includes an *Analyzer* to detect PII such as names or addresses, and an *Anonymizer* to replace or mask them.
- **FLAIR** [3] is a pre-trained NER model based on a bidirectional LSTM and *contextual string embeddings*, which capture word meaning from the surrounding context, improving disambiguation and sense adaptation.
- **NER_CoNLL2003_BERT** is a BERT-based model fine-tuned on the CoNLL-2003 dataset [27], offering strong generalization and robust entity extraction through bidirectional context representation.
- **GPT-3.5** [40] is a general purpose Transformer model enhanced with sparse attention and RLHF, enabling better alignment with human intent and better understanding of text.
- **GPT-4** [1] introduces refined attention mechanisms and fine-tuning for improved contextual performance. The lightweight **GPT-4o Mini** is also evaluated for low-latency, cost-efficient applications.

4 Experimental Framework

This section first introduces the databases used in this study, describing their structure, key characteristics, and relevance to the analysis. Following this, we detail the experimental protocol itself, outlining the methodologies applied, and the experimental setup.

4.1 Dataset: FairCVdb

FairCVdb [41,45,46] is a comprehensive dataset that aims to advance research on multimodal fairness within the realm of artificial intelligence. This dataset comprises 24,000 synthetic profiles generated by integrating identities sourced from the DiveFace Face Recognition database [36] and leveraging biographies. Each profile in FairCVdb includes:

- **Textual Information**: FairCVdb was created using data from the occupation database [10], with samples assigned to profiles based on gender. Each sample contains two brief biographies (one raw and one gender neutral), an occupation and a name covering ten occupations in four labor sectors. This diversity in occupations allows for a broader analysis of privacy in several fields.
- **Structured Information**: Seven attributes were derived from five categories that define candidate competencies: education, availability, previous experience, recommendations, and language proficiency. Gender and ethnicity were also included as demographic factors. Each resume received a blind score (excluding demographics) and biased scores, with sector-specific weightings. The candidates were labeled with an occupational group ($g \in$ O1: Health, O2:

[2] https://microsoft.github.io/presidio/.

Education and Knowledge, O3: Communication and Media, O4: Legal) based on their profession and assigned a human-evaluated score $y \in [0, 1]$ following human resource processes [41].

FairCVdb includes gender-biased scores generated to assess performance and fairness in automatic recruitment tools. By analyzing the performance of these scores, it is possible to investigate the disparities in hiring and evaluation processes, explore the implications of demographic attributes on employment outcomes, and propose strategies for enhancing fairness and privacy in multimodal AI systems.

4.2 Privacy Enhancing Module

The model N is responsible for the anonymization of the texts according to the desired entity. In the present research work, two distinct entities are differentiated: Location and Person. A total of six distinct models are utilized to observe how the choice of anonymizer can affect the final performance of the tasks.

All anonymizers receive raw information $\mathbf{t}$, which consists of textual data from each of the 24,000 resumes present in FairCVdb, and the model output retains the same contextual information from each resume, except that the selected entity is masked, resulting in $\mathbf{t}^-$.

The analysis was carried out on the short biographies available in the FairCVdb database, which contains a total of 1,416,693 words. The number of entities removed varies from 28,000 to 66,000 depending on the NER model and the type of entity. This information represents approximately between 4% and 9% of the total words included in the short biographies. The variation in the number of entities identified by the NER algorithms is mainly due to their different performances and different sensitivities. For example, compound nouns may be recognized as single or multiple entities based on the algorithm used. Although the NER algorithms assessed are not without errors, each of the six NER modules achieved performance exceeding 80% on the public *CoNLL*-2003 benchmark [50].

4.3 Occupancy Classification and Candidate Scoring

We used two distinct pre-trained Transformer models for the Text Comprehension Module in our experiments: *BERT* and *RoBERTa*.

- **BERT** [27] is a Transformer-based model that introduced bidirectional pre-training using *Masked Language Modeling (MLM)* and *Next Sentence Prediction (NSP)*. It captures left and right context simultaneously, enabling deep understanding of sentence structure and inter-sentence relationships.
- **RoBERTa** [32] improves on BERT by removing the NSP objective, training on a larger corpus and dynamically applying masking at each iteration. These adjustments enhance performance while keeping the original architecture.

We employ here the base versions of models, characterized by presenting a latent space embedding of dimensions 768. To maintain the integrity of their pre-trained weights, we freeze the Transformer layers throughout the training process. The feature vector $\mathbf{f}_i \in \mathbb{R}^{768}$ is generated by averaging the embeddings of the output tokens of the Transformer model, ensuring that the attention mask is applied to exclude padding tokens from the averaging process. Before feeding the bios into the models, we preprocess them by eliminating punctuation and using the respective tokenizer, adhering to a fixed sequence length of 256, since the longest bio in the training dataset comprises 197 words.

- For the **Occupancy Prediction task** we have trained a *Multilayer Feedforward Perceptron (MLP)* with three layers characterized by their parameters $\mathbf{w}_o$. The number of units in each layer is 300, 70, and 4, respectively. The output of $\mathbf{w}_o$ is a vector $\mathbf{g}_i$ corresponding to the probabilities of the four predefined occupancy sectors in FairCVdb.
- Subsequently, for the **Scoring Prediction task**, we concatenate the outputs of these layers with the candidate competencies $\mathbf{v}_i$ to predict the final score $\hat{y}$ through an output layer (1 unit) characterized by the parameters $\mathbf{w}_s$.

All trained layers ($\mathbf{w}_o$ and $\mathbf{w}_s$) use the sigmoid activation function, with a dropout rate set at 0.3 to alleviate the risk of overfitting. The layers were trained during 50 epochs, using the root mean square error (RMSE) as a loss function, with a batch size of 32. We employed the *AdamW* optimizer with parameters $\beta_1 = 0.9$, $\beta_2 = 0.999$, and $\epsilon = 2 \times 10^{-8}$, along with a learning rate of 1×10^{-3}. FairCVdb contains a total of 24,000 synthetic resumes and is divided into training and testing subsets for model evaluation. Specifically, 80% of the data, amounting to 19,200 resumes, is used for training purposes. The remaining 20%, comprising 4,800 resumes, is designated for testing, providing an independent set of data to assess the model's generalization capabilities.

5 Experiments and Results

5.1 Privacy-Aware Models: Impact of Anonymized Text

The experimental framework illustrates the functionality of a Recruitment Tool using anonymized data, ensuring the protection of sensitive information. This approach enables an effective evaluation of the tool's ability to classify candidates and assess gender bias in recruitment processes. Anonymized resumes are used to preserve privacy while retaining the realism of data.

The resume scoring tool is initially trained using textual data from non-anonymized resumes, establishing a baseline performance benchmark. This baseline serves as a reference point to assess how the model performs when all personal details are available, allowing for a direct comparison in later trials. Following this, the resume scoring tool is trained on anonymized data using six independent NER modules. Each NER module masked the same entity types (names and locations), and the goal is to determine whether the removal of

Table 1. Performance of Task 1 (Occupancy Prediction) for the six NER models. The results are categorized depending on the private information removed: locations (LOC) and persons (PER). This experiment incorporates the use of two pre-trained Transformers, which optimize content understanding and enhance the models' effectiveness.

		LOC				PER			
Transformer	Anonymization Module	O1	O2	O3	O4	O1	O2	O3	O4
RoBERTa	**None (Baseline)**	0.86	0.88	0.85	0.79	0.89	0.84	0.87	0.72
	GPT-4	0.92	0.94	0.87	0.80	0.91	0.87	0.87	0.79
	GPT-4o-mini	0.91	0.93	0.86	0.78	0.91	0.84	0.85	0.66
	GPT-3.5	0.83	0.82	0.86	0.66	0.82	0.86	0.89	0.71
	Presidio	0.91	0.89	0.88	0.79	0.93	0.85	0.88	0.69
	FLAIR	0.92	0.87	0.84	0.71	0.91	0.85	0.88	0.78
	NER CoNLL2003 BERT	0.93	0.84	0.81	0.72	0.85	0.86	0.88	0.76
BERT	**None (Baseline)**	0.87	0.85	0.87	0.79	0.89	0.86	0.87	0.79
	GPT-4	0.92	0.87	0.89	0.81	0.92	0.88	0.87	0.82
	GPT-4o-mini	0.91	0.86	0.90	0.77	0.90	0.88	0.88	0.78
	GPT-3.5	0.90	0.88	0.91	0.79	0.89	0.87	0.90	0.79
	Presidio	0.91	0.86	0.89	0.82	0.87	0.85	0.91	0.78
	FLAIR	0.90	0.87	0.88	0.80	0.90	0.88	0.88	0.80
	NER CoNLL2003 BERT	0.89	0.87	0.89	0.77	0.91	0.85	0.90	0.81

these entities affects the performance of the scoring tool relative to the established baseline (i.e., non-anonymized data).

Our experiments include two learning tasks (occupancy prediction and scoring prediction). The results obtained for the occupancy prediction are summarized in Table 1, which outlines the comparative performance metrics for each anonymization model, highlighting the extent to which the anonymization process impacts the performance and effectiveness of the resume scoring tool. As reflected in Table 1, the results obtained compared to the baseline are even slightly superior, especially in cases where the anonymization method has been conducted with a GPT-based model. This shows that these methods are capable of, at the very least, maintaining performance and, in some cases, enhancing privacy. These findings suggest that it is possible to strike a balance between protecting privacy and achieving good model performance. In the worst case, performance remains unchanged, while in the best case, privacy is increased without compromising accuracy, which is a significant benefit when dealing with sensitive data. Furthermore, the learning process of the models throughout these experiments is depicted in Fig. 3 for BERT and Fig. 4 for RoBERTa, where it becomes clear that the learning trajectory of the model is influenced both by the anonymization technique and the specific type of entity being masked. This figure provides valuable information on how the model adapts to varying levels of information removal, showing whether certain anonymization methods lead

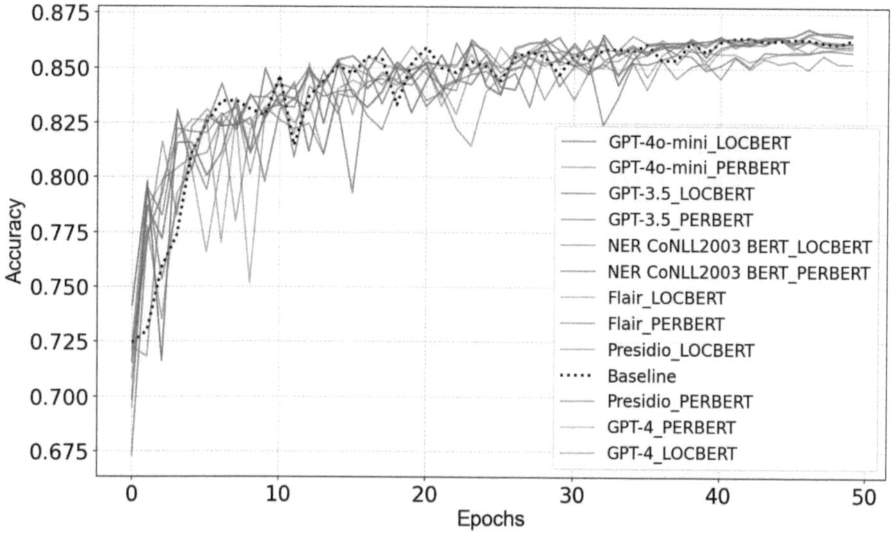

Fig. 3. Accuracy over epochs for the BERT-based occupancy prediction model. The Figure compares different implementations, including six anonymization modules over different two entities (PER and LOC). The baseline model, without anonymization module, is highlighted with a dotted black line. The results reflect performance across different configurations, with darker colors indicating specific models.

to better generalization or, conversely, hinder the model's ability to maintain consistent scoring performance. Ultimately, this experiment sheds light on the delicate balance between the preservation of individual privacy and the predictive power of machine learning models in sensitive domains.

5.2 Privacy- and Bias-Aware Models

In Sect. 5.1 it has been observed that anonymizing text related to PER (persons) and LOC (locations) does not impact the performance of the occupation prediction model. Building on this insight, the objective is to create a recruitment tool that prioritizes both privacy and fairness, ensuring that it is free of gender bias.

The approach involves several steps. First, entity anonymization will be implemented to mask sensitive personal data, such as names and locations. This ensures that no identification information influences the decision-making process. Second, the elimination of gender bias will target both explicit indicators (e.g., pronouns, titles) as in Sect. 5.1 and implicit markers (e.g., gendered phrases or role associations). To address this, we will employ the bias-aware learning framework introduced in [41] and further detailed in [42], which represents a method to systematically remove gender characteristics from the model's internal representations. This approach allows the model to make predictions that are independent of gender-specific information while preserving task-relevant linguistic

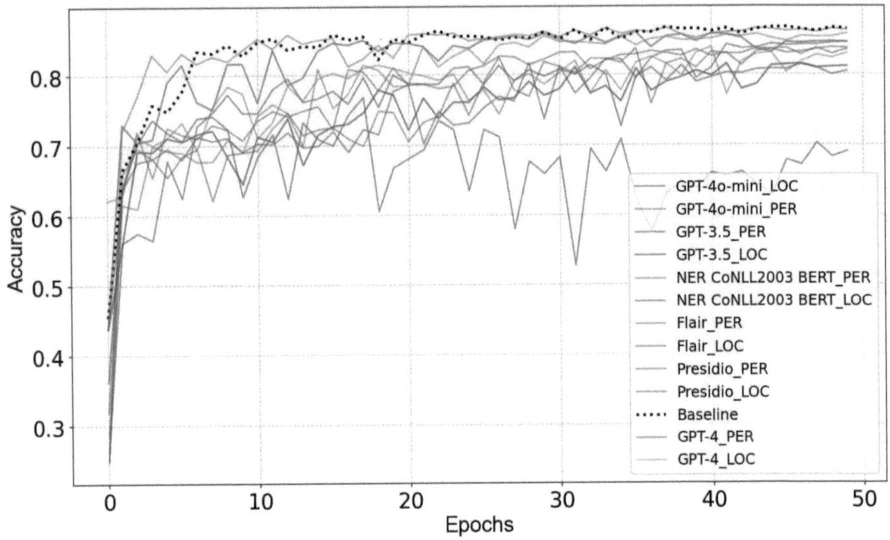

Fig. 4. Accuracy over epochs for the RoBERTa-based occupancy prediction model. The figure compares different implementations, including six anonymization modules over two different entities (PER and LOC). The baseline model, without anonymization module, is highlighted with a dotted black line. The results reflect performance across different configurations, with darker colors indicating specific models.

patterns, ensuring that gender does not influence predictions. In combination, these steps aim to produce a recruitment tool that is both privacy conscious and bias aware.

As can be seen in Table 2, the results demonstrate that the combined application of anonymization of entities (masking private information such as locations and persons) along with gender bias mitigation techniques does not degrade the performance of occupation prediction models. Rather, this dual approach often results in stable or improved predictive accuracy compared to baseline models without any anonymization or bias-aware learning. Importantly, the findings confirm that the implementation of privacy-preserving and fairness-enhancing mechanisms together can enhance the robustness and generalization capacity of the models without sacrificing performance.

The results shown before have been evaluated for the task of Occupancy Classification (see Sect. 4.3). We now evaluate our PBa-LLM approach (incorporating both NER-based privacy enhancement and gender bias reduction, similar to Table 2) for the other task studied in the present paper: Candidate Scoring. For this second task of candidate scoring, we follow the same protocol of [41,42], generating a top 100 list of candidates based on the estimated candidate scores $\hat{y}_i$ (see the right output in Fig. 2). Recall that we used the men-biased dataset from FairCVdb, so the top 100 list is expected to have more men than women for similar competencies given the gender-biased expert score annotations. Elim-

Table 2. Performance of Occupancy Prediction for the six NER models. The results are categorized depending on the private information removed: locations (LOC), persons (PER) and Gender (all rows except **None (Baseline)**). This experiment incorporates the use of two pre-trained Transformers, which optimize content understanding and enhance the models' effectiveness.

		LOC				PER			
Transformer	Anonymization Module	O1	O2	O3	O4	O1	O2	O3	O4
RoBERTa	**None (Baseline)**	0.86	0.88	0.85	0.79	0.89	0.84	0.87	0.72
	None (Bias-aware)	0.88	0.84	0.85	0.82	0.91	0.87	0.86	0.75
	GPT-4	0.91	0.95	0.84	0.82	0.89	0.87	0.85	0.75
	GPT-4o-mini	0.89	0.90	0.86	0.80	0.92	0.85	0.87	0.64
	GPT-3.5	0.82	0.85	0.85	0.65	0.90	0.87	0.90	0.73
	Presidio	0.90	0.90	0.89	0.78	0.91	0.83	0.84	0.71
	FLAIR	0.90	0.86	0.83	0.74	0.89	0.85	0.87	0.76
	NER CoNLL2003 BERT	0.94	0.85	0.83	0.69	0.86	0.89	0.91	0.80
BERT	**None (Baseline)**	0.87	0.85	0.87	0.79	0.89	0.86	0.87	0.79
	None (Bias-aware)	0.89	0.86	0.90	0.83	0.89	0.86	0.92	0.77
	GPT-4	0.94	0.87	0.87	0.82	0.94	0.92	0.91	0.79
	GPT-4o-mini	0.93	0.87	0.91	0.77	0.90	0.91	0.89	0.83
	GPT-3.5	0.88	0.90	0.89	0.75	0.88	0.85	0.92	0.81
	Presidio	0.89	0.88	0.93	0.81	0.85	0.89	0.90	0.74
	FLAIR	0.92	0.86	0.88	0.81	0.91	0.88	0.90	0.77
	NER CoNLL2003 BERT	0.91	0.88	0.90	0.75	0.93	0.86	0.91	0.82

inating personal data and gender cues in biographies using our PBa-LLMs, Table 3 shows that even with these gender-biased training data, one can generate privacy-preserving bias-free AI models (useful enough for NLP tasks like the two tasks studied here: Occupancy Classification and Candidate Scoring).

6 Conclusions

In this work, a privacy-preserving learning framework was proposed and evaluated using Named Entity Recognition (NER) to anonymize sensitive information in text while maintaining system performance. The framework demonstrated its effectiveness in automated recruitment tools using two Transformer-based language models and six anonymization algorithms on a dataset of 24,000 resumes. The experimental results revealed that anonymized data preserved the utility of the models for tasks such as occupancy and scoring predictions, while simultaneously enhancing privacy and mitigating bias. These findings underscore the potential of NER-based anonymization techniques as a scalable and domain-independent solution to ensure both ethical compliance and trust in AI-driven decision-making systems.

Furthermore, the study highlighted the robustness of the NER-based approach, which showed minimal degradation between different anonymization models. The results also revealed that, in certain scenarios, privacy-preserving techniques not only safeguarded sensitive information but also led to slight improvements in fairness, as seen in bias-aware recruitment tasks.

7 Future Works

A promising direction for future work involves extending the current NER-based anonymization framework to other sensitive domains beyond recruitment, such as healthcare [28], education [6,8], or legal affairs [33,43], where privacy concerns are critical [63]. These areas introduce different entity types and context-dependent privacy risks, necessitating the development of domain-specific NER models or hybrid anonymization strategies that combine rule-based and learned approaches. In addition, embedding adaptive mechanisms that automatically calibrate the strength of anonymization to the sensitivity of each domain would help balance privacy with data utility in real-time deployments.

Another important avenue is to expand the bias mitigation capabilities of the system beyond gender to address other social and demographic biases [58], such

Table 3. Gender proportions in the FairCVdb validation set for both LOC and PER labels were analyzed using candidate shortlists from resume scores $\hat{y}$ across different scoring tool configurations. This table reports the male and female representation to assess gender balance after different anonymization strategies.

Transformer	Anonymization Module	LOC		PER	
		Prop (M)	Prop (F)	Prop (M)	Prop (F)
RoBERTa	**None (Baseline)**	67.00%	33.00%	67.00%	33.00%
	None (Bias-aware)	47.80%	52.20%	47.80%	52.20%
	GPT-4	50.60%	49.40%	50.60%	49.40%
	GPT-4o-mini	50.80%	49.20%	50.90%	49.10%
	GPT-3.5	50.90%	49.10%	50.70%	49.30%
	Presidio	50.70%	49.30%	50.80%	49.20%
	FLAIR	50.75%	49.25%	50.85%	49.15%
	NER CoNLL2003 BERT	50.80%	49.20%	50.75%	49.25%
BERT	**None (Baseline)**	69.40%	30.60%	69.40%	30.60%
	None (Bias-aware)	48.60%	51.40%	48.60%	51.40%
	GPT-4	50.70%	49.30%	50.85%	49.15%
	GPT-4o-mini	50.80%	49.20%	50.95%	49.05%
	GPT-3.5	50.70%	49.30%	50.80%	49.20%
	Presidio	50.75%	49.25%	50.88%	49.12%
	FLAIR	50.75%	49.25%	50.92%	49.08%
	NER CoNLL2003 BERT	50.80%	49.20%	50.83%	49.17%

as ethnicity, age, disability, or socioeconomic status. Incorporating detection and debiasing techniques [12,42,52,57] tailored to these attributes would promote fairness on multiple axes and make the recruitment tool more inclusive and equitable across diverse candidate populations.

Acknowledgement. Supported by: BBforTAI (PID2021-127641OB-I00 MICINN/FEDER), HumanCAIC (TED2021-131787B-I00 MICINN), M2RAI (PID2024-160053OB-I00 MICIU/FEDER), Cátedra ENIA UAM-VERIDAS en IA-Responsable (NextGenerationEU PRTR TSI-100927-2023-2), and R&D Agreement DGGC/UAM/FUAM for Biometrics and Applied AI. G. Mancera is supported by FPI-PRE2022-104499 MICINN/FEDER. Work conducted in the ELLIS Unit Madrid.

References

1. Achiam, J., et al.: GPT-4 technical report. arXiv:2303.08774 (2023)
2. Acquisti, A., Brandimarte, L., Loewenstein, G.: Privacy and human behavior in the age of information. Science **347**(6221), 509–514 (2015)
3. Akbik, A., Bergmann, T., Blythe, D., Rasul, K., Schweter, S., Vollgraf, R.: FLAIR: an easy-to-use framework for state-of-the-art NLP. In: Proceedings of the Conference of the North American Chapter of the Association for Computational Linguistics (Demonstrations), pp. 54–59 (2019)
4. Al-Alawi, A.I., Naureen, M., AlAlawi, E.I., Al-Hadad, A.A.N.: The role of artificial intelligence in recruitment process decision-making. In: International Conference on Decision Aid Sciences and Application (DASA), pp. 197–203. IEEE (2021)
5. Babych, B., Hartley, A.: Improving machine translation quality with automatic named entity recognition. In: Proceedings of the 7th International EAMT Workshop on MT and Other Language Technology Toolsat EACL (2003)
6. Becerra, A., Daza, R., Cobos, R., Morales, A., Cukurova, M., Fierrez, J.: AI-based multimodal biometrics for detecting smartphone distractions: Application to online learning. In: EATEL European Conference on Technology-Enhanced Learning (ECTEL) (2025)
7. Cheng, L., Varshney, K.R., Liu, H.: Socially responsible AI algorithms: Issues, purposes, and challenges. J. Artif. Intell. Res. **71**, 1137–1181 (2021)
8. Daza, R., Becerra, A., Cobos, R., Fierrez, J., Morales, A.: A multimodal dataset for understanding the impact of mobile phones on remote online virtual education. Sci. Data (2025)
9. Daza, R., Morales, A., Tolosana, R., Gomez, L.F., Fierrez, J., Ortega-Garcia, J.: edBB-Demo: biometrics and behavior analysis for online educational platforms. In: AAAI Conference on Artificial Intelligence (AAAI), vol. 37, pp. 16422–1642 (Feb 2023)
10. De-Arteaga, M., Romanov, A., Wallach, H., Chayes, J., Borgs, C., Chouldechova, A., Geyik, S., Kenthapadi, K., Kalai, A.T.: Bias in bios: A case study of semantic representation bias in a high-stakes setting. In: Proceedings of the Conference on Fairness, Accountability, and Transparency. pp. 120–128 (2019)
11. DeAlcala, D., Morales, A., Mancera, G., Fierrez, J., Tolosana, R., Ortega-Garcia, J.: Is my data in your AI model? Membership inference test with application to face images. arXiv:2402.09225 (2024)

12. DeAlcala, D., Serna, I., Morales, A., Fierrez, J., Ortega-Garcia, J.: Measuring bias in AI models: an statistical approach introducing n-sigma. In: IEEE Conference on Computers, Software, and Applications (COMPSAC), pp. 1167–1172 (June 2023)
13. Diakopoulos, N.: Algorithmic accountability: journalistic investigation of computational power structures. Digit. J. **3**(3), 398–415 (2015)
14. Ding, J., Sun, H., Wang, X., Liu, X.: Entity-level sentiment analysis of issue comments. In: Proceedings of the 3rd International Workshop on Emotion Awareness in Software Engineering, pp. 7–13 (2018)
15. Europen Union: Artificial intelligence act. European Parliament: European Parliamentary Research Service 2024/1689 (2024)
16. Fernández, M., Cantador, I., López, V., Vallet, D., Castells, P., Motta, E.: Semantically enhanced information retrieval: an ontology-based approach. J. Web Semantics **9**(4), 434–452 (2011)
17. Fu, J., Huang, X.J., Liu, P.: Spanner: Named entity re-/recognition as span prediction. In: Proceedings of the 59th Annual Meeting of the ACL, pp. 7183–7195 (2021)
18. Geetha, R., Bhanu, S.R.D.: Recruitment through artificial intelligence: a conceptual study. Int. J. Mech. Eng. Technol. **9**(7), 63–70 (2018)
19. Ghafourian, M., et al.: Toward face biometric de-identification using adversarial examples. In: IEEE Conference on Computers, Software, and Applications (COMPSAC), pp. 723–728 (June 2023), also presented at AAAI Workshop on Artificial Intelligence for Cyber Security, AICS 2023
20. Ghafourian, M., Fierrez, J., Vera-Rodriguez, R., Tolosana, R., Morales, A.: SaFL: Sybil-aware federated learning with application to face recognition. In: IEEE International Conference on Image Processing Challenges and Workshops (ICIPcw), pp. 4050–4056 (2024)
21. Ghafourian, M., et al.: Blockchain and biometrics: Survey, GDPR elements, and future directions. arXiv:2302.10883 (2025)
22. Gomez-Barrero, M., Galbally, J., Morales, A., Fierrez, J.: Privacy-preserving comparison of variable-length data with application to biometric template protection. IEEE Access **5**, 8606–8619 (2017)
23. Hassanpour, A., Moradikia, M., Yang, B., Abdelhadi, A., Busch, C., Fierrez, J.: Differential privacy preservation in robust continual learning. IEEE Access **10** (Feb 2022)
24. Jiang, A.Q., et al.: Mistral 7b. arXiv:2310.06825 (2023)
25. Jurado, F., Rodriguez, P.: Sentiment analysis in monitoring software development processes: an exploratory case study on GitHub's project issues. J. Syst. Softw. **104**, 82–89 (2015)
26. Kaaniche, N., Laurent, M., Belguith, S.: Privacy enhancing technologies for solving the privacy-personalization paradox: taxonomy and survey. J. Netw. Comput. Appl. **171**, 102807 (2020)
27. Kenton, J.D.M.W.C., Toutanova, L.K.: Bert: pre-training of deep bidirectional transformers for language understanding. In: Proceedings of NACCL-HLT (2019)
28. Lacruz-Pleguezuelos, B., Bazán, G.X., et al.: AI4Food, a feasibility study for the implementation of automated devices in the nutritional advice and follow up within a weight loss intervention. Clin. Nutr. **48**, 80–89 (2025)
29. Lepri, B., Oliver, N., Pentland, A.: Ethical Machines: The Human-centric Use of Artificial Intelligence. IScience **24**(3) (2021)
30. Li, J., Sun, A., Han, J., Li, C.: A survey on deep learning for named entity recognition. IEEE Trans. Knowl. Data Eng. **34**(1), 50–70 (2022)

31. Li, L., Fan, Y., Tse, M., Lin, K.Y.: A review of applications in federated learning. Comput. Indust. Eng. **149**, 106854 (2020)
32. Liu, Y.: RoBERTa: A robustly optimized bert pretraining approach. arXiv:1907.11692 (2019)
33. Lopez-Duran, M., Fierrez, J., Morales, A., et al.: Benchmarking graph neural networks for document layout analysis in public affairs. In: IAPR International Conference on Document Analysis and Recognition Workshops (ICDARw) (September 2025)
34. Mancera, G., DeAlcala, D., Fierrez, J., Tolosana, R., Morales, A.: Is my text in your AI model? Gradient-based membership inference test applied to LLMs. arXiv preprint arXiv:2503.07384 (2025)
35. Mollá, D., Van Zaanen, M., Smith, D.: Named entity recognition for question answering. In: Proceedings of the Australasian Language Technology Association Workshop, pp. 51–58 (2006)
36. Morales, A., Fierrez, J., Vera-Rodriguez, R., Tolosana, R.: SensitiveNets: Learning agnostic representations with application to face images. IEEE Trans. Pattern Anal. Mach. Intell. **43**(6), 2158–2164 (2020)
37. Mumtaz, R.: Extracting Named Entities and Relations From Text for Populating Knowledge Graph. Ph.D. thesis, Capital University (2023)
38. Nasar, Z., Jaffry, S.W., Malik, M.K.: Named entity recognition and relation extraction: state-of-the-art. ACM Comput. Surv. (CSUR) **54**(1), 1–39 (2021)
39. Neumann, G.K., Grace, P., Burns, D., Surridge, M.: Pseudonymization risk analysis in distributed systems. J. Internet Serv. Appl. **10**(1), 1–16 (2019). https://doi.org/10.1186/s13174-018-0098-z
40. Ouyang, L., Wu, J., Jiang, X., Almeida, D., et al.: Training language models to follow instructions with human feedback. Proc. Adv. Neural Inform. Process. Syst. **35**, 27730–27744 (2022)
41. Peña, A., et al.: Human-centric multimodal machine learning: recent advances and testbed on AI-based recruitment. SN Comput. Sci. **4**(5), 434 (2023)
42. Peña, A., Fierrez, J., Morales, A., Mancera, G., Lopez, M., Tolosana, R.: Addressing bias in LLMs: Strategies and application to fair AI-based recruitment. In: AAAI/ACM Conference on AI, Ethics, and Society (AIES) (October 2025)
43. Peña, A., et al.: Continuous document layout analysis: Human-in-the-loop AI-based data curation, database, and evaluation in the domain of public affairs. Inform. Fusion **108**, 102398 (2024)
44. Peña, A., et al.: Leveraging large language models for topic classification in the domain of public affairs. In: IAPR Intl. Conf. on Document Analysis and Recognition Workshops (ICDARw). LNCS, vol. 14193, pp. 20–33. Springer (August 2023)
45. Peña, A., Serna, I., Morales, A., Fierrez, J.: Bias in multimodal AI: Testbed for fair automatic recruitment. In: IEEE/CVF Conference on Computer Vision and Pattern Recognition Workshops (CVPRw) (June 2020), also presented at ICML 2020 Workshop on Human-in-the-Loop Learning
46. Peña, A., Serna, I., Morales, A., Fierrez, J.: FairCVtest Demo: understanding bias in multimodal learning with a testbed in fair automatic recruitment. In: ACM Intl. Conf. on Multimodal Interaction (ICMI), pp. 760–761 (October 2020), also presented at European Conference on Computer Vision, ECCV 2020
47. Pfitzmann, A., Hansen, M.: A terminology for talking about privacy by data minimization: Anonymity, unlinkability, undetectability, unobservability, pseudonymity, and identity management (2010)

48. Ritter, A., Clark, S., Etzioni, O., et al.: Named entity recognition in tweets: an experimental study. In: Proceedings of the Conference on Empirical Methods in Natural Language Processing, pp. 1524–1534 (2011)
49. Sandhu, R.S., Samarati, P.: Access control: principle and practice. IEEE Commun. Mag. **32**(9), 40–48 (1994)
50. Sang, E.F., De Meulder, F.: Introduction to the CoNLL-2003 shared task: Language-independent named entity recognition. In: Proceedings of the Conference on Natural Language Learning at HLT-NAACL 2003, pp. 142–147 (2003)
51. Schwartz, P.M., Solove, D.J.: The PII problem: privacy and a new concept of personally identifiable information. NYUL rev. **86**, 1814 (2011)
52. Serna, I., DeAlcala, D., Morales, A., Fierrez, J., Ortega-Garcia, J.: IFBiD: Inference-free bias detection. In: AAAI Workshop on Artificial Intelligence Safety (SafeAI), CEUR-WS, vol. 3087 (February 2022)
53. Serna, I., Morales, A., Fierrez, J., Obradovich, N.: Sensitive loss: improving accuracy and fairness of face representations with discrimination-aware deep learning. Artif. Intell. **305**, 103682 (2022)
54. Serrano-Silva, Y.O., et al.: Automatic feature weighting for improving financial decision support systems. Decis. Support Syst. **107**, 78–87 (2018)
55. Staab, R., Vero, M., Balunović, M., Vechev, M.: Large language models are advanced anonymizers. arXiv preprint arXiv:2402.13846 (2024)
56. Stalla-Bourdillon, S., Knight, A.: Anonymous data v. personal data-false debate: an eu perspective on anonymization, pseudonymization and personal data. Wisconsin Int. Law J. **34**(2), 284–322 (2017)
57. Tello, J., et al.: Symbolic AI (LFIT) for XAI to handle biases. In: European Conference on AI Workshops (ECAIw). CEUR-WS, vol. 3523 (2023)
58. Terhorst, P., et al.: A comprehensive study on face recognition biases beyond demographics. IEEE Trans. Technol. Society **3**(1), 16–30 (2022)
59. The White House: Memorandum on Advancing the United States' Leadership in Artificial Intelligence. (2024). https://www.whitehouse.gov/briefing-room/presidential-actions/2024/10/24/
60. Touvron, H., et al.: LLaMA: Open and efficient foundation language models. arXiv:2302.13971 (2023)
61. Touvron, H., et al.: LLaMA 2: Open foundation and fine-tuned chat models. arXiv:2307.09288 (2023)
62. Vaswani, A.: Attention is all you need. In: Proceedings of Advances in Neural Information Processing Systems (2017)
63. Veldhuis, R., et al.: Privacy and Security Matters in Biometric Technologies. Springer (2025)
64. Voigt, P., Von dem Bussche, A.: The EU General Data Protection Regulation (GDPR) - A Practical Guide. Springer (2017)
65. Weston, L., et al.: Named entity recognition and normalization applied to large-scale information extraction from the materials science literature. J. Chem. Inf. Model. **59**(9), 3692–3702 (2019)

Benchmarking Graph Neural Networks for Document Layout Analysis in Public Affairs

Miguel Lopez-Duran(✉), Julian Fierrez, Aythami Morales, Ruben Tolosana, Oscar Delgado-Mohatar, and Alvaro Ortigosa

School of Engineering, University Autónoma de Madrid (UAM), Madrid 28049, Spain
{miguel.lopezd,julian.fierrez,aythami.morales,ruben.tolosana, oscar.delgado,alvaro.ortigosa}@uam.es

Abstract. The automatic analysis of document layouts in digital-born PDF documents remains a challenging problem due to the heterogeneous arrangement of textual and nontextual elements and the imprecision of the textual metadata in the Portable Document Format. In this work, we benchmark Graph Neural Network (GNN) architectures for the task of fine-grained layout classification of text blocks from digital native documents. We introduce two graph construction structures: a k-closest-neighbor graph and a fully connected graph, and generate node features via pre-trained text and vision models, thus avoiding manual feature engineering. Three experimental frameworks are evaluated: single-modality (text or visual), concatenated multimodal, and dual-branch multimodal. We evaluated four foundational GNN models and compared them with the baseline. Our experiments are specifically conducted on a rich dataset of public affairs documents that includes more than 20 sources (e.g., regional and national-level official gazettes), 37K PDF documents, with 441K pages in total. Our results demonstrate that GraphSAGE operating on the k-closest-neighbor graph in a dual-branch configuration achieves the highest per-class and overall accuracy, outperforming the baseline in some sources. These findings confirm the importance of local layout relationships and multimodal fusion exploited through GNNs for the analysis of native digital document layouts.

Keywords: Graph Neural Networks · Document Layout Analysis · Digital document classification

1 Introduction

The increasing volume of digital documents has created a need for accurate, efficient, and scalable algorithms capable of automatically processing the information within the documents. A large proportion of these documents are stored in the Portable Document Format (PDF), a standard originally developed by Adobe and later standardized in 2008 [14]. The widespread adoption of PDF is

L. Jin et al. (Eds.): ICDAR 2025 Workshops, LNCS 16225, pp. 21–35, 2026.
https://doi.org/10.1007/978-3-032-09368-4_2

largely due to its ability to present documents independently of the software, hardware, or operating system used, as well as its support for encryption, compression, digital signatures, and editing.

As a result of these advantages, PDF has become the standard format for document management in public administrations and private companies. However, the automatic processing of PDF documents remains a challenging task. Extracting and analyzing relevant information requires understanding the structure and relationships between the various elements of the document, especially text blocks, since they provide different information depending on their semantic role within the document.

Document Layout Analysis (DLA) addresses this challenge by aiming to detect and classify the fundamental components of a document. This task is essential for the automatic processing of digital documents, as it constitutes a preliminary step in the extraction of knowledge and enables further processing with other techniques such as Natural Language Processing (NLP) algorithms [3,15,16].

As mentioned above, understanding the interactions between the different layout components within a document is essential to accurately determine its semantic structure. As demonstrated by numerous studies in various domains such as social networks [21], or transaction networks [20], graphs represent a particularly effective tool for modeling these relationships and have been extensively studied in recent years [19]. Among the techniques for processing graph-structured data, Graph Neural Networks (GNNs) are one of the most prominent approaches [35]. GNNs have been used for various tasks in document understanding such as text summarization [37] or document object detection [1].

In this work, we explore the use of GNN-based models combined with multiple graph representations. We use official digital native documents from Spanish public administrations to classify the textual components of each document into specific semantic categories such as title or body. Our approach integrates both text and visual information [29] in order to enhance the classification of each layout element through specialized models. Furthermore, we evaluated the performance of different GNN architectures with various experimental frameworks to assess the contribution of each component of our method and evaluate its effectiveness.

The main contributions of this work are summarized as follows:

- We propose a multimodal graph-based methodology for general Document Layout Analysis, and apply it to Spanish digital native official document preprocessing and representation, leveraging pre-trained models to generate node embeddings, and thus avoiding the need for handcrafted features.
- A benchmark for Document Layout Analysis (DLA) is presented on digitally conceived documents using GNNs. In this benchmark, we explore two different document graph structures and evaluate the performance of four baseline GNN architectures. We evaluated our approach with three different frameworks to assess the need for each component of our models.

- Our proposed approaches are compared against a strong baseline in the DLA of public affairs documents from Spanish administrations. The experiments are carried out on a dataset of public affairs documents that includes 20 sources, 37K PDF documents, with 441K pages in total.[1] Our models achieve a higher classification accuracy per class for some classes in most document sources and improve the overall accuracy in one third of them.

The paper is organized as follows. In Sect. 2 we review the state of the art on multimodal DLA and graph-based DLA methods. Section 3 provides the details of the document graph construction and the proposed approaches. The dataset used and the results of our experiments are reported in Sect. 4. Finally, the conclusions and future directions of the work are drawn in Sect. 5.

2 Related Works

2.1 Document Layout Analysis

The literature on DLA typically distinguishes between two categories of PDF documents: (1) *digitally conceived documents*, which are generated directly from digital sources, and (2) *document images*, which are scanned versions of physical or digital documents (possibly including handwritten text and graphics).

Traditionally, DLA tools targeting digital-born documents have focused primarily on text extraction. These tools are constrained by the structure of the PDF format, which often lacks semantic metadata, making the identification of layout elements such as tables particularly challenging. Bast *et al.* [2] introduced a benchmark for evaluating text extraction methods in digital PDFs, compiling a dataset of more than 10K *arXiv* articles annotated by parsing the associated TeX source files. Zhong *et al.* [40] proposed an automated annotation approach for digital PDF articles by aligning the output of the PDFMiner library with the XML representations of scientific publications. All of these approaches focus on the detection and extraction of the layout components, especially tables, but they lack classification of the semantics of the layout components detected.

2.2 Multimodal Layout Analysis

Early work on layout-related tasks relied only on textual or visual information, failing to take advantage of the strong complementarity between these two modalities. Recent work has shown the effectiveness of combining visual, textual, and other types of features such as structural information to improve model expressiveness and performance.

For example, Gu *et al.* [10] extract textual, visual, and layout features to address document understanding tasks using a multimodal transformer architecture. Similarly, Cheng *et al.* [4] proposed a vision-grid transformer that integrates visual and textual inputs, leveraging multimodal fusion [7,29] and pre-training

[1] https://github.com/BiDAlab/PALdb.

strategies to improve downstream performance. In [38], Zhang *et al.* outline the limitations of previous multimodal DLA approaches and propose a dual multimodal fusion strategy for DLA.

2.3 GNN-Based Document Layout Analysis

GNNs have shown strong performance in a variety of document-related tasks [36]. Their main strength lies in their ability to model and exploit relational information between entities, which is especially useful in structured document environments.

In the context of DLA, GNNs have been successfully applied to tasks such as table detection and extraction [9,30], document classification [39], and multimodal document understanding [32].

However, most existing approaches focus on scanned document images or perform classification at the document level. In contrast, our goal is to classify individual text blocks within digital-born PDF documents from Spanish official sources, leveraging both semantic, visual, and structural information. One advantage of using digital-born documents is the fact that the layout objects are exactly placed, so there is no need of Optical Character Recognition (OCR) assessment, which is not true for scanned documents. To the best of our knowledge, this is the first work to apply a GNN-based method for DLA to official Spanish digital-born documents.

3 Graph Construction and Proposed Frameworks

In this section, we describe the methodology followed to represent document layouts as graph structures and the approaches adopted to compare various GNN models against the baseline methods.

3.1 Graph Construction and Initial Features

A graph $G = (V, E)$ is a mathematical structure consisting of a set of nodes or vertices, denoted V, and a set of edges or links, denoted E, which define the connections between the nodes.

Given a document D comprising P pages, we represent each page P_i using two distinct graphs: $G_i^K = (V_i, E_i^K)$ and $G_i^C = (V_i, E_i^C)$. In both cases, the set of nodes V_i corresponds to the layout objects within the page.

The construction of edges differs between the two graph structures. For G_i^K, which we refer to as the k-closest graph, edges are established by connecting each node to its k closest neighbors, where proximity is determined based on the Euclidean distance between the centroids of the layout objects within the page (see Fig. 1(b)). In all of our experiments, we set $k = 4$ in order to focus on local dependencies while keeping the diameter of the graph small, to easily connect distant nodes. This approach is motivated by the hypothesis that the category of a layout object is significantly influenced by its neighboring elements. In contrast,

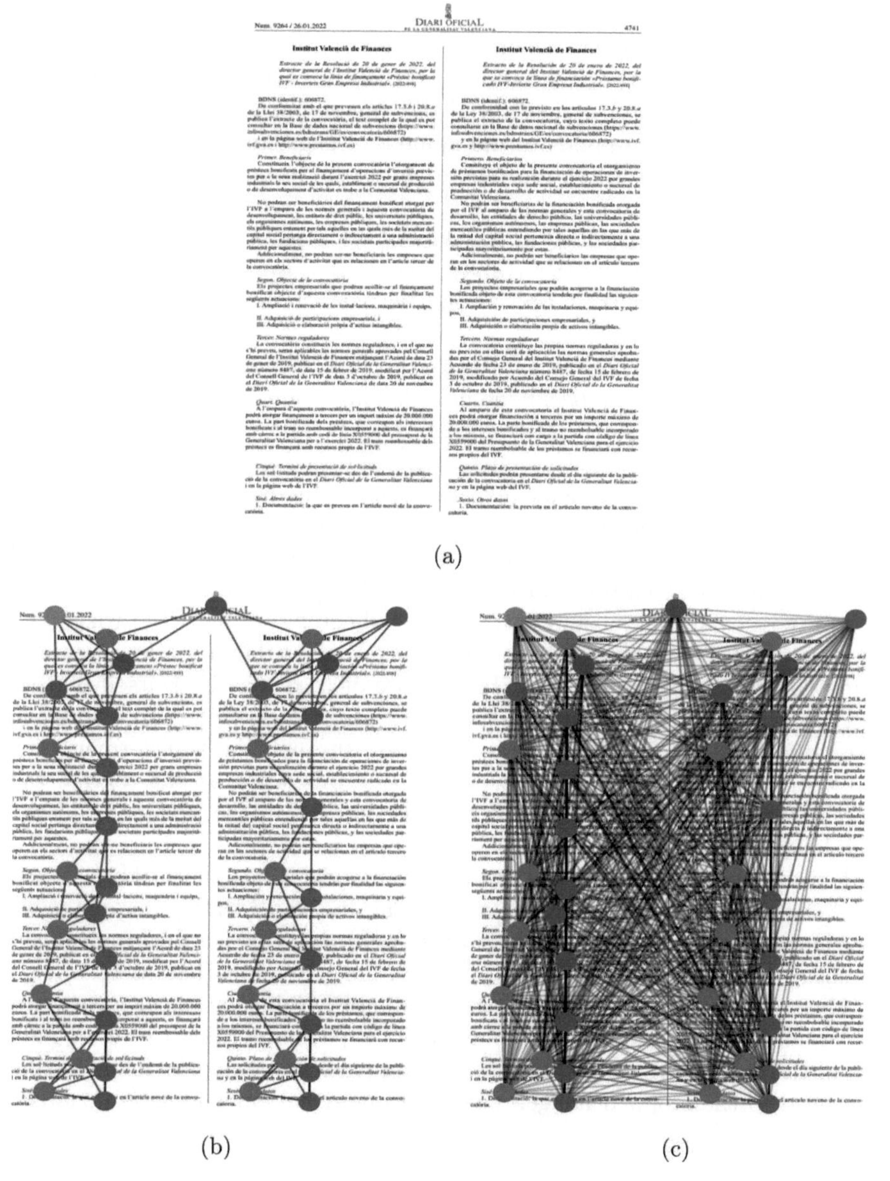

(a)

(b)

(c)

Fig. 1. Illustration of the graph construction process applied to a document from the *Diario Oficial de la Generalitat Valenciana*. Original document page (a), k-closest graph (b), and complete graph (c) representation of the page.

G_i^C, which is called the complete graph, is designed as a fully connected graph, where each node is linked to all other nodes within the graph (see Fig. 1(c)).

This structure is intended to minimize the introduction of expert-driven biases, allowing the model to generalize effectively across various document layouts.

After constructing the graphs for each page, each document is represented as two sequences of graph structures of the same type, denoted $\{G_i^j\}_{i=1}^P$, where $j \in \{K, C\}$ and P corresponds to the number of pages in the document.

Once the document graph representation is established, initial node features are generated to use them as input for the models. Rather than relying on the hand-crafted features provided in the dataset we used (see Sect. 4.1), which are partially incomplete for images, tables, and URLs, we employ two pretrained models to effectively extract textual and visual information from each layout object.

For text embedding, based on previous work by Peña et al. [27], we employ *RoBERTa-base* [11] as the backbone model. This model is available in the *HuggingFace* repository[2]. *RoBERTa* models [23] are variants of *BERT* [16] that incorporate optimized hyperparameter selection and an improved pretraining strategy. *RoBERTa-base*, specifically, is a Spanish-language version of these models, pretrained on a corpus comprising 570 GB of clean Spanish text.

Given a layout object, we process its textual information as follows. If the object corresponds to a text block, we apply the RoBERTa-base model directly to its raw textual content and extract the embedding of the [CLS] token as its representation. For images, due to the absence of textual information, we use the string "0" as input to the model. In the case of links, the URL itself is passed as a string. For tables, we extract their contents, flatten them into a single-line text representation, process the resulting text through the transformer model, and use the final classification token ([CLS]) embedding as the textual table representation. After applying this process, we obtain the text embedding of each layout object, denoted as $\mathbf{x}_T$.

For visual embedding, we employ a *ResNet* model [13] as the backbone. Residual Networks (*ResNet*) leverage residual learning and identity mappings with shortcut connections to effectively capture representative features from images. To ensure efficient and lightweight preprocessing, we select ResNet-18.

Given a layout object and its corresponding PDF document page, we extract the visual representation by cropping the PDF page according to the bounding box of the object. We then apply the vision model excluding the final classification head to the cropped image. The resulting visual embedding is denoted as $\mathbf{x}_V$.

3.2 Proposed Frameworks

To assess the performance of various GNN models, we propose three distinct frameworks in which GNNs serve as the backbone. Although images, tables, and links are not used directly during training, their embeddings are updated and retained within the graphs. This decision is based on the hypothesis that these

[2] https://huggingface.co/PlanTL-GOB-ES.

elements contribute structural information relevant to the semantic interpretation of their neighboring text blocks.

All proposed frameworks are designed for a supervised node classification task, where each text block is categorized into one of four classes: Identifier, Summary, Title, or Body. The models consist of one or two branches in which node features are processed through a GNN backbone model. To evaluate the performance and suitability of these frameworks, and determine whether specialized GNN models can be developed for this task, we select four fundamental GNN architectures:

- **Graph Convolutional Network (GCN)** [17]: One of the first GNN architectures, designed to learn representative node and graph features in a semi-supervised manner.
- **Graph Attention Network (GAT)** [34]: One of the most widely used graph architectures employs an attention mechanism to learn the relative importance of neighboring nodes for the downstream task.
- **GraphSAGE** [12]: An inductive architecture that updates the node embeddings by efficiently aggregating features from the local neighborhood of each node.
- **Topology Adaptive Graph Convolutional Network (TAGCN)** [6]: A GNN variant that adaptively processes graphs to learn across different topologies and effectively captures long-range dependencies.

The three proposed frameworks in which we will test the four previous GNNs differ in the type of features utilized (textual and vision) and in the number of processing branches. The defined frameworks are as follows: a single branch using only one type of feature, a single branch incorporating both textual and vision features, and a dual-branch architecture leveraging both feature types. The specific classification strategy varies depending on the framework.

Single Feature Type, Single Branch. In this framework, each document page node is processed through the GNN model using only one type of feature (textual or visual), while discarding the other. A fully connected layer is then applied to classify each layout object on the page, represented as:

$$\mathbb{P}_1(\omega|\mathbf{x}_f),\ f \in \{T, V\}\,. \tag{1}$$

Both Feature Types, Single Branch. In this framework, the initial embedding of the initial features of each document page node is the concatenation of both the text and the vision features, denoted as $\mathbf{x} = (\mathbf{x}_T \parallel \mathbf{x}_V)$, which is passed through the GNN model, allowing the model to learn both representations simultaneously. A fully connected layer is applied to classify each layout object on the page, resulting in:

$$\mathbb{P}_2(\omega|\mathbf{x}). \tag{2}$$

Both Feature Types, Dual Branch. In this framework, each feature type is processed independently through dedicated models, creating specialized textual and visual representations. The outputs of both models are then concatenated and passed through two fully connected layers to classify each layout object on the page, which results in:

$$\mathbb{P}_3(\omega|\mathbf{x}_T, \mathbf{x}_V), \tag{3}$$

where $\omega \in \{ID, Title, Summary, Body\}$ in all settings.

An overview of the different frameworks is shown in Fig. 2.

In all frameworks, we perform a train/test split at the document level and execute the forward pass at the page level to mitigate intra-document biases, as observed in [26]. The selected GNN architectures are applied in a sequence of layers, each followed by batch normalization, an Exponential Linear Unit (ELU) activation function, and a dropout layer to enhance performance.

4 Experiments and Results

In this section, we describe the dataset we used to evaluate the models. The experimental setup and the results of our approaches are also presented. The primary objective of our experiments is to assess the effectiveness of multimodality and the usefulness of GNNs in the layout classification task for official documents.

4.1 Public Affairs Layout (PAL) Dataset

The Public Affairs Layout (PAL) dataset [26] contains over 37.9K PDF document layouts extracted from a total of 24 official Spanish gazettes.[3] The documents were annotated using a human-in-the-loop approach, assigning each layout object one of the following labels: Image, Table, Link, Identifier, Title, Summary, or Body. Although the first three categories can be directly retrieved from the internal PDF metadata, the remaining four are initially classified as generic text blocks. This limitation highlights the need to develop algorithms and models capable of accurately distinguishing semantic differences between these text-based categories.

The PAL database is divided into two subsets: a training set, originally used to train the Random Forest models introduced in [26] to classify the text blocks, and a validation set, initially used for model evaluation and subsequently verified by a human annotator. In all of our experiments, we exclusively use the validation set. This decision is motivated by the fact that the training set may contain labeling inaccuracies, as it was not externally validated. For detailed statistics on the validation set, we refer to the original publication. Some examples of documents and layouts from the dataset are presented in Fig. 3.

[3] https://github.com/BiDAlab/PALdb.

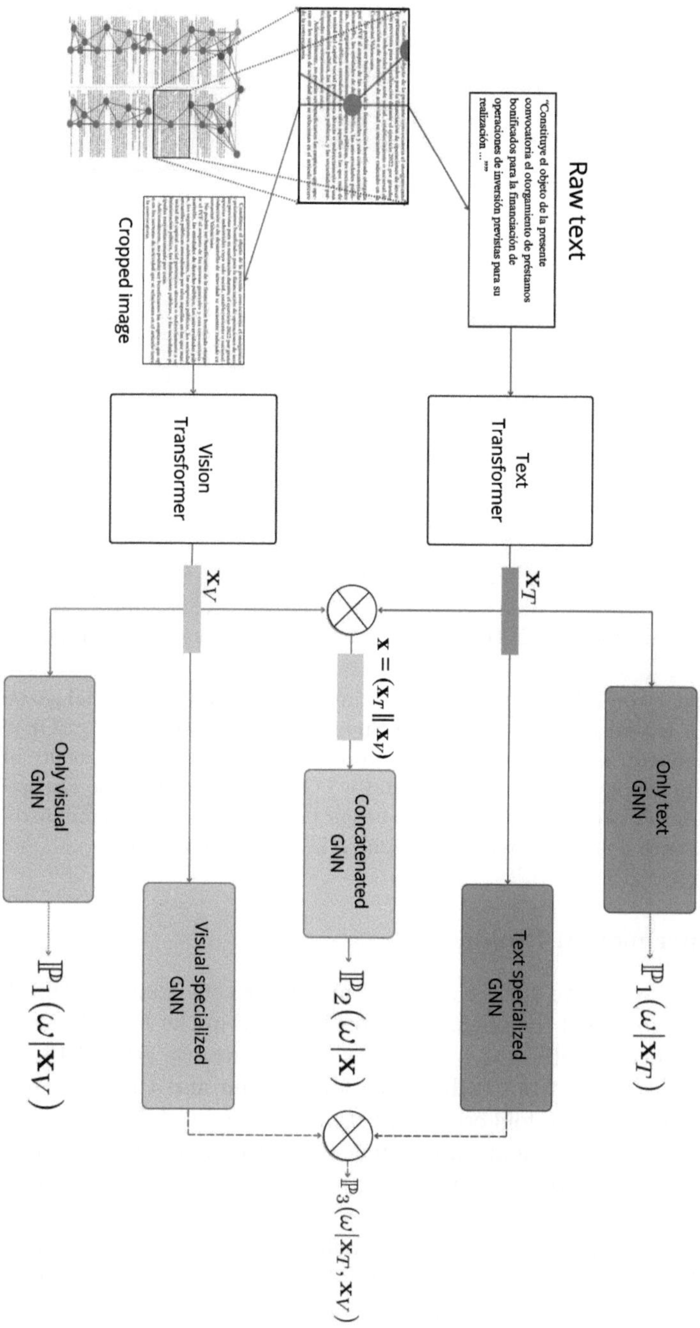

Fig. 2. Overview of the three experimental frameworks. After passing each object through the transformers, four different models are trained independently to perform three different classifications of the text block object.

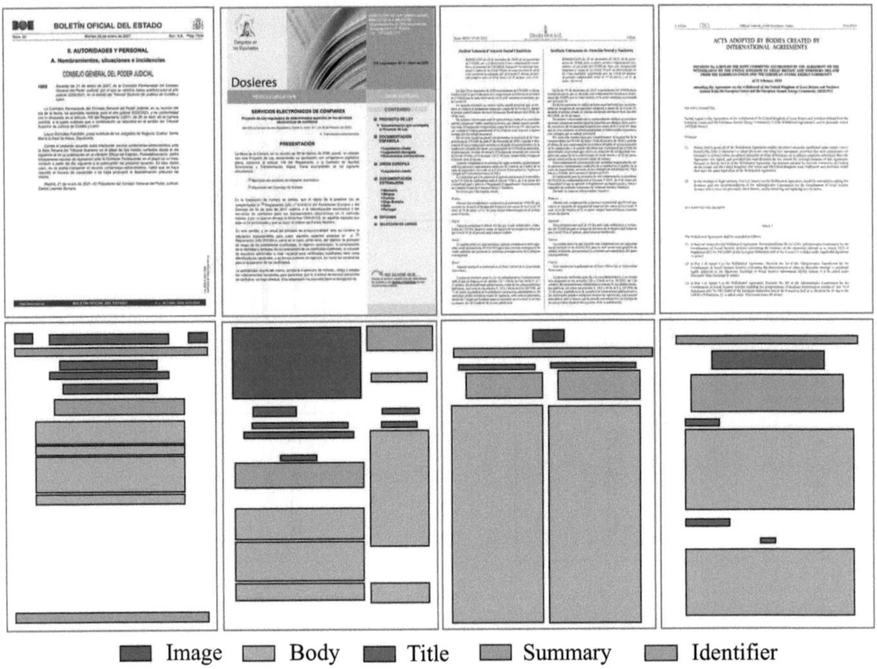

Fig. 3. Visual examples of page document images from different official gazettes (upper row). From left to right we present: i) Spanish official gazette (i.e., BOE); ii) public affairs document with the results of a vote in the Spanish Parliament; iii) Spanish regional official gazette (i.e. DOGV); and iv) Official Journal of the European Union. The bottom row presents illustrations of their layout components simplified by colored block (i.e., each color represents a document layout component).

4.2 Experimental Protocol

In all experiments, we train a separate model for each source, where the source corresponds to one of the official gazettes included in the PAL database. Following the protocol established by Peña *et al.* [26], we use 80% of the documents in the validation set from each source for training and the remaining 20% for testing. We employ a k-fold cross-validation strategy with $k = 5$ and random initializations. Given the significant imbalance in the number of documents across sources, the batch size is manually adjusted per source to ensure stable and consistent training.

All models were trained for 350 epochs with a weighted cross-entropy loss, and the stochastic gradient descent (SGD) optimizer with a learning rate of 10^{-3} and a momentum of 0.9 and a dropout of 0.1.

4.3 Results

Table 1. Overall and per-class mean classification accuracy across all folds and the 24 sources compared against the baseline. We report the best performing models for Only Text (OT) and Only Visual (OV) features, and Concatenated (Conc.) framework. We report the six best performing models for dual setting, where the first model refers to the text specialization and the second model to the visual specialization. G = GCN, S = GraphSAGE, T = TAGCN, RF = Random Forest.

Model	Backbone	Type of graph	Classification accuracy (%)				
			Overall	ID	Title	Summary	Body
[26]	RF	-	97.48	99.81	93.70	95.81	98.25
OT	GraphSAGE	k-closest	95.44	99.01	93.20	92.13	95.52
OV	GraphSAGE	k-closest	75.97	81.76	65.03	41.47	81.29
Conc.	GraphSAGE	Complete	95.03	98.15	92.12	81.99	96.09
Dual	S+S	k-closest	95.85	98.86	92.42	89.97	96.42
	S+S	Complete	95.98	99.32	93.36	82.63	96.92
	S+G	k-closest	95.81	98.91	92.88	91.03	96.29
	S+T	k-closest	95.10	98.56	91.16	90.15	96.03
	S+T	Complete	95.84	99.27	93.04	83.06	96.56

The general results for the proposed frameworks are presented in Table 1, which summarizes the graph structures and models that perform the best, together with a comparison to the baseline. Given the large number of document sources and experimental settings, we report the average classification accuracy across all folds and sources. For the first two frameworks (Only Text or Only Visual, and Concatenated Features), we report the best-performing model, while for the dual-branch framework, we report the top five model combinations, specifying the graph type and GNN backbones used.

Our analysis reveals that the k-closest graph structure consistently yields better results, outperforming the complete graph structure in five of the eight top-performing configurations. Additionally, GraphSAGE emerges as the most effective backbone, present in all eight configurations, either for text or visual features in the dual-branch setting. These findings support our initial hypothesis regarding the importance of local layout relationships in determining the semantic category of each text block. Furthermore, the dual-branch framework outperforms the others in four of the five top configurations, confirming the benefit of processing text and visual features separately before classification.

However, these results may present an overly optimistic view, as they are based on selecting the best configuration for each framework independently and reporting averaged scores across all folds and sources. To provide a more detailed and realistic evaluation, we assess the performance of a fixed configuration in five representative sources. Based on the previous results, we selected the k-closest graph structure and the dual-branch framework with GraphSAGE, as it was the configuration that performed the best on average. The selected sources correspond to those with the largest number of documents in the validation set. The results presented in Table 2 highlight the fields in which our models outperform the baseline method proposed by Peña *et al.* [26].

Table 2. Overall and per-class classification accuracy for k-closest graph and the best model selection containing GraphSAGE for the five selected sources. We report the accuracy in terms of $mean_{std}$ across five folds of cross validation. The first model refers to the text specialization and the second model to the visual specialization. S = GraphSAGE, T = TAGCN. The BOE (Boletín Oficial del Estado) source is the Official State Bulletin where the Spanish government publishes laws, regulations, and official acts that must be made public. The remaining sources correspond to the official gazettes of other regions within the Spanish state. BOPA = Boletín Oficial del Principado de Asturias, BOPV = Boletín Oficial del Páis Vasco, BOME = Boletín Oficial de Melilla, BOAB = Boletín Oficial del Ayuntamiento de Barcelona.

Source	Backbone	Classification accuracy (%)				
		Overall	ID	Title	Summary	Body
BOE	S+S	$97.93_{2.03}$	$99,52_{0,76}$	$\mathbf{97,77_{1,84}}$	$98,45_{1,27}$	$97,45_{3,56}$
BOPA	S+T	$95,91_{0,92}$	100_0	$95,50_{1,81}$	$93,43_{7,66}$	$95,73_{1,26}$
BOPV	S+S	$\mathbf{98,40_{0,81}}$	$99,72_{0,56}$	$\mathbf{98,59_{1,75}}$	$\mathbf{99,09_{1,82}}$	$97,85_{0,85}$
BOME	S+T	$96,08_{0,87}$	100_0	$92,54_{4,80}$	$90,51_{8,63}$	$97,07_{1,81}$
BOAB	S+S	$97,83_{1,59}$	100_0	$\mathbf{95,83_{2,76}}$	$98,92_{0,89}$	$98,01_{1,60}$

We observe that the selected model and graph structure consistently perform well across the evaluated sources, particularly with the use of GraphSAGE as the backbone for both textual and vision models. The results demonstrate strong performance in all five selected sources, with our approach exceeding the baseline in the classification of specific classes in four of the selected sources and achieving a higher overall accuracy in one of the cases.

5 Conclusions and Future Work

In this work, we introduced several experimental frameworks to evaluate the performance of different graph-based models and graph structures for the semantic classification of text blocks in digital-born PDF documents from the public affairs domain.

We represent each document page as a graph, where nodes correspond to layout objects and edges are defined using two alternative structures: a k-closest neighbor graph, where each node is connected to its k closest neighbors on the page, and a complete graph, where all nodes are connected to each other. Thus, each document is modeled as a sequence of graphs, and classification is performed at the text block level. To avoid manual feature engineering, we initialize node embeddings using pretrained models for both text and vision features.

We proposed and evaluated three experimental frameworks: one that uses text or vision features independently, a second that concatenates both types of feature in a single model, and a third that trains separate models for each modality and subsequently fuses their outputs for classification. Our experiments in the PAL dataset [26] demonstrate strong performance, particularly for the dual-branch framework with the GraphSAGE architecture, which outperforms the baseline in per-class accuracy for several sources and overall accuracy in one third of them, while maintaining consistency between different subsets of the data.

For future work, several research directions remain open. These include enhancing the representation of text and vision features, incorporating additional information such as relative position within the page (an approach that has proven effective in scanned document settings [8]) and exploring more advanced graph models to capture deeper structural relationships. Furthermore, the application of sequential Graph Neural Networks, which have already been used for spatiotemporal forecasting [22], or alternative graph structures, such as visibility graphs [31], could further improve the semantic classification of layout elements in digital-born PDF documents. Furthermore, we also plan to explore DLA in combination with Named Entity Recognition (NER) and Entity Linking (EL) technologies to enhance the information extracted from the document's text, a proposal that has already been applied to privacy-preserving text classification tasks [18,25] and form understanding [5]. Finally, addressing biases [28,33] and facilitating audit tools [24] in this research line on DLA/NER/EL is also part of our research agenda.

Acknowledgments. Supported by HumanCAIC (TED2021-131787B-I00 MICINN), BBforTAI (PID2021-127641OB-I00 MICINN/FEDER), M2RAI (PID2024-160053OB-I00 MICIU/FEDER), Cátedra ENIA UAM-VERIDAS en IA Responsable (NextGenerationEU PRTR TSI-100927-2023-2), and Research Agreement DGGC/UAM/FUAM for Biometrics and Applied AI. Morales is also supported by the Madrid Government in the line of Excellence for University Teaching Staff (V PRICIT). Work conducted within the ELLIS Unit Madrid.

Disclosure of Interests. The authors declare no conflict of interests.

References

1. Banerjee, A., Biswas, S., Lladós, J., Pal, U.: GraphKD: exploring Knowledge distillation towards document object detection with structured graph creation. In: Proceedings of ICDAR, pp. 354–373. Springer (2024)
2. Bast, H., Korzen, C.: A benchmark and evaluation for text extraction from PDF. In: ACM/IEEE Joint Conferenceon Digital Libraries (JCDL), pp. 1–10 (2017)
3. Brown, T., et al.: Language models are few-shot learners. Adv. Neural. Inf. Process. Syst. **33**, 1877–1901 (2020)

4. Da, C., Luo, C., Zheng, Q., Yao, C.: Vision grid transformer for document layout analysis. In: Proceedings of ICCV, pp. 19462–19472 (2023)
5. Dang, T.A.N., Hoang, D.T., Tran, Q.B., Pan, C.W., Nguyen, T.D.: End-to-end hierarchical relation extraction for generic form understanding. In: 2020 25th International Conference on Pattern Recognition (ICPR), pp. 5238–5245. IEEE (2021)
6. Du, J., Zhang, S., Wu, G., Moura, J.M., Kar, S.: Topology adaptive graph convolutional networks. arXiv preprint arXiv:1710.10370 (2017)
7. Fierrez, J., Morales, A., Vera-Rodriguez, R., Camacho, D.: Multiple classifiers in biometrics. Part 2: Trends and challenges. Inform. Fusion **44**, 103–112 (2018)
8. Gemelli, A., Biswas, S., Civitelli, E., Lladós, J., Marinai, S.: Doc2graph: a task agnostic document understanding framework based on graph neural networks. In: European Conference on Computer Vision, pp. 329–344. Springer (2022)
9. Gemelli, A., Vivoli, E., Marinai, S.: Graph neural networks and representation embedding for table extraction in PDF documents. In: 2022 26th International Conference on Pattern Recognition (ICPR), pp. 1719–1726. IEEE (2022)
10. Gu, Z., et al.: XYLayoutLM: Towards layout-aware multimodal networks for visually-rich document understanding. In: Proceedings of the IEEE/CVF Conference on Computer Vision and Pattern Recognition, pp. 4583–4592 (2022)
11. Gutiérrez Fandiño, A., et al.: MarIA: Spanish Language Models. Procesamiento del Lenguaje Natural **68** (2022)
12. Hamilton, W., Ying, Z., Leskovec, J.: Inductive representation learning on large graphs. In: Advances in Neural Information Processing Systems, vol. 30 (2017)
13. He, K., Zhang, X., Ren, S., Sun, J.: Deep residual learning for image recognition. In: Proceedings of CVPR, pp. 770–778 (2016)
14. ISO: Document management—Portable document format—Part 1: PDF 1.7. Standard (2008)
15. Josi, F., Wartena, C., Heid, U.: Preparing legal documents for NLP analysis: Improving the classification of text elements by using page features. In: Computer Science & Information Technology (CS & IT), pp. 17–29. AIRCC Publishing Corporation (2022)
16. Kenton, J.D.M.W.C., Toutanova, L.K.: BERT: Pre-training of deep bidirectional transformers for language understanding. In: Proceedings of NAACL-HLT. vol. 1. Minneapolis, Minnesota (2019)
17. Kipf, T.N., Welling, M.: Semi-supervised classification with graph convolutional networks. arXiv preprint arXiv:1609.02907 (2016)
18. Kutbi, M.: Named entity recognition utilized to enhance text classification while preserving privacy. IEEE Access **11**, 117576–117581 (2023)
19. Leus, G., Marques, A.G., Moura, J.M., Ortega, A., Shuman, D.I.: Graph signal processing: history, development, impact, and outlook. IEEE Signal Process. Mag. **40**(4), 49–60 (2023)
20. Li, S., Gou, G., Liu, C., Hou, C., Li, Z., Xiong, G.: TTAGN: temporal transaction aggregation graph network for Ethereum phishing scams detection. In: Proceedings of the ACM Web Conference 2022, pp. 661–669 (2022)
21. Li, X., Sun, L., Ling, M., Peng, Y.: A survey of graph neural network based recommendation in social networks. Neurocomputing **549**, 126441 (2023)
22. Liang, G., Tiwari, P., et al.: Dynamic causal explanation based diffusion-variational graph neural network for spatiotemporal forecasting. IEEE Trans. on Neural Networks and Learning Systems (2024)
23. Liu, Y., et al.: RoBERTa: A robustly optimized bert pretraining approach. arXiv preprint arXiv:1907.11692 (2019)

24. Mancera, G., DeAlcala, D., Fierrez, J., Tolosana, R., Morales, A.: Is my text in your AI model? Gradient-based membership inference test applied to LLMs. arXiv preprint arXiv:2503.07384 (2025)
25. Mancera, G., et al.: PBa-LLM: Privacy- and bias-aware NLP using named-entity recognition (NER). In: IAPR International Conference on Document Analysis and Recognition Workshops (ICDARw) (September 2025)
26. Peña, A., et al.: Continuous document layout analysis: human-in-the-loop AI-based data curation, database, and evaluation in the domain of public affairs. Inform. Fusion **108**, 102398 (2024)
27. Peña, A., et al.: Leveraging large language models for topic classification in the domain of public affairs. In: International Conference on Document Analysis and Recognition, pp. 20–33. Springer (2023)
28. Peña, A., Fierrez, J., Morales, A., Mancera, G., Lopez, M., Tolosana, R.: Addressing bias in LLMs: strategies and application to fair AI-based recruitment. In: AAAI/ACM Conference on AI, Ethics, and Society (AIES) (October 2025)
29. Peña, A., et al.: Human-centric multimodal machine learning: recent advances and testbed on AI-based recruitment. SN Comput. Sci. **4**(5), 434 (2023)
30. Qasim, S.R., Mahmood, H., Shafait, F.: Rethinking table recognition using graph neural networks. In: 2019 International Conference on Document Analysis and Recognition (ICDAR), pp. 142–147. IEEE (2019)
31. Riba, P., Goldmann, L., Terrades, O.R., Rusticus, D., Fornés, A., Lladós, J.: Table detection in business document images by message passing networks. Pattern Recogn. **127**, 108641 (2022)
32. Tang, Z., et al.: Unifying vision, text, and layout for universal document processing. In: Proceedings of the IEEE/CVF Conference on Computer Vision and Pattern Recognition, pp. 19254–19264 (2023)
33. Tello, J., et al.: Symbolic AI (LFIT) for XAI to handle biases. In: European Conference on AI Workshops (ECAIw). CEUR-WS, vol. 3523 (2023)
34. Veličković, P., Cucurull, G., Casanova, A., Romero, A., Liò, P., Bengio, Y.: Graph Attention Networks. In: ICLR (2018)
35. Waikhom, L., Patgiri, R.: A survey of graph neural networks in various learning paradigms: methods, applications, and challenges. Artif. Intell. Rev. **56**(7), 6295–6364 (2023)
36. Wu, L., et al.: Graph neural networks for natural language processing: a survey. Found. Trends® Mach. Learn. **16**(2), 119–328 (2023)
37. Xie, Q., Tiwari, P., Ananiadou, S.: Knowledge-enhanced graph topic transformer for explainable biomedical text summarization. IEEE J. Biomed. Health Inform. **28**(4), 1836–1847 (2023)
38. Zhang, N., et al.: M2Doc: a multi-modal fusion approach for document layout analysis. In: Proceedings of the AAAI Conference on Artificial Intelligence. vol. 38, pp. 7233–7241 (2024)
39. Zhang, Y., Yu, X., Cui, Z., Wu, S., Wen, Z., Wang, L.: Every document owns its structure: inductive text classification via graph neural networks. In: Proceedings of ACL, pp. 334–339 (2020)
40. Zhong, X., Tang, J., Yepes, A.J.: PubLayNet: largest dataset ever for document layout analysis. In: 2019 International Conference on Document Analysis and Recognition (ICDAR), pp. 1015–1022. IEEE (2019)

Improving Handwritten Text Recognition via 3D Attention and Multi-scale Training

Zi-Rui Wang(✉)

Chongqing University of Posts and Telecommunications, Chongqing, China
cs211@mail.ustc.edu.cn

Abstract. The segmentation-free research efforts for addressing handwritten text recognition can be divided into three categories: connectionist temporal classification (CTC), hidden Markov model and encoderdecoder methods. In this paper, inspired by the above three modeling methods, we propose a new recognition network by using a novel three-dimensional (3D) attention module and global-local context information. Based on the feature maps of the last convolutional layer, a series of 3D blocks with different resolutions are split. Then, these 3D blocks are fed into the 3D attention module to generate sequential visual features. Finally, by fusing the visual features and the corresponding global-local context features, a well-designed representation can be obtained. Main canonical neural units including attention mechanisms, fully-connected layers, recurrent units and convolutional layers are efficiently organized into a network and can be jointly trained by the CTC loss and the cross-entropy loss. Experiments on the latest Chinese handwritten text datasets (the SCUT-HCCDoc and the SCUT-EPT) and one English handwritten text dataset (the IAM) show that the proposed method can achieve comparable results with the state-of-the-art methods. The code is available at https://github.com/Wukong90/3DAttention-MultiScaleTraining-for-HTR.

Keywords: Handwritten text recognition · Segmentation-free recognition · 3D attention module · Multi-scale training · Global-local context information

1 Introduction

In text recognition tasks, there are three typical segmentation-free methods, i.e., hidden Markov model (HMM) [1,2], connectionist temporal classification (CTC) [3–8] and encoderdecoder (ED) framework [9–12]. As shown in Fig. 1, in the HMM-based method, each character is modeled by an HMM and a text line can be represented by cascaded HMMs. A series of frames extracted from an original image by a left-to-right sliding window are assigned to the underlying states. Then, a neural network is used to estimate the posterior probabilities of the states, while the outputs of networks in CTC and ED-based approaches are

L. Jin et al. (Eds.): ICDAR 2025 Workshops, LNCS 16225, pp. 36–52, 2026.
https://doi.org/10.1007/978-3-032-09368-4_3

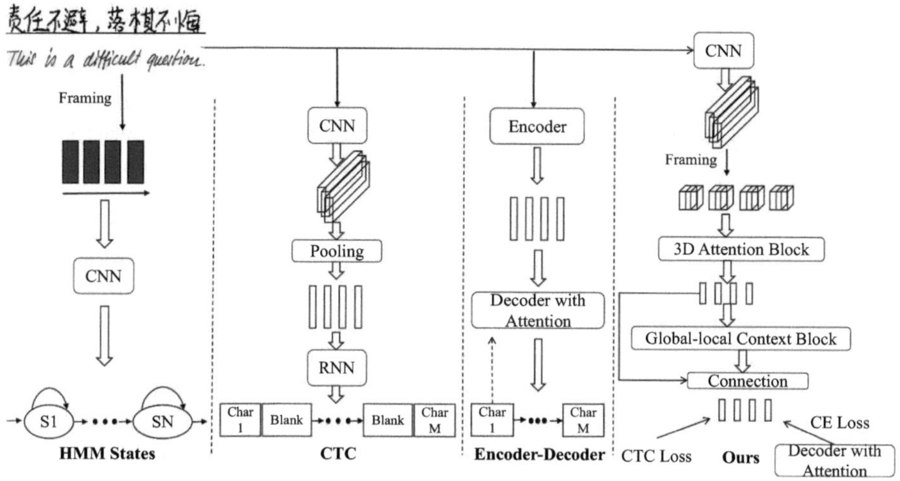

Fig. 1. The typical segmentation-free methods and the proposed network. The abbreviations CNN, RNN, Char, S denote convolutional neural network, recurrent neural network, character and state, respectively.

character classes. In the CTC loss, a special character "blank" and a sophisticated rule are designed to split different characters, and the forward-backward algorithm can efficiently compute the probability of the underlying character sequence. In an encoder-decoder network, an image is usually fed into the encoder to generate the corresponding middle features. Based on the middle representations, the decoder is used to locate and predict the character sequence via attention mechanisms. The common cross-entropy loss can be directly used to adjust the parameters of the ED network.

It is reasonable to explicitly model the 2D information of characters. However, it is very difficult to expand 1D HMM to 2D HMM [13] due to the computational complexity. Even for the CTC, ED-based approaches [6,10,14,15], early networks usually simply depend on the local receptive field of convolutional layers or recurrent units and gradually shrink the height of feature maps to small pixels via stacked pooling layers, which may lead to information loss. Although some attempts [11,16–18] explicitly or implicitly use 2D information in scene text recognition and character recognition, there is no obvious evidence showing that these methods can be directly used in long handwritten text recognition. Obviously, the context information is important for the text recognition. In [19–21], the visual features and the linguistic knowledge are simultaneously integrated into a network. Actually, based on the visual features extracted from a certain image, the corresponding context information can also be naturally obtained by using recurrent units [5,8,16,22] or transformers [17,23,24].

For the handwritten text recognition task, recently, Hoang et al. [7] combined the radical-level CTC loss and the character-level loss while Ngo et al. [12] constructed a joint decoder to integrate the visual feature and the linguistic context feature. More recently, Peng et al. [25] built a full convolution network to simultaneously achieve the purpose of the character location, the character bounding boxes prediction and the character prediction. Lin et al. [26] propose a mobile text recognizer via searching the lightweight neural units. Furthermore, Peng et al. [27] detect and recognize characters in page-level handwritten text while Coquenet et al. [23] use a fully convolutional network as the encoder with a transformer decoder for document recognition.

In this paper, inspired by the above three modeling ways, we propose a new recognition network by using a novel 3D attention module and global-local context information. The 3D attention module is employed to explicitly extract 2D information of block features with different resolutions. In detail, the 3D attention module is ingeniously decoupled into a 2D self-attention operation and a 1D attention-based aggregation operation. Based on the outputs (visual features) of the 3D attention module, we further extract the corresponding global-local context information via the self-attention and the recurrent unit, respectively. Finally, by fusing the visual features and the corresponding global-local context features, a well-designed representation for the recognition task can be obtained. In summary, the main contributions of this paper are as follows:

1. Inspired by the typical segmentation-free approaches, we improve the text recognition network by using a novel 3D attention module and global-local context information.
2. Main canonical neural units including attention mechanisms, fully-connected layers, recurrent units and convolutional layers are ingeniously organized into a network.
3. We propose the multi-scale training approach that includes extracting 3D blocks with different resolutions and simultaneously adopting the CTC and the CE losses.
4. Compared with the state-of-the-art methods, the proposed network can achieve comparable results on all datasets. We conduct a comprehensive analysis to verify the effects of the 3D attention module and the different features.

The remainder of this paper is organized as follows: Sect. 2 elaborates on the details of the proposed method. Section 3 reports the experimental results and analysis. Finally, we discuss the advantages of the proposed method and conclude the paper.

2 Methodology

As shown in Fig. 2, the proposed network includes three parts, i.e., the convolutional neural network (CNN), the 3D attention module and the features fusion block. In the CNN, the hybrid attention module (HAM) [8] is used and each convolutional layer is equipped with the batch normalization [28]. For the outputs of the CNN, we use a multi-scale framing strategy to extract different solutions. In this section, we elaborate on the details of the 3D attention module and the global-local context information. Moreover, the multi-scale training pipeline is also shown.

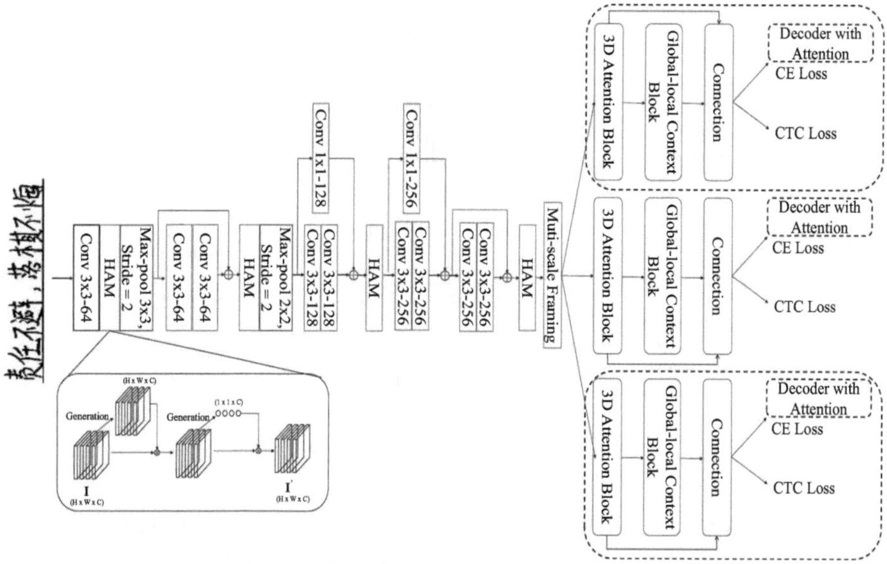

Fig. 2. The parts within the dashed lines are only used during the training stage. The details of the 3D attention, global-local context, decoder with attention, CE loss and CTC loss are illustrated in the following sections.

2.1 3D Attention

As shown in Fig. 3, assuming the front CNN has the output tensor $\mathbf{O} \in R^{C \times W \times H}$ with each feature map $\mathbf{O}_c \in R^{W \times H}$, each 3D block represented by a sliding window $\mathbf{f} \in R^{C \times S \times H}$ from left to right, with the window shift of S pixels, is scanned across the feature maps. Firstly, each location p in the plane $S \times H$ is added the corresponding positional encoding:

$$\mathbf{f}_p[0 : C/2] = \mathbf{f}_p[0 : C/2] + \mathbf{p}_{\mathrm{w}} \tag{1}$$

$$\mathbf{f}_p[C/2\text{: }(\mathrm{C}-1)] = \mathbf{f}_p[C/2\text{: }(\mathrm{C}-1)] + \mathbf{p}_{\mathrm{h}} \tag{2}$$

where $\mathbf{p}_{\mathrm{w}}$ and $\mathbf{p}_{\mathrm{h}}$ are sinusoidal positional encoding over height and width, respectively, as defined in [29]:

$$\mathbf{p}_{p,2i} = \sin(pe^{2i(-\ln(10000)/C)}) \tag{3}$$

$$\mathbf{p}_{p,2i+1} = \cos(pe^{2i(-\ln(10000)/C)}) \tag{4}$$

here i is the index along hidden dimensions. And then, we reshape each 3D block from $C \times S \times H$ to $SH \times C$ and conduct a self-attention operation for these sequential vectors $\mathbf{f}_p$ as Eq. 5:

$$\mathbf{F}^{'} = \mathrm{softmax}(\frac{\mathbf{F}\mathbf{F}^{T}}{s})\mathbf{F} \tag{5}$$

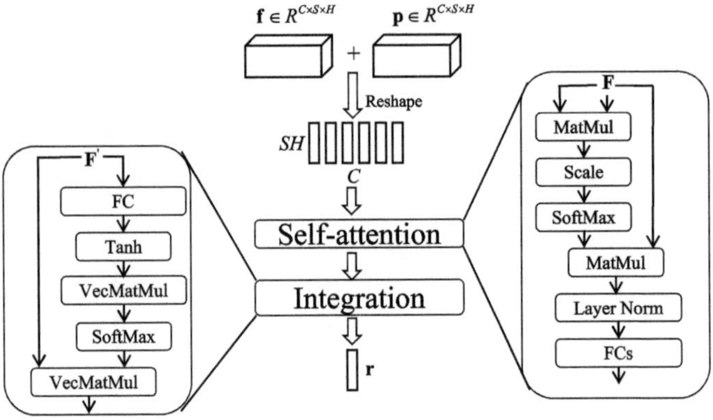

Fig. 3. The proposed 3D attention block.

The matrices $\mathbf{F}$, $\mathbf{F}^{'}$ pack all vectors $\mathbf{f}_p$ and the corresponding $\mathbf{f}_p^{'}$,respectively. Finally, a representation vector $\mathbf{r}$ for all $\mathbf{f}_p^{'}$ can be obtained:

$$e_p = \mathbf{w}^{\mathrm{T}} \tanh\left(\mathbf{W}\mathbf{f}_p^{'} + \mathbf{b}\right) \tag{6}$$

$$\alpha_p = \frac{e_p}{\sum_i e_i} \tag{7}$$

$$\mathbf{r} = \sum_p \alpha_p \mathbf{f}_p^{'} \tag{8}$$

Similar to the transformer encoder, the vectors $\mathbf{f}_p^{'}$ are fed into a layer normalization (Layer Norm) followed by two fully connected layers (FCs) before computing Eq. 6–8.

2.2 Features Fusion

Through the 3D attention module, a series of 3D feature maps can be transformed into the corresponding visual features $\mathbf{r}_t(t = 0...T)$. As shown in Fig. 4-(a), for the visual feature $\mathbf{r}_t$, the global context feature $\mathbf{l}_t$ and the local context $\mathbf{s}_t$ can be extracted by using the self-attention mechanism and the recurrent units, respectively. We directly use the computation method of the $\mathbf{f}^{'}$ in Fig. 3 to obtain the global context information $\mathbf{l}_t$. For the local context features, the standard LSTM (Fig. 4-(b)) is used. $\mathbf{r}_t$ represents the input at time t and $\mathbf{s}_t$ is the corresponding output, $\mathbf{i}$, $\mathbf{g}$, $\mathbf{o}$ and $\mathbf{c}$ are the outputs of input gate, forget gate, output gate and cell vectors, respectively. Finally, the visual feature $\mathbf{r}_t$, the global feature $\mathbf{l}_t$ and the local feature $\mathbf{s}_t$ are simply contacted together:

$$\mathbf{v}_t = \mathbf{r}_t \oplus \mathbf{l}_t \oplus \mathbf{s}_t \tag{9}$$

The well-defined representations $\mathbf{v}$ can be directly fed into a classification layer and also be used as the input of the decoder during the joint training of the CTC and the CE losses.

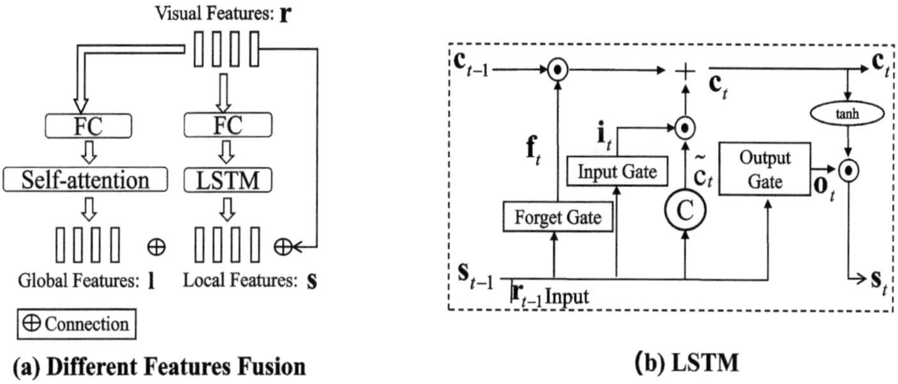

Fig. 4. The different features (visual features, global-local features) are combined before the classification layer and the details of the LSTM are illustrated.

2.3 Multi-scale Training

The proposed multi-scale training includes two parts. Firstly, as shown in Fig. 2, in the training stage, multiple feature sequences with different frame lengths are extracted. Although parallel branches including the same 3D attention block and global-local context block are used, we only retained the corresponding branch of the frame length 3 during the inference stage in our experiments.

Secondly, we use the joint training of the CTC and the CE losses. Given the feature sequence $\mathbf{V}$ of a text line image, the text recognition task is to find the corresponding underlying n-character sequence $\mathbf{C} = \{C_1, C_2, ..., C_n\}$, i.e., compute the posterior probability $p(\mathbf{C}|\mathbf{V})$. The CTC can be regarded as a loss function of neural networks:

$$L_{\mathrm{CTC}}(\mathbf{\Theta}) = -\log(\sum_{\boldsymbol{\pi}:\varphi(\boldsymbol{\pi})=\mathbf{C}} p(\boldsymbol{\pi}|\mathbf{V}, \mathbf{\Theta})) \tag{10}$$

where $\mathbf{\Theta}$ represents the parameters of the recognition network and $\boldsymbol{\pi}$ is the predicted character sequence under the constraint of $\varphi(\boldsymbol{\pi}) = \mathbf{C}$. The function φ only retains one of the consecutive adjacent repeating characters and removes the 'blank' characters. The CTC loss (Eq. 10) can be efficiently computed by using the forward-backward algorithm [3]. For the auxiliary decoder in Fig. 5, the CE loss directly predicts the probability $p(C_t|\mathbf{x}_t)$:

$$L_{\mathrm{CE}}(\mathbf{\Theta}, \mathbf{\Gamma}) = \prod_t p(C_t|\mathbf{x}_t) \tag{11}$$

$\mathbf{\Gamma}$ is the parameters of the decoder. Based on the feature sequence $\mathbf{V}$, the vector $\mathbf{x}_t$ at the time step t can be computed using the following attention method:

$$\mathbf{q}_t = \tanh(\mathbf{W}_{\mathrm{q}}\mathbf{h}_{t-1} + \mathbf{b}_{\mathrm{q}}) \tag{12}$$

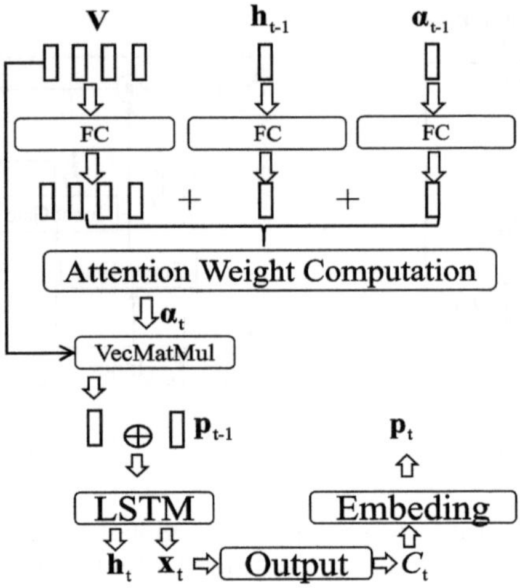

Fig. 5. The decoder with attention. The Attention Weight Computation module is similar to the Integration module in Fig. 3.

$$\mathbf{k}_i = \tanh(\mathbf{W}_\mathrm{k}\mathbf{v}_i + \mathbf{b}_\mathrm{k}) \tag{13}$$

$$\mathbf{a}_t = \tanh(\mathbf{W}_\mathrm{a}\boldsymbol{\alpha}_{t-1}) \tag{14}$$

$$e_{t,i} = \mathbf{w}^\mathrm{T}\tanh(\mathbf{q}_t + \mathbf{k}_i + \mathbf{a}_t) \tag{15}$$

$$\alpha_{t,i} = \frac{e_{t,i}}{\sum\limits_i e_{t,i}} \tag{16}$$

$$\mathbf{x}_t = \mathrm{LSTM}((\sum_i \alpha_{t,i}\mathbf{v}_i) \oplus \mathbf{p}_{t-1}) \tag{17}$$

where $\mathbf{h}_{t-1}$,$\mathbf{p}_{t-1}$are the hidden state of the LSTM and the decoded embedding vector at the time step $t-1$, respectively.

Finally, Algorithm1 describes the joint training of the CTC and the CE Losses.

3 Experiments

3.1 Datasets and Evaluation Metrics

The proposed network is validated on two Chinese handwritten text datasets(the SCUT-HCCDoc [31] and the SCUT-EPT [32]) and one English handwritten text dataset: the IAM [33].

All the images of SCUT-HCCDoc were obtained by Internet search. According to certain rules, only 12,253 images were preserved from the initial 100 thousand candidate images. These images were randomly split into the training and test sets with a ratio of 4:1. After splitting, the training set contains 9,801

Algorithm 1. The joint training pipeline of the CTC and the CE losses.

Require:

The randomly initialized parameter sets $\{\boldsymbol{\Theta}, \boldsymbol{\Gamma}\}$;
The loss functions L_{CTC}, L_{CE} and the coressponding weight coefficients λ_1, λ_2 in the training stage of the main recognition network (MRN($\boldsymbol{\Theta}$)), and the auxiliary decoder (Dec($\boldsymbol{\Gamma}$)), respectively.

1: Optimize the MRN parameter set $\boldsymbol{\Theta}$ by using the Adam algorithm [30].
$\boldsymbol{\Theta} = \mathrm{Adam}(\boldsymbol{\Theta}, \mathrm{L_{CTC}})$

2: Jointly train the parameter sets $\boldsymbol{\Theta}, \boldsymbol{\Gamma}$ based on the CTC and the CE losses.
$L = \lambda_1 L_{\mathrm{CTC}} + \lambda_2 L_{\mathrm{CE}}$
$\boldsymbol{\Theta}, \boldsymbol{\Gamma} = \mathrm{Adam}(\boldsymbol{\Theta}, \boldsymbol{\Gamma}, \mathrm{L})$

3: **return** The MRN($\boldsymbol{\Theta}$).

Table 1. The number of text lines, characters and classes in the SCUT-HCCDoc and the SCUT-EPT.

Type	SCUT-HCCDoc		SCUT-EPT	
	Training set	Test set	Training set	Test set
Text lines	93,254	23,484	40,000	10,000
Characters	925,200	230,019	1,018,432	248,730
Classes	5,922	4,435	4,058	3,236

images with 93,254 text instances and 925,200 characters. The test set contains 2,452 images with 23,484 text instances and 230,019 characters.

The Chinese handwritten text data set SCUT-EPT contains 4,250 categories of Chinese characters and symbols, totaling 1,267,162 characters, and it was split into 40,000 training text lines and 10,000 test text lines. These samples were collected from examination papers and were written by 2,986 students, and the training and test sets do not include the same writers. There are a total of 4,250 classes, but the number of classes in the training set is just 4,058. Therefore, some classes in the testing set do not appear in the training set and can not be correctly recognized.

Table 2. The standard partition of the IAM dataset.

Set name	Text lines	Writers
Train	6,161	283
Validation1	900	46
Validation2	940	43
Test	1,861	128
Total	9,862	500

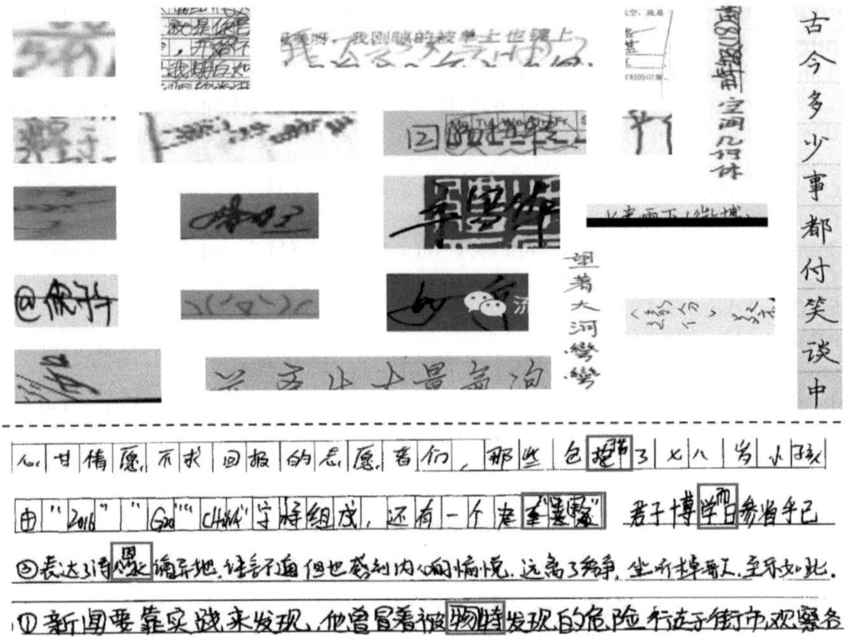

Fig. 6. Some removed images in the SCUT-HCCDoc dataset are shown above the dashed line. They are very low-quality (illegible text, random graffiti, high overlap of text and background, and incomplete characters) or vertical writing text. The lower part of the dashed line shows typical removed images in the SCUT-EPT dataset. These images include abnormal structures, i.e., characters swapping and overlap, and more details can be found in [34].

For the English handwritten text recognition, we evaluate the performance of our method on the IAM dataset. The IAM dataset contains a total of 9,862 text lines written by 500 writers. It provides one training set, one test set and two validation sets. The text lines of all data sets are mutually exclusive, thus, each writer has contributed to one set only (Fig. 6).

Table 1 lists the number of text lines, characters and classes in both Chinese datasets. In the SCUT-EPT dataset[1], we removed 681 training text images including abnormal structures, i.e., characters swapping and overlap. Similarly, in the SCUT-HCCDoc dataset[2], some very low-quality (illegible text, random graffiti, high overlap of text and background, and incomplete characters) or vertical writing text images were removed and only 91,261 text instances were used

[1] https://github.com/HCIILAB/SCUT-EPT_Dataset_Release. Prof. Lianwen Jin (eelwjin@scut.edu.cn) at South China University of Technology (SCUT) is the owner of this dataset.

[2] https://github.com/HCIILAB/SCUT-HCCDoc_Dataset_Release. Prof. Lianwen Jin (eelwjin@scut.edu.cn) at SCUT is the owner of this dataset. Thanks to his team for building two useful datasets.

in the training set. All test images in both datasets were evaluated. Table 2 lists the standard partition of the English dataset[3]. All training images, validation images and test images were used in our experiments and the best model was selected for testing according to the results on the validation sets. The character accuracy rate (AR) was used to evaluate the performance of the text recognition model. The evaluation criterion is defined as follows:

$$1 - \frac{N_{\mathrm{s}} + N_{\mathrm{i}} + N_{\mathrm{d}}}{N} \tag{18}$$

where N is the total number of samples in the evaluation set. N_{s}, N_{i} and N_{d} denote the number of substitution errors, insertion errors and deletion errors, respectively.

3.2 Experimental Results and Analysis

During the training stage, common data argument methods including text location shift, image blur, grey level and contrast changes, and linear transformation were used. The Adam optimization algorithm with default parameters was adopted, we adjusted the learning rate according to training steps until the model converged into a small range. At certain epochs, the learning rate was multiplied by 0.5. The deep learning platform Pytorch [35] and an NVIDIA RTX A6000 with 48GB memory were used. Firstly, we conducted ablation experiments to verify the effectiveness of the proposed network.

Ablation Experiments. In this section, we examine several important factors for recognition performance. These factors include the frame resolution, the 3D attention module, the global-local context information and the multi-scale training.

We only compare the results of different frame lengths and all sliding windows have the same height. Table 3 shows the different results of different frame resolutions. We can observe that frame length 3 is the optimal configuration for all experiments, which means the corresponding window size is suitable for most characters. For a longer or shorter feature sequence, it is difficult to discriminate different regions of characters. For example, when we used a small frame length, the Chinese recognition results obviously decreased from 88.8% to 88.06%, 76.67% to 75.13%, respectively, while the English recognition performance only changed from 94.17% to 94.11%, which may be owing to a smaller width for most English characters. Furthermore, we can obtain consistent improvements by integrating multiple resolutions, i.e., the ARs achieve 89.16%, 76.96% and 94.37%, respectively. As shown in Fig. 2, in the training stage, although three 3D attention blocks and global-local context blocks were used in parallel, we only retained the corresponding branch of the frame length 3 during the inference stage.

[3] https://fki.tic.heia-fr.ch/databases/iam-handwriting-database. Accessed 2021.

Table 3. The different results of different frame resolutions.

Frame length	SCUT-HCCDoc	SCUT-EPT	IAM
2	88.06	75.13	94.11
3	88.80	76.67	94.17
4	87.21	76.36	93.62
Multi-scale	**89.16**	**76.96**	**94.37**

Based on the optimal frame length and single-scale training, Table 4 lists the results with/without the 3D attention module and the global-local context information. If we did not adopt the 3D attention module in experiments, a pooling operation was directly used to reduce the height of the last convolutional feature maps to 1. The 3D attention module can steadily improve the Chinese recognition performance. On the SCUT-HCCDoc data set, the AR is improved from 88.51% to 88.8%. On the SCUT-EPT data set, the AR increases from 75.00% to 76.67%. Although it seems that the English recognition without the 3D attention module has a higher AR (94.31% vs. 94.17%), the network based on the multiple-resolutions training still obtains the best AR. As the conclusions demonstrated by many previous researches, there exists an obvious gap with/without using the context information for the sequence task. The results dramatically decrease from 88.8% to 85.36%, 76.67% to 72.28% and 94.17% to 87.0%, respectively. It is interesting to observe that the pure CNN-based models without the 3D attention block and the context information can also achieve 85.76%, 72.53% and 85.93%, respectively.

Table 4. The recognition performance with/without different blocks.

3D attention module	Global-local context information	SCUT-HCCDoc	SCUT-EPT	IAM
✓	✓	**88.80**	**76.67**	94.17
✓	×	85.36	72.28	87.00
×	✓	88.51	75.00	**94.31**
×	×	85.76	72.53	85.93

In Table 5, we compare the results of different training strategies, i.e., the weights are only optimized by the CTC loss or jointly trained by the CTC loss and the CE loss. Although the Chinese decoding results of the auxiliary network are not good, the main recognition network still benefits from the joint training. The best network can achieve 89.34%, 77.15% and 94.41%, respectively. Moreover, we also list the storage and the running time comparisons of different methods. In order to make a fair comparison, all experiments were evaluated on the same machine and the experimental configurations on the same dataset were consistent. In Table 5, the decoding time of the CTC is defined as 1. Generally speaking, considering the auxiliary decoder with attention needed and the cyclic decoding way used during the inference stage, it is reasonable that the ED method needs more time consumption and storage. As shown in Fig 2, for the joint training strategy of CTC and CE, only the faster CTC method is used in the test stage, i.e., it has the same time consumption with the pure CTC training strategy.

Table 5. The comparisons of accuracy rate, storage and relative speed.

Training strategy	SCUT-HCCDoc			SCUT-EPT			IAM		
	AR	Speed	Storage	AR	Speed	Storage	AR	Speed	Storage
CTC	89.16	1.0	41.92MB	76.67	1.0	35.62MB	94.37	1.0	22.04MB
CE	85.53	3.10	46.89MB	72.58	6.24	41.50MB	93.94	2.63	30.15MB
CTC + CE	**89.34**	1.0	41.92MB	**77.15**	1.0	35.62MB	**94.41**	1.0	22.04MB

Table 6. Performance comparison of our proposed method and other state-of-the-art methods. For the English handwritten recognition task, all listed results do not use any lexicons.

Dataset	Method	Without external data	With external data
SCUT-HCCDoc	Zhang et al. [31]	78.65	79.84
	Peng et al. [25]	-	**90.85**
	Lin et al. [26]	-	88.10
	Ours	**89.34**	-
SCUT-EPT	Zhu et al. [32]	75.37	75.97
	Hoang et al. [7]	76.61	**77.61**
	Ngo et al. [12]	76.85	-
	Ours	**77.15**	-
IAM	Dutta et al. [6]	-	94.3
	Chowdhury et al. [10]	91.9	-
	Ours	**94.41**	-

Overall Comparison. Finally, Table 6 shows an overall comparison of our proposed method and other state-of-the-art methods on different test sets. Without using external data, our proposed network can make a new milestone. Compared with the best networks trained by using external data, the proposed network can also achieve comparable results on all datasets. For the listed results on the IAM dataset, the standard partition and no word lexicons were used. Please note that there exists an unofficial partition (RWTH partition) of the IAM dataset. It is widely used by many researchers, their results are not compared in this paper.

Visualization Analysis. In Fig. 7, Fig. 8 and Fig. 9, the attention weights in the 3D block are shown on the original images. It is interesting to observe that even for long-text images and some small symbols, most attention weights (red parts) are still located on the handwriting characters, which demonstrates the proposed attention mechanism is reasonable.

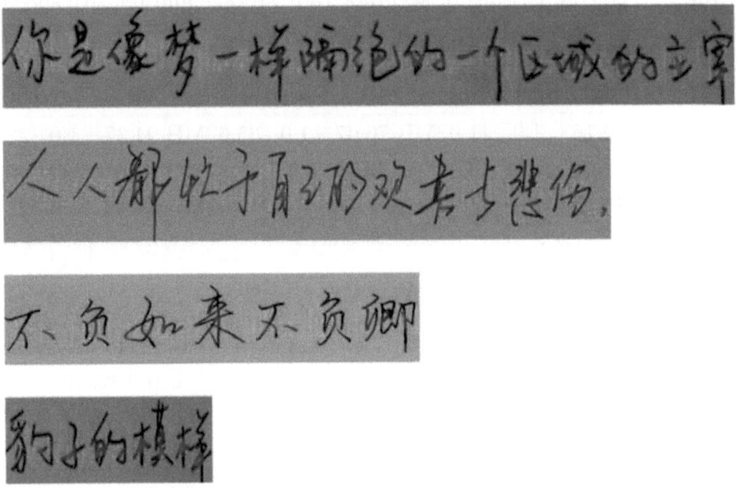

Fig. 7. The typical samples of the attention visualization on the SCUT-HCCDoc images [31].

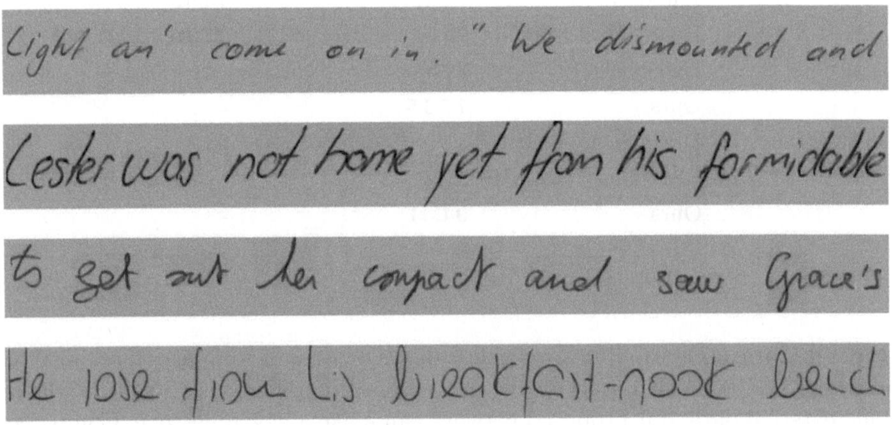

Fig. 8. The visualization of the 3D attention weights on the IAM images [33].

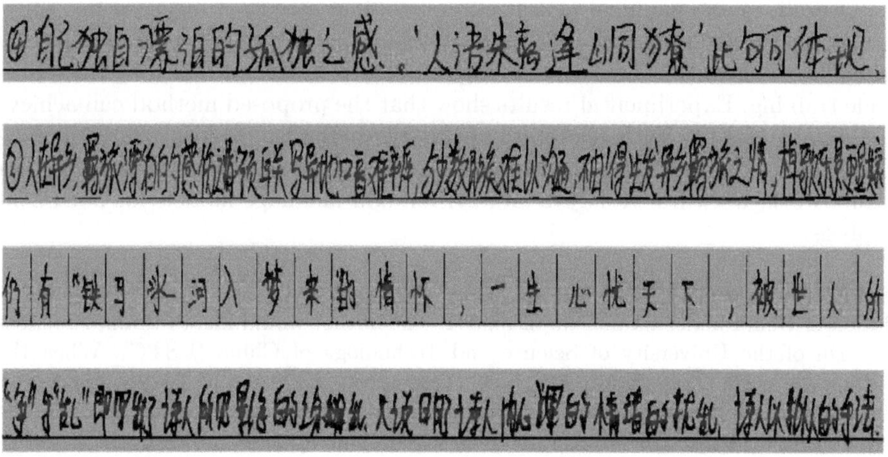

Fig. 9. The typical samples of the attention visualization on the SCUT-EPT images [32].

4 Discussion and Conclusion

From the above experiments, we can observe that the proposed network has achieved good results on multiple handwritten text datasets. One important factor is that the proposed network can absorb the advantages of three typical segmentation-free sequence recognition algorithms. Compared with the HMM, which relies on artificially designed features and the generative model (GMM-HMM) for character modeling, we do not need to perform complex character modeling and inference. However, we have integrated the state-level feature extraction method based on sliding windows in HMM into the intermediate process of the network, and ingeniously extracted multi-scale block features for the training of the network, which potentially increases the training data and enables the network to better handle characters of different sizes. Compared with the encoder-decoder method based on the attention mechanism, which is prone to attention drift when applied to long handwritten text recognition, the proposed 3D attention mechanism is conducted on the sliding window to generate better local visual features. Whether from intuition or from our experimental results, the proposed 3D attention can better take into account the 2D natural properties of characters. Without too many tricks, we use the most popular neural units to implement the 3D attention. The extracted sequence visual features are used to obtain global and local context features through the self-attention mechanism and the LSTMs, respectively. Obviously, the representation after combining visual features and contextual features is suitable for the text recognition task. Finally, by cascading CTC loss and CE loss, the whole network can be optimized well. As shown in Fig. 2, in this paper, during the test stage, we only use one of the network branches and only use the faster CTC method, without the cyclic decoding way of common ED methods.

Overall, main canonical neural units including attention mechanisms, fully-connected layers, recurrent units and convolutional layers are ingeniously organized into a network. The network weights are effectively optimized by the multi-scale training. Experimental results show that the proposed method can achieve comparable results with the state-of-the-art methods. Besides, the visualization analysis demonstrates the proposed attention mechanism is reasonable. For future work, we will investigate large vision and language models for document analysis.

Acknowledgments. This work was supported by the National Natural Science Foundation of China under Grant No. 62106031. The author would like to thank Professor Jun Du of the University of Science and Technology of China (USTC). When the author was a Ph.D. candidate at USTC (September 2015 to June 2020), we successfully applied the HMM and the deep neural network to the text recognition task. The supervisor (Prof. Jun Du) always said, can we combine three mainly segmentation-free methods? Although the HMM is not explicitly used in this paper, the question inspired the author to delve into this problem while self-isolated at home. The author also would like to thank Yan Wang at USTC for applying the code of extracting text lines from original SCUT-HCCDoc documents.

References

1. Espana-Boquera, S., Castro-Bleda, M.J., Gorbe-Moya, J., Zamora-Martinez, F.: Improving offline handwritten text recognition with hybrid HMM/ANN models. IEEE Trans. Pattern Anal. Mach. Intell. **33**(4), 767–779 (2010)
2. Wang, Z.-R., Jun, D., Wang, J.-M.: Writer-aware CNN for parsimonious hmm-based offline handwritten Chinese text recognition. Pattern Recogn. **100**, 107102 (2020)
3. Graves, A., Fernández, S., Gomez, F., Schmidhuber, J.: Connectionist temporal classification: labelling unsegmented sequence data with recurrent neural networks. In: Proceedings of the 23rd International Conference on Machine Learning, pp. 369–376 (2006)
4. Messina, R., Louradour, J.: Segmentation-free handwritten Chinese text recognition with LSTM-RNN. In: 2015 13th International Conference on Document Analysis and Recognition (ICDAR), pp. 171–175. IEEE (2015)
5. Shi, B., Bai, X., Yao, C.: An end-to-end trainable neural network for image-based sequence recognition and its application to scene text recognition. IEEE Trans. Pattern Anal. Mach. Intell. **39**(11), 2298–2304 (2016)
6. Dutta, K., Krishnan, P., Mathew, M., Jawahar, C.V.: Improving CNN-RNN hybrid networks for handwriting recognition. In: 2018 16th International Conference on Frontiers in Handwriting Recognition (ICFHR), pp. 80–85. IEEE (2018)
7. Hoang, H.T., Peng, C.J., Tran, H.V., Le, H., Nguyen, H.H.: LodeNet: a holistic approach to offline handwritten Chinese and Japanese text line recognition. In: 2020 25th International Conference on Pattern Recognition (ICPR), pp, 4813–4820. IEEE (2021)
8. Wang, Z.-R., Jun, D.: Fast writer adaptation with style extractor network for handwritten text recognition. Neural Netw. **147**, 42–52 (2022)

9. Bahdanau, D., Cho, K., Bengio, Y.: Neural machine translation by jointly learning to align and translate. arXiv preprint arXiv:1409.0473 (2014)
10. Chowdhury, A., Vig, L.: An efficient end-to-end neural model for handwritten text recognition. arXiv preprint arXiv:1807.07965 (2018)
11. Zhang, J., Jun, D., Dai, L.: Radical analysis network for learning hierarchies of Chinese characters. Pattern Recogn. **103**, 107305 (2020)
12. Ngo, T.T., Nguyen, H.T., Ly, N.T., Nakagawa, M.: Recurrent neural network transducer for Japanese and Chinese offline handwritten text recognition. In: Barney Smith, E.H., Pal, U. (eds.) ICDAR 2021. LNCS, vol. 12917, pp. 364–376. Springer, Cham (2021). https://doi.org/10.1007/978-3-030-86159-9_26
13. Ma, J., Wang, Z., Du, J.: An open-source library of 2D-GMM-HMM based on Kaldi Toolkit and its application to handwritten Chinese character recognition. In: Peng, Y., Hu, S.-M., Gabbouj, M., Zhou, K., Elad, M., Xu, K. (eds.) ICIG 2021. LNCS, vol. 12888, pp. 235–244. Springer, Cham (2021). https://doi.org/10.1007/978-3-030-87355-4_20
14. Graves, A., Schmidhuber, J.: Offline handwriting recognition with multidimensional recurrent neural networks. In: Advances in Neural Information Processing Systems, vol. 21 (2008)
15. Puigcerver, J.: Are multidimensional recurrent layers really necessary for handwritten text recognition? In: 2017 14th IAPR International Conference on Document Analysis and Recognition (ICDAR), vol. 1, pp. 67–72. IEEE (2017)
16. Cheng, Z., Bai, F., Xu, Y., Zheng, G., Pu, S., Zhou, S.: Focusing attention: towards accurate text recognition in natural images. In: Proceedings of the IEEE International Conference on Computer Vision, pp. 5076–5084 (2017)
17. Lee, J., Park, S., Baek, J., Oh, S.J., Kim, S., Lee, H.: On recognizing texts of arbitrary shapes with 2d self-attention. In: Proceedings of the IEEE/CVF Conference on Computer Vision and Pattern Recognition Workshops, pp. 546–547 (2020)
18. Yu, H., Wang, X., Li, B., Xue, X.: Chinese text recognition with a pre-trained clip-like model through image-ids aligning. In: Proceedings of the IEEE/CVF International Conference on Computer Vision, pp. 11943–11952 (2023)
19. Qiao, Z., Zhou, Y., Yang, D., Zhou, Y., Wang, W.: Seed: Semantics enhanced encoder-decoder framework for scene text recognition. In: Proceedings of the IEEE/CVF Conference on Computer Vision and Pattern Recognition, pp. 13528–13537 (2020)
20. Wang, Y., Xie, H., Fang, S., Wang, J., Zhu, S., Zhang, Y.: From two to one: a new scene text recognizer with visual language modeling network. In: Proceedings of the IEEE/CVF International Conference on Computer Vision, pp. 14194–14203 (2021)
21. Fang, S., Xie, H., Wang, Y., Mao, Z., Zhang, Y.: Read like humans: autonomous, bidirectional and iterative language modeling for scene text recognition. In: Proceedings of the IEEE/CVF Conference on Computer Vision and Pattern Recognition, pp. 7098–7107 (2021)
22. Shi, B., Yang, M., Wang, X., Lyu, P., Yao, C., Bai, X.: Aster: an attentional scene text recognizer with flexible rectification. IEEE Trans. Pattern Anal. Mach. Intell. **41**(9), 2035–2048 (2018)
23. Coquenet, D., Chatelain, C., Paquet, T.: Dan: a segmentation-free document attention network for handwritten document recognition. IEEE Trans. Pattern Anal. Mach. Intell. **45**(7), 8227–8243 (2023)
24. Tan, Y.L., Kong, A.WK., Kim, J.J.: Pure transformer with integrated experts for scene text recognition. In: European Conference on Computer Vision, pp. 481–497. Springer (2022). https://doi.org/10.1007/978-3-031-19815-1_28

25. Peng, D., et al.: Recognition of handwritten Chinese text by segmentation: a segment-annotation-free approach. IEEE Trans. Multimedia **25**, 2368–2381 (2022)
26. Lin, W., et al.: Building a mobile text recognizer via truncated SVD-based knowledge distillation-guided NAS. In: BMVC, p. 375 (2023)
27. Peng, D., Jin, L., Liu, Y., Luo, C., Lai, S.: PageNet: towards end-to-end weakly supervised page-level handwritten Chinese text recognition. Int. J. Comput. Vis. **130**(11), 2623–2645 (2022)
28. Ioffe, S., Szegedy, C.: Batch normalization: accelerating deep network training by reducing internal covariate shift. In: International Conference on Machine Learning, pp. 448–456. PMLR (2015)
29. Vaswani, A., et al.: Attention is all you need. In: Advances in Neural Information Processing Systems, vol. 30 (2017)
30. Kinga, D., Adam, J.B.: Adam: a method for stochastic optimization. arXiv preprint arXiv:1412.6980 (2014)
31. Zhang, H., Liang, L., Jin, L.: SCUT-HCCDOC: a new benchmark dataset of handwritten Chinese text in unconstrained camera-captured documents. Pattern Recogn. **108**, 107559 (2020)
32. Zhu, Y., Xie, Z., Jin, L., Chen, X., Huang, Y., Zhang, M.: SCUT-EPT: new dataset and benchmark for offline Chinese text recognition in examination paper. IEEE Access **7**, 370–382 (2018)
33. U-V Marti and Horst Bunke: The IAM-database: an English sentence database for offline handwriting recognition. Int. J. Doc. Anal. Recogn. **5**, 39–46 (2002)
34. Wang, Z.-R.: Leveraging structure knowledge and deep models for the detection of abnormal handwritten text. In: Lin, Z., et al. (eds.) Chinese Conference on Pattern Recognition and Computer Vision (PRCV), pp. 187–200. Springer (2024). https://doi.org/10.1007/978-981-97-8511-7_14
35. Paszke, A., et al.: PyTorch: an imperative style, high-performance deep learning library. In: Advances in Neural Information Processing Systems, vol. 32 (2019)

Masked Self-supervised Pre-training for Text Recognition Transformers on Large-Scale Datasets

Martin Kišš(✉) and Michal Hradiš

Faculty of Information Technology, Brno University of Technology, Brno, Czech Republic
{ikiss,hradis}@fit.vutbr.cz

Abstract. Self-supervised learning has emerged as a powerful approach for leveraging large-scale unlabeled data to improve model performance in various domains. In this paper, we explore masked self-supervised pre-training for text recognition transformers. Specifically, we propose two modifications to the pre-training phase: progressively increasing the masking probability, and modifying the loss function to incorporate both masked and non-masked patches. We conduct extensive experiments using a dataset of 50M unlabeled text lines for pre-training and four differently sized annotated datasets for fine-tuning. Furthermore, we compare our pre-trained models against those trained with transfer learning, demonstrating the effectiveness of the self-supervised pre-training. In particular, pre-training consistently improves the character error rate of models, in some cases up to 30 % relatively. It is also on par with transfer learning but without relying on extra annotated text lines.

Keywords: Self-supervised pre-training · Transformers · OCR · HTR

1 Introduction

Large models have become fundamental in machine learning, with notable examples including large language models [11,23,36], the ASR model Whisper [41], and generative models like Stable Diffusion [19,42] and DALLE-3 [10]. Such large models require huge amounts of training data to generalize well. In text recognition, and especially in handwritten text recognition, acquiring large datasets containing millions of text lines is costly and time-consuming. On the other hand, there are machine learning paradigms that utilize unlabeled data to overcome this issue, namely semi-supervised and self-supervised learning.

Approaches to semi-supervised learning most commonly use pseudo-labeling, also called self-training, [27,29,32,48] and consistency regularization [1,9,31]. In text recognition, pseudo-labeling uses unlabeled text lines annotated with machine transcriptions to improve an existing system, while consistency regularization is based on the idea that a text line should be transcribed the same under

L. Jin et al. (Eds.): ICDAR 2025 Workshops, LNCS 16225, pp. 53–70, 2026.
https://doi.org/10.1007/978-3-032-09368-4_4

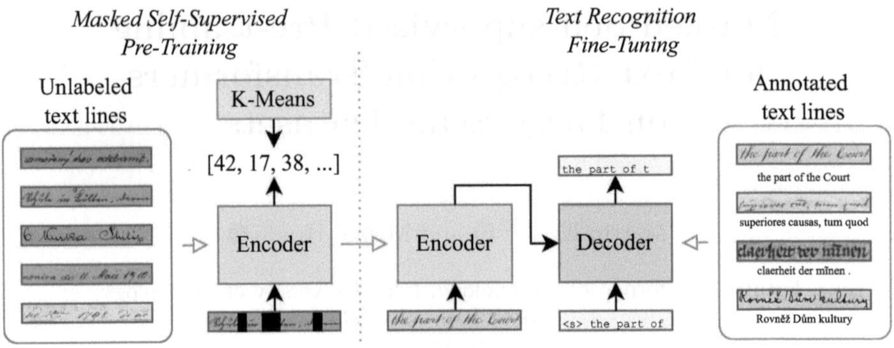

Fig. 1. Overview of our self-supervised pre-training. First, we pre-train encoder using fitted K-Means and masked label prediction. Then, we use the encoder in encoder-decoder model and we fine-tune it on an annotated dataset.

different perturbations. The self-supervised learning is typically used to pre-train a model on unlabeled data and it is then fine-tuned for a downstream task, like text recognition, speech recognition, or image classification. Self-supervised approaches are mainly based on masked label prediction [3,6,16,25,26,40] and joint-embedding learning [7,8,15]. In the masked label prediction, random parts of the input data are masked and the model is trained to predict correct labels of the masked parts. The joint-embedding learning uses two views of the input data that are processed by a model which is trained to produce similar outputs for the corresponding outputs while distancing the unrelated ones.

In this paper, we investigate the self-supervised pre-training, specifically the masked label prediction utilizing K-Means clustering [26,28] (see Fig. 1). In contrast to the paper by Kišš et al. [28], we explore the effect of self-supervised encoder pre-training on Transformer-based encoder-decoder models instead of CTC-trained models. We propose two modifications to the pre-training phase – progressive masking probability and non-masked patches loss – which we experimentally evaluate. We also experiment with two fine-tuning strategies and six differently sized models. Within the experiments, we use dataset of 50M unlabeled text lines for the pre-training and four datasets of different sizes for fine-tuning – three standard, publicly available datasets (the Bentham collection [45], the Bullinger dataset [43], and the CATMuS Medieval dataset [17]) and an in-house handwritten dataset called PERO-HWR. The sizes of the fine-tuning datasets range from 9k to over 1M annotated text lines. We compare our pre-trained models also with models trained using transfer learning, which is usually considered a very strong baseline [28].

The contributions of the paper are as follows:

1. We investigate two proposed modifications of the masked label prediction-based self-supervised pre-training on a large-scale unlabeled dataset.
2. We evaluate two fine-tuning strategies on four differently sized annotated datasets.

3. We experiment with six differently sized pre-trained models, we compare them to models trained using transfer learning and we also compare them with the state-of-the-art results.

2 Related Work

In recent years, self-supervised learning has become a prominent approach in machine learning, enabling large Transformer-based models to be trained without relying on extensive labeled datasets. This paradigm encourages models to learn meaningful representations by generating supervision signals from the input data. Self-supervised learning has been successfully applied to various domains, including language modeling [4,18], image modeling [6,7,13,15,50], and automatic speech recognition [14,16,26,34].

A common technique in self-supervised learning is masked label prediction [6,26,27], where parts of the input or intermediate representation are masked, and the model is trained to infer the missing information based on its context. In this setup, the labels are not the actual annotations from the downstream task but are derived from the input data itself, ensuring that similar patterns receive consistent labels. By leveraging this method, the model learns dependencies between the masked elements and their surrounding context, ultimately improving its ability to represent data. The masking-based modeling approach was first introduced in the pre-training of Transformer-based language model BERT [18].

The second commonly used technique is joint-embedding learning, where the model processes two different views of the input data and optimizes their outputs to be similar for corresponding pairs while pushing apart all other outputs. This approach relies on specialized loss functions, such as VICReg [7,8] or NT-Xent [15], to achieve effective representation learning.

2.1 Speech Recognition

One of the earliest self-supervised ASR approaches is wav2vec 2.0 [5] by Baevski et al., which processes raw audio through a convolutional encoder to produce a latent representation. Randomly masked parts are refined by a Transformer encoder, while product quantization is applied. The model is trained via contrastive loss, distinguishing masked representations from quantized positives and negatives. HuBERT [26] by Hsu et al. follows a similar pipeline but generates training targets by clustering Mel-frequency cepstral coefficients (MFCCs) with K-Means. WavLM [14] builds on HuBERT, introducing noise augmentation for speaker recognition, separation, and diarization. Another masked training approach, BEST-RQ [16], generates labels using a frozen randomly initialized projection and codebook. Similarly, in NEST-RQ [25] the labels are generated using a frozen randomly initialized projection and codebook, but the model is trained using next token prediction. In UniWav [34], a student encoder is trained with masked audio modeling and pseudo-labels obtained using a teacher encoder model. The student is trained using standard backpropagation and the teacher is updated with an exponential moving average of the student encoder weights. The model also contains a decoder trained using Flow Matching.

2.2 Image Modeling

Self-supervised learning in image modeling includes approaches like DINO [13, 37], iBOT [50], SwAV [12], BEiT [6,40,47], and SimCLR [15]. SimCLR [15] uses two views of an image processed by a model which is trained to maximize agreement between the two outputs. BEiT [6] employs masked image modeling to pre-train a Transformer, later fine-tuned for classification and segmentation. It initially uses a Discrete Variational AutoEncoder (dVAE) for masked training labels, which is replaced in BEiT v2 [40] and BEiT-3 [47] by the VQ-KD model. DiT [33] applies a similar strategy to document images. In DINO [13,37], a teacher model creates targets for training a student model. The student is then trained using standard backpropagation, while the teacher is updated with an exponential moving average of the student weights. A similar approach to DINO is data2vec 2.0 [4], where models are pre-trained in a very similar way on speech, images, and text.

2.3 Text Recognition

Self-supervised pre-training for text recognition primarily focuses on contrastive learning for scene text recognition [2,24,49] and learning degradation-invariant models [44]. SeqCLR [2] processes two augmented views of a word image using noise contrastive estimation loss, while DiG [49] employs InfoNCE loss and adds a reconstruction head to recover the masked original image. CCD [24] extends these methods with character-based segmentation, improving contrastive learning under geometric transformations. In MaskOCR [35], the encoder is pre-trained as a Context AutoEncoder and the decoder is trained predicting masked characters in synthetic text lines. Text-DIAE [44] trains an AutoEncoder to reconstruct degraded images, with its encoder later adapted for handwritten and scene text recognition. Penarrubia et al. [39] used rotation, flipping, jigsawing, and sorting to pretrain a CNN for text recognition model. In SSM [22], using augmentations two views of an image are combined into a single image and the model is pre-trained via reconstructing the original views.

3 Masked Self-supervised Pre-training

Our masked self-supervised pre-training is based on the masked label prediction where random patches in the input text line are masked and the model is trained to predict the class for these masked patches. In particular, our method follows the Feature Quantization [28] method which utilizes existing feature extractor to encode patches of text lines into feature vectors that are used to fit a K-Means model. Then, the K-Means model is used to obtain discrete labels for feature

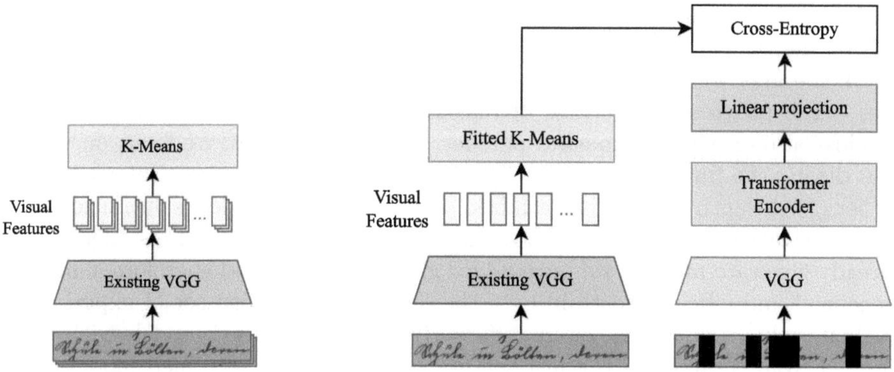

(a) Fitting K-Means (b) Pre-training

Fig. 2. Details of the masked self-supervised pre-training. We process text lines using existing VGG-like model to extract visual features and we fit a K-Means on them (Fig. 2a). In the pre-training, we employ a VGG-like model and the fitted K-Means to generate discrete labels for a given text line. Random patches of the text line are masked, and the encoder is trained to predict labels of the masked parts. (Figure 2b).

vectors obtained using the feature extractor. The entire pre-training schema is visualized in Fig. 2.

We propose two modifications to the pre-training phase which both originate in HuBERT [26]. Originally, the masking was done with a constant probability during the whole pre-training. We propose to progressively increase the masking probability during training. The reasoning behind this approach is that as a larger portion of the input image is masked, the task becomes more challenging for the model due to reduced available information. This increased difficulty is expected to drive the model toward improved performance and it also improves utilization of the input data. The second proposed modification changes the loss function which is originally calculated only on the masked patches. We propose to compute the loss function also on the non-masked patches, again to make better use of the information in the input data and to make the training more efficient, as the model learns on the entire text line, not only on a small subset. Both of these modifications showed improvement in HuBERT [26].

4 Experiments

We experimentally demonstrate the benefits of the masked self-supervised pre-training method when training encoder-decoder Transformers. We perform our experiments on four various and differently sized datasets to show the robustness of the pre-training. In the experiments, we follow the traditional self-supervised scheme, where the encoder is first pre-trained and then the whole model is fine-tuned on the downstream task using standard supervised training. For the evaluation and comparison of the trained models, we use the character error rate (CER) metric.

In the first set of experiments, we evaluate the proposed pre-training strategies – the progressive increase of masking probability and the calculation of the loss also on the non-masked patches. In the second set, we focus on strategies during the fine-tuning phase. Specifically, we compare two strategies where either (1) the entire model is optimized or (2) only the decoder is optimized first. The motivation for the second strategy is that the pre-trained encoder should already generate meaningful outputs for given text lines and therefore it may be beneficial and efficient to train only the decoder. The third set of experiments evaluates the performance of six differently sized pre-trained models with those trained from scratch. In the last set, we compare the pre-trained models with models fine-tuned from existing OCR models, i.e. trained using transfer learning, which is a strong baseline in this kind of setup [28].

4.1 Datasets

In our experiments, we use two types of text line datasets. The first type is unlabeled and it is used during the pre-training phase. The second type is annotated with line-level transcriptions and it is used during the fine-tuning phase.

Unlabeled Dataset. The unlabeled dataset is composed of 30M text lines detected in 408k pages from 3k historical documents provided by archive in Opava[1] and 20M text lines detected in 247k pages from obtained from our PERO-OCR[2] web application. In both cases, the documents contain mainly Czech and German handwritten pages from the late 19th century to the present. The documents are mostly from chronicles, parish records, land registers, and similar. The collection from the OCR application also contains some printed documents. To detect text lines in these documents, we used ParseNet-based [30] network trained on our in-house layout dataset with annotated text lines.

Fine-tuning Datasets. For the fine-tuning phase, we use four differently sized datasets: three standard, publicly available datasets and an internal handwritten dataset called *PERO-HWR*. The three standard datasets are: the ICFHR 2014 Bentham collection [45], The Bullinger Dataset [43], and the CATMuS Medieval dataset [17]. The Bentham collection contains pages written in historical English by Jeremy Bentham and a few other authors. The Bullinger dataset contains text lines from letter correspondences written in German and Latin. The letters were either written by Heinrich Bullinger or addressed to him. Unfortunately, we were not able to obtain the dataset as it is described in the original research paper because some parts with training text lines are not in the data source[3] The downloaded dataset also contains more text lines in the validation and the test

[1] https://www.archives.cz/.
[2] https://pero-ocr.fit.vutbr.cz/.
[3] https://github.com/pstroe/bullinger-htr.

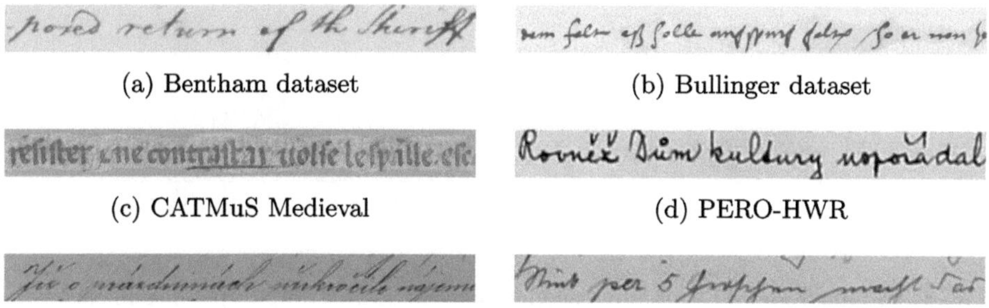

(a) Bentham dataset (b) Bullinger dataset

(c) CATMuS Medieval (d) PERO-HWR

(e) Unlabeled dataset

Fig. 3. Examples of text lines from datasets.

Table 1. Sizes of the used datasets in the number of lines.

Dataset	Training	Validation	Test
Unlabeled	50 573 532	–	2 000
Bentham	9 194	1 415	860
Bullinger	97 442	16 567	16 567
CATMuS Medieval	146 885	8 317	8 317
PERO-HWR	1 149 995	7 383	–

sets than they report in the paper. We pre-processed the obtained dataset using our text line detector resulting in text lines with more consistent baseline placement and text size. The CATMuS Medieval dataset contains pages of historical documents spanning from the 8th to the 16th century written in 10 different languages. We use the general split of the CATMuS Medieval v1.0 dataset. The PERO-HWR dataset has similar properties to the unlabeled dataset – it contains primarily Czech and German handwritten text lines from the late 19th century to the presetnt. To a lesser extent, it also contains text lines from other existing datasets, such as ICDAR2017 READ dataset [46], the Parzival dataset [21], or the Saint-Gall dataset [20]. The sizes of all datasets are summarized in Table 1 and their examples are shown in Fig. 3.

4.2 Preparation of the Self-supervised Dataset

For the self-supervised dataset, we use the text lines from the unlabeled dataset. The text lines are processed using an existing VGG-like model producing visual features. We obtained the model from a previously train OCR on previous version of the PERO-HWR dataset. We fit a K-Means model with 4 096 classes on visual features of 15k randomly selected text lines, and we then produce discrete labels for the whole dataset. To qualitatively evaluate the resulting labels, we visualized frequent trigrams in the dataset, as proposed by Kišš et al. [28]. Example visualizations are depicted in Fig. 4. Visualizations show that visually

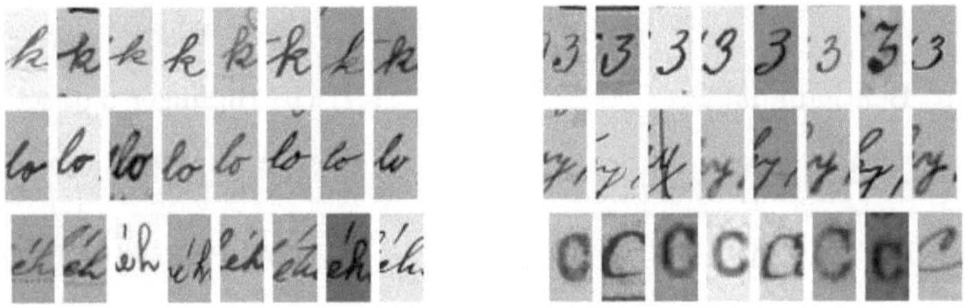

Fig. 4. Visualization of identically encoded patches. Each group of patches in each row represent patches with the same trigrams (label triplets).

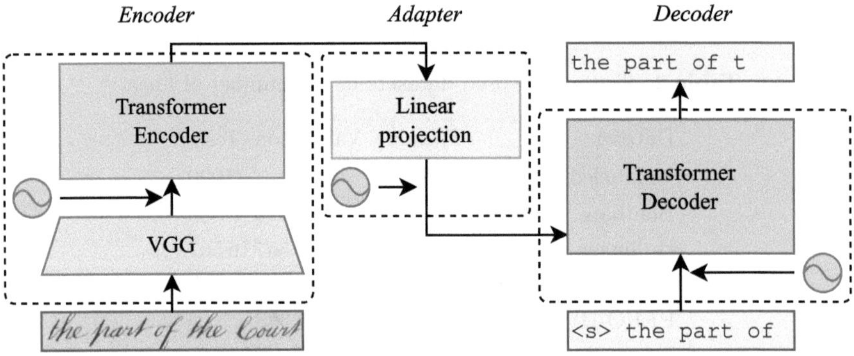

Fig. 5. Schema of our models consisting of an encoder, a decoder, and an adapter that transforms the encoder outputs for the decoder. The purple circles represent 1D sine-based positional encoding.

similar patches of text lines are encoded in the same trigrams (label triplets), which is essential from a pre-training perspective.

4.3 Models

In the experiments, we use encoder-decoder Transformer architecture operating on arbitrarily long text lines with a normalized height of 48 pixels. The visualization of the entire model is shown in Fig. 5. In the encoder, we use a VGG-like convolutional neural network, which transforms the input into a sequence of feature vectors with a horizontal subsampling factor of 8. The feature vectors are further processed with the standard Transformer encoder layers. The decoder is a standard Transformer decoder consisting of an embedding layer translating the transcription characters into vectors, standard Transformer decoder layers, and an output projection producing categorical distribution over used alphabet. The decoder is trained to generate the output sequence in an autoregressive manner. In both, the encoder and the decoder, we add 1D positional information to the vectors before the Transformer layers using sine-based positional encoding.

Since we experiment with encoders and decoders of different dimensionality, we also have a small block, which we call an adapter (the middle part in Fig. 5),

Table 2. Definitions of the encoders and decoders.

Name	Layers	Heads	Dim.	Params
Encoder E6	6	8	512	25 M
Encoder E12	12	16	1 024	158 M
Decoder D2	2	4	320	3.6 M
Decoder D6	6	8	512	26 M
Decoder D10	10	12	768	95 M

that transforms the encoder output into a representation with correct dimensionality for the decoder. The adapter is a linear layer with input and output channels equal to the dimensionality of the encoder and decoder, respectively. After the linear layer, we also add 1D sine-based positional encoding. In our preliminary experiments, this has proven to be a good practice when training a text recognizer with a pre-trained encoder, as the encoder does not necessarily retain positional information in its outputs. Without this positional information, the decoder is prone to skipping or repeating sequences in its output. On the other hand, it does not negatively affect performance of models without pre-trained encoder.

As mentioned before, we use encoders and decoders of different sizes. In particular, we define two encoders, marked as E6 and E12, and three decoders, marked as D2, D6, and D10. In all encoders and decoders, the MLP expansion ratio in the attention layers is 4. Details about the properties of the encoders and the decoders are summarized in Table 2.

4.4 Training Details

Pre-training phase. In the pre-training phase, we optimize the encoders (the left part in Fig. 5) with batch size 128 for 500k iterations. The initial learning rate is set to 2×10^{-4} and it is halved after 100k and 300k iterations. The optimized model consists of the convolutional encoder, Transformer encoder layers, and a linear projection layer, which transforms the output of the Transformer encoder layers into the classes produced by the K-Means model. This projection layer is discarded from the model once the pre-training phase is finished. Since the training targets are discrete labels, we train the encoders using standard cross-entropy loss. For the pre-training, we use the code available in `pero-pretraining` github repository[4].

Fine-tuning phase. In the fine-tuning phase, we train the models with batch size 64 using the standard cross-entropy loss. We train the models for a different number of iterations depending on the dataset: 100k on the Bentham dataset, 500k on the CATMuS Medieval dataset and the Bullinger dataset, and 1M iterations on the PERO-HWR dataset. The initial learning rate depends on the size

[4] https://github.com/DCGM/pero-pretraining.

of the model: we use 5×10^{-5} for the encoder E6 and 1×10^{-5} for the encoder E12. When training on the Bentham dataset, the learning rate is halved after 30k iterations, on the Bullinger dataset and the CATMuS medieval dataset after 250k iterations, and on the PERO-HWR dataset after 500k iterations.

In our experiment, we also evaluate fine-tuning strategy in which we first train only the adapter and the decoder. We train these two parts for the first 20 % of iterations – we refer to this as the decoder stage – and then we train the entire model for the rest of the iterations. Otherwise, the training setup remains the same.

In both the pre-training and fine-tuning phases, we train the models using bfloat16. During the pre-training phase we do not use any augmentations as these would change the appearance of text lines and the self-supervised labels could be potentially wrong. In the fine-tuning phase, we use augmentations consisting of color changing, adding noise, gamma adjusting, motion and defocus blurring, geometric transformations, and masking as proposed in AT-ST [27].

Transfer Learning. In the transfer learning experiments, we train the models for 150k iterations with learning rate 2.5×10^{-5}. Since we only perform these experiments on the Bullinger and the CATMuS datasets with encoder E6 and decoder D6, we only consider this single setup. Otherwise, the training setup remains the same as in the fine-tuning phase.

4.5 Pre-training Experiments

The first set of experiments is focused on the pre-training phase. Kišš et al. [28] randomly mask 20 % of the text lines. In our experiments, we use the same strategy, and we also propose two modifications: progressive masking probability and training on non-masked patches. In the first case, we increase the masking probability during training, in particular from the 100k-th iteration, we mask 33 % of the text line, and from the 300k-th iteration, we mask 50 % of the text line. In the latter modification, we calculate the cross-entropy loss also on patches which were not masked and we add this to the loss calculated on the masked patches with weight 0.1.

We evaluate the pre-training strategies by fine-tuning a model consisting of the smaller pre-trained encoder E6 and randomly initialized adapter and decoder D6 on the four text recognition datasets, measuring the character error rate of the fine-tuned models on the respective validation sets. The results of the fine-tuned models are summarized in Table 3. The results show that the pre-training strategy with the progressive masking probability performs the best – it consistently achieves the lowest CER on all datasets, except the PERO-HWR dataset where the results are very close for each strategy. On the other hand, computing the loss on non-masked patches does not lead to any performance gains. Throughout the rest of the experiments, we use only encoders pre-trained using the progressive masking probability strategy.

Table 3. Results of models (E6+D6) with encoders pre-trained using the proposed strategies. We report the results in CER [%] for validation sets of the respective fine-tuning datasets after the fine-tuning phase.

Progressive masking prob.	Non-masked patches loss	Dataset			
		Bentham	Bullinger	CATMuS	PERO-HWR
		3.41	7.39	3.84	1.94
✓		**3.13**	**7.21**	**3.45**	1.94
	✓	3.57	7.29	3.49	**1.91**
✓	✓	3.24	**7.21**	3.82	1.94

4.6 Fine-Tuning Strategy Experiments

In this set of experiments, we evaluate two fine-tuning strategies. The first strategy was already used in the previous set of experiments and it optimizes the entire model. The second strategy first optimizes only the decoder and the adapter in the so-called decoder stage (first 20 % of the training iterations) and later optimizes the entire model for the rest of the iterations. As mentioned in the beginning of this section, the motivation behind this strategy is based on the idea that the pre-trained encoder should already be capable of producing meaningful outputs for given text lines, so it could be useful to train only the rest of the model, which is randomly initialized in the beginning of the training. We used the same combination of encoder and decoder (E6 + D6) as in the previous set of experiments and we also report the character error rates on the respective validation sets. The results of these experiments are in Table 4 and they show that training only the decoder in the decoder stage is not beneficial. This also suggests that the encoder does not produce output directly suitable for the decoder that would allow simple downstream fine-tuning for text recognition.

4.7 Model Size Experiments

The third set of experiments is focused on the different sizes of encoders and decoders. We use the E6 and E12 encoders pre-trained using the progressive masking probability strategy from the first set of experiments. As decoders we use the D2, D6, and D10 decoders, and we optimize the entire encoder-decoder model as resulted from the second set of experiments. A visual comparison of the performance of the models is shown in Fig. 6. It shows that the self-supervised pre-training helps to train better-performing models, especially on the smaller dataset. It also turns out that training smaller models on these datasets is a better choice than training large models, even with pre-training. On the other hand, the results for the PERO-HWR dataset show rather small improvements for the models with encoder E6. There is even quite large increase of CER for the decoders D2 and D6 with the encoder E12. It is necessary to note that we experienced instabilities during training of these models, which could probably be overcome by better hyper-parameter selection. However, for the biggest model

Table 4. Results of models (E6+D6) with pre-trained encoders fine-tuned with the investigated fine-tuning strategies. We report the results in CER [%] for validation sets of the respective fine-tuning datasets after the decoder stage of training (upper part) and for the best-performing checkpoint after the entire training (lower part).

Strategy	Dataset			
	Bentham	Bullinger	CATMuS	PERO-HWR
After 20 % iterations (decoder stage)				
Decoder	17.71	15.30	9.81	4.21
Full	3.64	8.14	3.81	2.56
Best performance after entire training				
Decoder	3.70	7.88	4.16	**1.93**
Full	**3.13**	**7.21**	**3.45**	1.94

Table 5. Results of the best-performing models from the experiments with model sizes. We report the results in CER [%] for validation sets of the respective fine-tuning datasets. Under the name of each dataset, the configuration of the best-performing model is specified.

Training method	Dataset			
	Bentham (E6+D2)	Bullinger (E6+D2)	CATMuS (E6+D2)	PERO-HWR (E6+D10)
From scratch	4.52	7.68	3.72	2.00
Self-supervised pre-training	**3.12**	**7.15**	**3.45**	**1.93**

(encoder E12 + decoder D10) the pre-trained model again performed better than the one trained from scratch. The particular CER values of the best-performing models on each dataset are in Table 5

4.8 Transfer Learning Experiments

In this set of experiments, we used the E6+D6 model trained from scratch on the PERO-HWR dataset and we fine-tuned it on the Bullinger and CATMuS Medieval datasets to compare the self-supervised pre-training and transfer learning. We did not run this experiment on the Bentham dataset because it is part of the PERO-HWR dataset, but unfortunately it does not follow the original splits, i.e. text lines from the original validation and test subsets are in the training subset of the PERO-HWR dataset. Therefore, a fair comparison would not be possible. The results are in Table 6 and they show that both transfer learning and self-supervised pre-training are beneficial when compared to training from scratch. Moreover, the results of the self-supervisely pre-trained models achieve comparable or even slightly better results when compared to the transfer learning, which is usually considered a strong baseline [28].

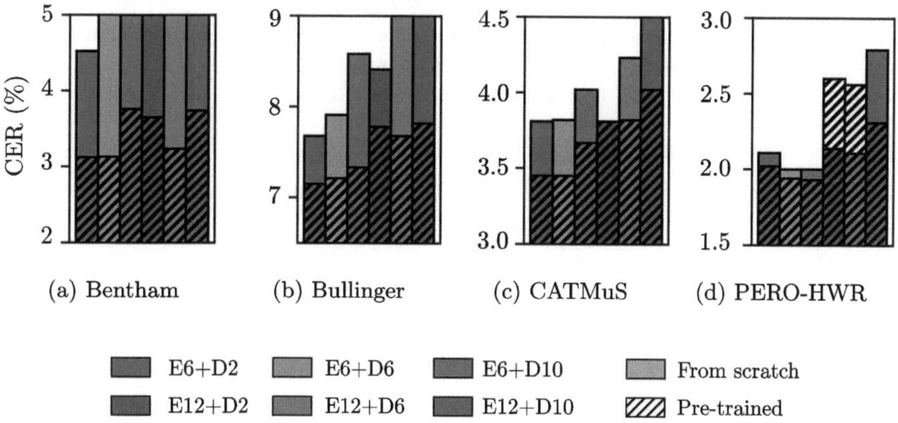

Fig. 6. Results of pre-trained models and models trained from scratch on validations sets of the fine-tuning datasets in CER [%]. The different colors represent different models. Colored bars are results of the models trained from scratch and the hatched parts present results of pre-trained models. Note the different scales on each y-axis.

Table 6. Results of the models (E6+D6) trained from scratch, using transfer learning, and using self-supervised pre-training. We report the results in CER [%] for validation sets of the Bullinger and the CATMuS datasets.

Training method	Dataset	
	Bullinger	CATMuS
From scratch	7.91	3.90
Transfer learning	**7.20**	3.59
Self-supervised pre-training	7.21	**3.45**

4.9 Final Comparison

In Table 7, we present the final results on test sets of the Bentham, the Bullinger, and the CATMuS Medieval datasets. In particular, we present the results of (1) the best-performing models (E6+D2) trained from scratch and using self-supervised pre-training, (2) models E6+D6 used through the experiments including transfer learning, and (3) state-of-the-art results for the Bentham dataset [38] and the CATMuS Medieval dataset [17] – we do not report state-of-the-art results for the Bullinger dataset since we do not have the exact validation and test sets for fair comparison, as mentioned in the beginning of this section. In the case of state-of-the-art results for the Bentham dataset, we report CER obtained by fine-tuning pre-trained TrOCR model on the Bentham, thus not really comparable with models trained from scratch, but comparable rather with the self-supervisedly pre-trained models. The state-of-the-art results on the CATMuS

Table 7. Final results on test sets of the respective datasets. We report the results in CER [%] for the best-performing models (upper part), models of configuration E6+D6 used through the experiments including transfer learning, which are thus mutually comparable in terms of size (middle part), and state-of-the-art results (lower part).

Training method	Dataset		
	Bentham	Bullinger	CATMuS
Best models (E6+D2)			
From scratch	3.86	7.62	3.72
Self-supervised pre-training	2.51	7.08	3.49
Comparable models (E6+D6)			
From scratch	9.54	7.81	3.94
Transfer learning	–	7.11	3.63
Self-supervised pre-training	2.47	7.12	3.49
State-of-the-art results	2.70	–	4.70

Medieval dataset are obtained by training only on that dataset and thus these results are directly comparable to our models trained from scratch. In each case, we selected the model with the lowest CER on the respective validation set and we report CER for the respective test set.

5 Conclusions

In this paper, we explored masked self-supervised pre-training for text recognition transformers, leveraging large-scale unlabeled data to improve recognition performance. We proposed two modifications to the pre-training phase: progressively increasing the masking probability during training and incorporating non-masked patches into the loss computation. Our experimental results demonstrate that the progressive masking modification leads to improved model performance after fine-tuning on the text recognition task. Through extensive experiments on six differently sized models and four fine-tuning datasets, we showed that self-supervised pre-training on a large-scale unlabeled dataset provides a strong foundation for text recognition, particularly in scenarios where labeled data is limited. Furthermore, our approach proved competitive with transfer learning, which is considered a strong baseline.

Future work may explore additional enhancements to the masked pre-training resulting in learning better representations. Another option is investigating self-supervised pre-training with different models which could further improve the text recognition accuracy. Additionaly, it could be beneficial either to pre-train the entire encoder-decoder model or to use pre-trained decoders.

Acknowledgments. This work has been supported by the Ministry of Culture Czech Republic in NAKI III project semANT - Semantic Document Exploration (DH23P03OVV060) and by the Ministry of Education, Youth and Sports of the Czech Republic through the e-INFRA CZ (ID:90254).

Disclosure of Interests. The authors have no competing interests to declare that are relevant to the content of this article.

References

1. Aberdam, A., Ganz, R., Mazor, S., Litman, R.: Multimodal semi-supervised learning for text recognition (2022)
2. Aberdam, A., et al.: Sequence-to-sequence contrastive learning for text recognition. In: Proceedings of the IEEE/CVF Conference on Computer Vision and Pattern Recognition, pp. 15302–15312 (2021)
3. Assran, M., et al.: Self-supervised learning from images with a joint-embedding predictive architecture
4. Baevski, A., Babu, A., Hsu, W.N., Auli, M.: Efficient self-supervised learning with contextualized target representations for vision, speech and language. In: Proceedings of the 40th International Conference on Machine Learning, pp. 1416–1429. PMLR (2023). ISSN: 2640-3498
5. Baevski, A., Zhou, Y., Mohamed, A., Auli, M.: wav2vec 2.0: a framework for self-supervised learning of speech representations. In: Advances in Neural Information Processing Systems, vol. 33, pp. 12449–12460. Curran Associates, Inc. (2020)
6. Bao, H., Dong, L., Piao, S., Wei, F.: BEiT: BERT pre-training of image transformers
7. Bardes, A., Ponce, J., LeCun, Y.: VICReg: variance-invariance-covariance regularization for self-supervised learning
8. Bardes, A., Ponce, J., LeCun, Y.: VICRegL: self-supervised learning of local visual features. Adv. Neural. Inf. Process. Syst. **35**, 8799–8810 (2022)
9. Berthelot, D., Carlini, N., Goodfellow, I., Papernot, N., Oliver, A., Raffel, C.A.: MixMatch: a holistic approach to semi-supervised learning. In: Advances in Neural Information Processing Systems, vol. 32. Curran Associates, Inc. (2019)
10. Betker, J., et al.: Improving image generation with better captions (2020)
11. Brown, T., et al.: Language models are few-shot learners. In: Advances in Neural Information Processing Systems, vol. 33, pp. 1877–1901. Curran Associates, Inc. (2020)
12. Caron, M., Misra, I., Mairal, J., Goyal, P., Bojanowski, P., Joulin, A.: Unsupervised learning of visual features by contrasting cluster assignments. In: Advances in Neural Information Processing Systems, vol. 33, pp. 9912–9924. Curran Associates, Inc. (2020)
13. Caron, M., Touvron, H., Misra, I., Jégou, H., Mairal, J., Bojanowski, P., Joulin, A.: Emerging properties in self-supervised vision transformers. In: Proceedings of the IEEE/CVF International Conference on Computer Vision, pp. 9650–9660 (2021)
14. Chen, S., et al.: WavLM: large-scale self-supervised pre-training for full stack speech processing. IEEE J. Sel. Top. Sig. Process. **16**(6), 1505–1518 (2022).
15. Chen, T., Kornblith, S., Norouzi, M., Hinton, G.: A simple framework for contrastive learning of visual representations

16. Chiu, C.C., Qin, J., Zhang, Y., Yu, J., Wu, Y.: Self-supervised learning with random-projection quantizer for speech recognition. In: Proceedings of the 39th International Conference on Machine Learning, pp. 3915–3924. PMLR (2022). ISSN: 2640-3498
17. Clérice, T., et al.: CATMuS medieval: a multilingual large-scale cross-century dataset in Latin script for handwritten text recognition and beyond. In: Barney Smith, E.H., Liwicki, M., Peng, L. (eds.) Document Analysis and Recognition - ICDAR 2024, pp. 174–194. Springer Nature Switzerland (2024). https://doi.org/10.1007/978-3-031-70543-4_11
18. Devlin, J., Chang, M.W., Lee, K., Toutanova, K.: BERT: pre-training of deep bidirectional transformers for language understanding. CoRR abs/1810.04805 (2018)
19. Esser, P., et al.: Scaling rectified flow transformers for high-resolution image synthesis
20. Fischer, A., Frinken, V., Fornés, A., Bunke, H.: Transcription alignment of Latin manuscripts using hidden Markov models. In: Proceedings of the 2011 Workshop on Historical Document Imaging and Processing, pp. 29–36. HIP 2011, Association for Computing Machinery (2011)
21. Fischer, A., Keller, A., Frinken, V., Bunke, H.: Lexicon-free handwritten word spotting using character HMMs. Pattern Recogn. Lett. **33**(7), 934–942 (2012)
22. Gao, Z., et al.: Self-supervised pre-training with symmetric superimposition modeling for scene text recognition
23. Grattafiori, A., Dubey, A., Jauhri, A., et al.: The llama 3 herd of models
24. Guan, T., Shen, W., Yang, X., Feng, Q., Jiang, Z., Yang, X.: Self-supervised character-to-character distillation for text recognition. In: Proceedings of the IEEE/CVF International Conference on Computer Vision, pp. 19473–19484 (2023)
25. Han, M., et al.: NEST-RQ: next token prediction for speech self-supervised pre-training
26. Hsu, W.N., Bolte, B., Tsai, Y.H.H., Lakhotia, K., Salakhutdinov, R., Mohamed, A.: HuBERT: self-supervised speech representation learning by masked prediction of hidden units. IEEE/ACM Transactions on Audio, Speech, and Language Processing **29**, 3451–3460 (2021)
27. Kišš, M., Beneš, K., Hradiš, M.: AT-ST: self-training adaptation strategy for OCR in domains with limited transcriptions. In: Lladós, J., Lopresti, D., Uchida, S. (eds.) ICDAR 2021. LNCS, vol. 12824, pp. 463–477. Springer, Cham (2021). https://doi.org/10.1007/978-3-030-86337-1_31
28. Kišš, M., Hradiš, M.: Self-supervised pre-training of text recognizers. In: Barney Smith, E.H., Liwicki, M., Peng, L. (eds.) Document Analysis and Recognition - ICDAR 2024, pp. 218–235. Springer Nature Switzerland (2024). https://doi.org/10.1007/978-3-031-70546-5_13
29. Kišš, M., Hradiš, M., Beneš, K., Buchal, P., Kula, M.: SoftCTC–semi-supervised learning for text recognition using soft pseudo-labels. Int. J. Document Anal. Recognit. (IJDAR) (2023). https://doi.org/10.1007/s10032-023-00452-9
30. Kodym, O., Hradiš, M.: Page layout analysis system for unconstrained historic documents. In: Lladós, J., Lopresti, D., Uchida, S. (eds.) Document Analysis and Recognition – ICDAR 2021, pp. 492–506. Springer International Publishing, Lecture Notes in Computer Science (2021). https://doi.org/10.1007/978-3-030-86331-9_32
31. Kurakin, A., et al.: ReMixMatch: semi-supervised learning with distribution matching and augmentation anchoring. In: ICLR (2020)

32. Lee, D.H.: Pseudo-label : The simple and efficient semi-supervised learning method for deep neural networks. ICML 2013 Workshop : Challenges in Representation Learning (WREPL) (2013)
33. Li, J., Xu, Y., Lv, T., Cui, L., Zhang, C., Wei, F.: DiT: Self-supervised pre-training for document image transformer. In: Proceedings of the 30th ACM International Conference on Multimedia, pp. 3530–3539. MM 2022, Association for Computing Machinery (2022)
34. Liu, A.H., et al.: UniWav: towards unified pre-training for speech representation learning and generation. In: The Thirteenth International Conference on Learning Representations (2024)
35. Lyu, P., et al.: MaskOCR: text recognition with masked encoder-decoder pretraining
36. OpenAI, Achiam, J., Adler, S., Agarwal, S., et al.: GPT-4 technical report
37. Oquab, M., et al.: DINOv2: learning robust visual features without supervision
38. Parres, D., Anitei, D., Paredes, R., Sánchez, J.A., Benedí, J.M.: Speed-up pre-trained vision encoder–decoder transformers by leveraging lightweight mixer layers for text recognition. In: Sfikas, G., Retsinas, G. (eds.) Document Analysis Systems. pp. 277–294. Springer Nature Switzerland (2024). https://doi.org/10.1007/978-3-031-70442-0_17
39. Penarrubia, C., Garrido-Munoz, C., Valero-Mas, J.J., Calvo-Zaragoza, J.: Spatial context-based self-supervised learning for handwritten text recognition
40. Peng, Z., Dong, L., Bao, H., Ye, Q., Wei, F.: BEiT v2: masked image modeling with vector-quantized visual tokenizers
41. Radford, A., Kim, J.W., Xu, T., Brockman, G., McLeavey, C., Sutskever, I.: Robust speech recognition via large-scale weak supervision (2022)
42. Rombach, R., Blattmann, A., Lorenz, D., Esser, P., Ommer, B.: High-resolution image synthesis with latent diffusion models. In: Proceedings of the IEEE/CVF Conference on Computer Vision and Pattern Recognition (CVPR), pp. 10684–10695 (2022)
43. Scius-Bertrand, A., Ströbel, P., Volk, M., Hodel, T., Fischer, A.: The bullinger dataset: a writer adaptation challenge. In: Fink, G.A., Jain, R., Kise, K., Zanibbi, R. (eds.) Document Analysis and Recognition - ICDAR 2023, pp. 397–410. Springer Nature Switzerland (2023). https://doi.org/10.1007/978-3-031-41676-7_23
44. Souibgui, M.A., et al.: Text-DIAE: a self-supervised degradation invariant autoencoder for text recognition and document enhancement. In: Proceedings of the Thirty-Seventh AAAI Conference on Artificial Intelligence and Thirty-Fifth Conference on Innovative Applications of Artificial Intelligence and Thirteenth Symposium on Educational Advances in Artificial Intelligence. AAAI 2023/IAAI 2023/EAAI 2023, vol. 37, pp. 2330–2338. AAAI Press (2023)
45. Sánchez, J.A., Romero, V., Toselli, A.H., Vidal, E.: ICFHR2014 competition on Handwritten Text Recognition on Transcriptorium Datasets (HTRtS). In: 2014 14th International Conference on Frontiers in Handwriting Recognition, pp. 785–790 (2014). iSSN: 2167-6445
46. Sánchez, J.A., Romero, V., Toselli, A.H., Villegas, M., Vidal, E.: ICDAR2017 competition on handwritten text recognition on the READ dataset. In: 2017 14th IAPR International Conference on Document Analysis and Recognition (ICDAR), vol. 01, pp. 1383–1388 (2017). ISSN: 2379-2140
47. Wang, W., et al.: Image as a foreign language: BEiT pretraining for vision and vision-language tasks. In: Proceedings of the IEEE/CVF Conference on Computer Vision and Pattern Recognition, pp. 19175–19186 (2023)

48. Xie, Q., Luong, M.T., Hovy, E., Le, Q.V.: Self-training with noisy student improves ImageNet classification. In: Proceedings of the IEEE/CVF Conference on Computer Vision and Pattern Recognition (CVPR) (2020)
49. Yang, M., et al.: Reading and writing: discriminative and generative modeling for self-supervised text recognition. In: Proceedings of the 30th ACM International Conference on Multimedia, pp. 4214–4223. MM 2022, Association for Computing Machinery (2022)
50. Zhou, J., et al.: iBOT: image BERT pre-training with online tokenizer

Text Prompt to Image Generation for Classification of Similar and Non-similar Scene Images to Improve Text Spotting Performance

Surajit Mukherjee[1], Shivakumara Palaiahnakote[2(✉)], Sukalpa Chanda[3], Umapada Pal[1], and Tong Lu[4]

[1] Computer Vision and Pattern Recognition, Indian Statistical Institute, Kolkata, India
umapada@isical.ac.in

[2] School of Science, Engineering and Environment, University of Salford, Manchester, UK
s.palaiahnakote@salford.ac.uk

[3] Østfold University College, Halden, Norway
sukalpa@ieee.org

[4] National Key Lab for Novel Software Technology, Nanjing University, Nanjing, China
lutong@nju.edu.cn

Abstract. Text spotting has received attention due to several real-world applications. Most existing spotting methods use textual information to solve text-spotting challenges and ignore non-text information (background part). Therefore, the current text spotting methods lack robustness and generalization. Here, we propose an approach that introduces a multimodal deep learning framework for classifying similar (interrelationship between scene text and non-text) and non-similar (no relationship between scene text and non-text) images to improve the text spotting performance. For an input scene text image, DeepSolo, which is a state-of-the-art text spotting method, has been used to segment text from non-text regions. The text regions are fed to the Google OCR for recognition results. Consequently, the recognition results are fed to the generative model (Fooocus) as a prompt for generating an image. It is expected that the generated image should be similar to the non-text region of the input image. This is because we believe that text and background have a strong correlation. The visual features from the generated image and input images are extracted using the combination of ViT and EfficientNet. The labels are generated for the input image and the generated images using the Google Vision API. The BERT is explored to extract textual features from the labels. The visual and the textual (semantic embedding) are combined using a bilinear fusion to classify similar and non-similar images. The proposed classification is validated and compared with the existing methods to demonstrate its promising performance. The state-of-the-art text spotting methods are tested before and after classification to show the effectiveness of the classification. This is the first work to classify similar and non-similar images using a diffusion model to improve text spotting performance.

Keywords: Text detection · Text spotting · DeepSolo · Transformer · Bilinear fusion · Similarity · Non-similarity

L. Jin et al. (Eds.): ICDAR 2025 Workshops, LNCS 16225, pp. 71–91, 2026.
https://doi.org/10.1007/978-3-032-09368-4_5

1 Introduction

Text spotting has been a long-studied research topic with significant real-world applications such as person and vehicle identification, scene understanding, and visual question answering. Despite years of investigation, current state-of-the-art methods still struggle with challenges arising from complex backgrounds and unstructured, unpredictable text. Traditional algorithms have gradually been replaced by deep learning models capable of detecting and segmenting text in complex images. For example, methods like CRAFT (Baek et al., 2019) and PSENet (Wang et al., 2019) focused on pixel-level text localization. In EAST (Zhou et al., 2017), the proposed algorithm is a regression method that directly predicts bounding boxes for text. Recent models such as DeepSolo (Ye et al., 2023) employ Transformer decoders to simultaneously localize text regions and recognize content, thereby streamlining the pipeline and enhancing performance.

Commercial solutions like Google Vision have demonstrated high accuracy in text extraction (Thammarak et al., 2022), and the Student-Teacher-Assistant (STA) network leverages dual CLIP models for text recognition in varied domains, including scene, underwater, and drone images (Purkayastha et al., 2025). Most of those existing approaches focus on the characteristics of text regions while largely ignoring non-text regions. However, a strong correlation can exist between scene text and non-text regions, which remains under-explored. This work proposes to exploit that relationship by categorizing images into two types, namely, Similar Images, where there is a correlation between scene text and non-text regions. Non-Similar Images-such a correlation is absent. Such classification is crucial because it guides the choice of text-spotting methods. For similar images, techniques that extract contextual features using text and non-text regions are preferable, whereas for non-similar images, methods focusing on text characteristics alone may yield better results.

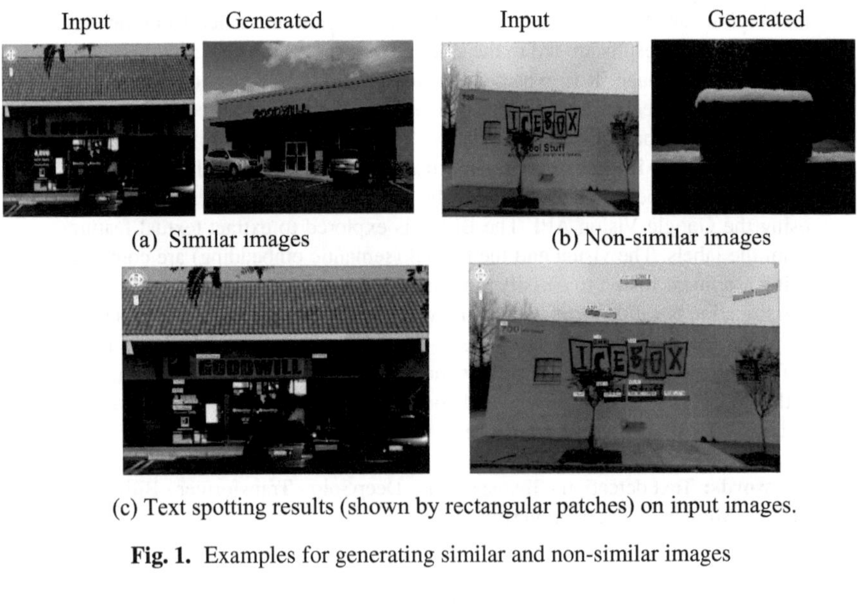

(a) Similar images (b) Non-similar images

(c) Text spotting results (shown by rectangular patches) on input images.

Fig. 1. Examples for generating similar and non-similar images

This intuition is illustrated in Fig. 1, where we can see sample similar and non-similar images in Fig. 1(a) and Fig. 1(b), respectively. It is observed from Fig. 1(a) and Fig. 1(b) that the generated image using the text prompt is almost the same as the input image in (a) while the generated image is different in (b). The effects of the similar and non-similar images can be seen in Fig. 1(c), where the state-of-the-art DeepSolo model (Ye et al., 2023) works well for similar images and does not work well for non-similar images. This shows that the DeepSolo model is good for the image where a correlation between scene text and non-text region exists. To achieve the best result for non-similar images, one should choose methods that focus on textual information for text spotting.

Thus, this work presents a new model for the classification of similar and non-similar images such that appropriate models can be chosen for improving text spotting performance. Inspired by the state-of-the-art performance of the DeepSolo model (Ye et al., 2023), we use the same model for segmenting text from non-text regions in the input images. The recognition results of segmented text are used as a prompt for generating images using the Fooocus diffusion generative model. The labels are obtained for input and generated images using Google Vision API. The features are extracted from input and generated images and the label of input and generated images using the combination of Vision Transformer (ViT) (Dosovitskiy et al., 2020) and EfficientNet (Tan et al., 2019) and BERT (Devlin et al., 2019), respectively. The features are fused using a bilinear fusion approach to classify similar and non-similar images. The main contributions of this work are:

(i) Introducing the use of text prompts and diffusion-generated images for classifying similar and non-similar images in text spotting. (ii) Integrating state-of-the-art models (ViT, EfficientNet, and BERT) for extracting robust visual and textual features. (iii) Proposing a bilinear fusion approach to effectively combine multiple modalities for improved classification accuracy.

The remainder of the paper is structured as follows. The methods of text-to-image generation, image-to-image generation, and the methods of multimodal approaches for image classification are reviewed in Sect. 2. Section 3 presents the architecture of the classification of similar and non-similar images. Experimental results and analysis are discussed in Sect. 4 and Sect. 5 summarizes the findings and inferences of the proposed work.

2 Related Works

This section reviews the methods of text prompt to image generation, image-to-image generation, and multimodal for general image classification.

Text-Guided Image Generation: Text-guided image generation has emerged as a promising technique for synthesizing realistic images conditioned on textual input. Diffusion-based models such as Stable Diffusion (Rombach et al., 2022) use latent representations and cross-attention mechanisms to produce high-quality images from text prompts. However, these models sometimes struggle with rendering legible text. To address this, Chen et al. (2023) developed a two-stage model, TextDiffuser that first predicts text layouts before generating images, ensuring better integration of textual information. Likewise, Zhang et al. (2024) introduced diff-text, a training-free method that guides image synthesis with localized attention and edge-map priors, achieving enhanced text fidelity. Tools like Fooocus further illustrate the potential of diffusion models in practical applications. These methods provide insight into generating synthetic images that reflect the semantic content of detected text, a key component of our multimodal framework. Although the diffusion models are elegant, none of these methods have been used to generate scene images to improve text spotting performance. Therefore, it is not certain that the existing models are robust for text spotting.

Image-to-Image Generation: Image-to-image translation approaches show potential in applications such as picture restoration, style transfer, and text-to-image generation. Isola et al. (2017) presented pix2pix, a conditional adversarial network for applications that generate realistic images from sketches. The model, CycleGAN, which was proposed by Zhu et al. (2017a), has completed unpaired image-to-image translation tasks, demonstrating strong domain adaptation skills. Zhu et al. (2017b) presented BicycleGAN, which further extends GAN-based approaches by combining conditional GANs with latent variable inference to generate a variety of outputs for a given input.

In addition, Choi et al. (2018) presented StarGAN, which is a unified architecture for multi-domain image-to-image translation, allowing translation among multiple domains using a single generator and discriminator. Moreover, Huang et al. (2018) developed MUNIT (Multimodal Unsupervised Image-to-Image Translation) disentangles content and style to facilitate flexible and diverse image translations in an unsupervised manner. These GAN-based methods collectively provide a robust set of tools for generating images that preserve key structural elements while incorporating novel stylistic or semantic modifications. There are image-to-image approaches for generating images. However, these approaches are not suitable for text spotting because our objective is to use scene text in the images as a prompt for generating images.

Multimodal Image Classification: Integrating information from multiple modalities has been effective in significantly boosting model performance on complex tasks (Li and Tang, 2024). In modern digital content, textual data often accompanies images (Kiela et al., 2019), motivating the development of multimodal classification models that can jointly analyze visual and textual features. Chanchal et al. (2021) introduced an end-to-end framework that predicts personality using multimodal fusion of facial, ambient, audio, and text features, achieving robust performance on Chalearn-17. These deep models were able to better capture the temporal dynamics and spatial details inherent in multimodal inputs, leading to improved classification performance.

Several surveys document this trend: Baltrušaitis et al. (2018) introduced an early taxonomy of multimodal learning, and Barua et al. (2023) provide a recent comprehensive review of multimodal applications across domains. These works highlight that leveraging complementary modalities not only improves accuracy but also enhances robustness and generalizability. These methods extract visual and textual features for classifying images. However, these are good for specific classification, but not similar and non-similar images of scene text images.

In summary, it is noted from the summary of the method of different categories that none of the methods used generated images for classifying similar and non-similar images for improving text spotting performance. Therefore, this study makes a difference compared to the existing methods as it gives a message for improving text spotting performance by choosing an appropriate model based on the nature of the images.

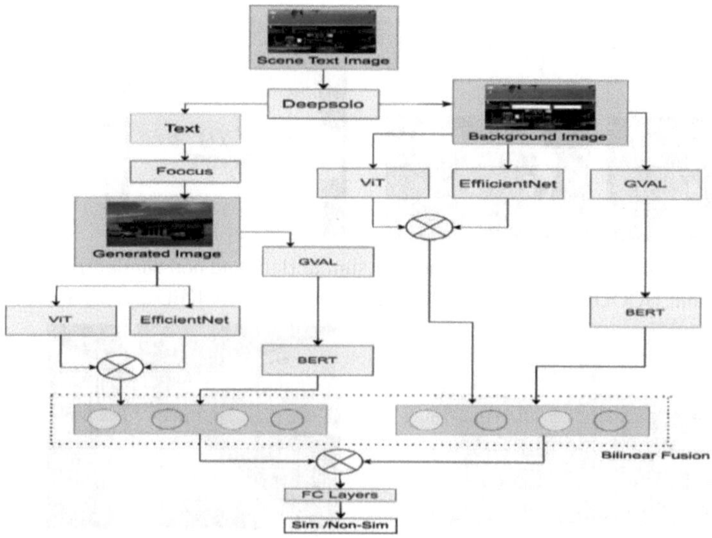

Fig. 2. Overview of the proposed method. The combination of ViT and EfficientNet is proposed for visual feature extraction, and BERT is used for textual feature extraction. The labels are obtained using the Google Vision API (GVAL). Bilinear approach for fusing two different modalities (visual and textual features).

3 The Proposed Method

As noted in the previous section, most of the existing methods ignore non-text regions for text spotting. This work aims to exploit the relationship between scene text and non-text regions to classify a given input image as similar or non-similar images. With this classification, one can choose an appropriate text-spotting method for addressing the challenges of text spotting. However, the scope of the proposed work is to classify the similar (correlation between scene text and non-text region) and non-similar images (no correlation between scene text and non-text region) for a given scene image as input. Since our objective is classification, inspired by the state-of-the-art performance of the DeepSolo text spotting models, we use the same for segmenting text from non-text in the input image. Motivated by the power of diffusion models for generating images using text prompts, we use segmented text (recognition results) as a prompt for generating an image. It is noted from the literature that the Vision Transformer and EfficientNet, and BERT are good for extracting visual and textual features, respectively. This observation motivated us to explore the combination of ViT and EfficientNet for extracting visual features from input images and generated images. BERT has been used for extracting textual features from the labels generated by using Google OCR for the input and generated images.

The proposed work introduces a bilinear fusion approach for integrating the visual and textual features to classify similar and non-similar images. The complete pipeline of the proposed method is shown in Fig. 2.

(a) Input and its text segmentation for similar (left) and non-similar (right) images.
Recognition: GOODWIL STORE 3000 Over New Itoms HOCKED and SUBWAY

(b) Generated images corresponding to input images of (a) using a text prompt

Fig. 3. Text extraction, recognition results and generated images

3.1 Text Extraction and Image Generation

An input text image, containing both text and background elements, is processed using the Deep SOLO model. This model specializes in instance segmentation, enabling it to detect individual text regions within the image. Consequently, two separate images are generated: one displaying only the segmented text regions, and the other showing the background with the text regions removed.

Next, the image containing only the text regions extracted from the previous step is processed using Google OCR. This model analyzes the segmented text areas to recognize and convert the visual text into machine-readable content. The recognition result of the text is fed to the Fooocus model (Zhang et al., 2024), which is a diffusion-generative model for generating an image. This is a state-of-the-art diffusion model for generating images using text prompts. The Fooocus model is a latent diffusion model that is trained on a large dataset like LAION. As a result, the performance is consistent and reliable. In addition, it works well for different situations and complex backgrounds. More details can be found in (Zhang et al., 2024). This model effectively translates textual information into a new visual representation, producing an image that reflects the content described by the recognized text. The effectiveness of this step is illustrated in Fig. 3, where (a) shows input, text segmentation, and recognition results, and (b) shows generated images corresponding to similar and non-similar images. It is noted from Fig. 3 that the generated image (left image in (b)) is almost similar to the input image in (left image in (a)) and the generated image (right image in (b)) is entirely different from the input image (right image in (a)).

3.2 Feature Extraction and Classification

To compare the real and the generated images, we extract visual and textual features using deep learning models. In this paper, deep visual embeddings are extracted from images using two distinct feature extractors. Let I be the Real and I' be the generated image. First, a pre-trained EfficientNet model is employed to capture local features, such as fine-grained textures and spatial details, producing a 1280-dimensional feature vector as defined in Eq. (1) and (2).

$$f_{v1} = \text{EfficientNet}(I) \in R^{1280}, \tag{1}$$

$$f_{v1'} = \text{EfficientNet}(I') \in R^{1280} \tag{2}$$

where f_{v1} represents the feature vectors extracted from the real image after processing through EfficientNet. And $f_{v1'}$ is the feature vectors generated from generated image after passing through EfficientNet. Next, we use a Vision Transformer (ViT-B/16) to extract global features, generating a 768-dimensional feature vector as defined in Eq. (3) and (4).

$$f_{v2} = \text{ViT-B/16}(I) \in R^{768}, \tag{3}$$

$$f_{v2'} = \text{ViT-B/16}(I') \in R^{768} \tag{4}$$

where f_{v2} is the feature vectors are generated from a real image after passing through ViT and $f_{v2'}$ is the feature vectors are obtained from the generated image after passing through Vit. Finally, we concatenate these two vectors to construct the final visual feature representation as defined in Eqs. (5) and (6).

$$f_v = [f_{v1}, f_{v2}] \in R^{2048}, \tag{5}$$

$$f'_v = [f'_{v1}, f'_{v2}] \in R^{2048} \tag{6}$$

For extracting textual features, the real and generated images are passed through Google Vision API to obtain labels. Google Vision API processes the images to generate descriptive labels as shown in Eqs. (7) and (8).

$$l = \text{GoogleVisionAPI}(I) \in R^{d_l}, \tag{7}$$

$$l' = \text{GoogleVisionAPI}(I') \in R^{d_l} \tag{8}$$

Here, d_l represents the dimensionality of the label representation output by the Google Vision API. The generated labels l and l' are fed into a pre-trained BERT model to refine the representation and generate the textual embeddings. Thus, the complete process can be summarized as defined in Eqs. (9) and (10).

$$f_t = \mathrm{BERT}(l) \in R^{[illegible]}, \tag{9}$$

$$f_t' = \mathrm{BERT}(l') \in R^{[illegible]} \tag{10}$$

Scene image-1 (segmented image of Fig.2(a) (left side): 'Logo', 'Advertising', 'Sign', Cars, Parking, 'Daytime',

Scene Image-2 (segmented image of Fig.2(a) (right side): 'Advertising', 'Palm Trees'

(a) Labels obtained for input scene images using Google Vision API.

Generated image-1 (left image of Fig.2(b)): Commercial building', 'Logo', 'Luxury vehicle'……

Generated image -2 (right image of Fig.2(b): 'Transport', 'Train', 'Public transport', ….

(b) Labels obtained or generated images using Google Vision API.

Fig. 4. Labels are generated for input and generated images using Google Vision API

Sample label generation is presented in Fig. 4, where (a) shows labels for similar images while (b) shows labels for non-similar images. One can notice from Fig. 4 that there is a strong correlation between scene text and non-text regions in similar images, while there is no correlation for non-similar images. This is the cue for classifying similar and non-similar images. Next, two bilinear fusion mechanisms are then employed to integrate the textual embeddings $f_{\mathrm{t}}, f_{\mathrm{t'}}$ and the visual embeddings $f_{\mathrm{v}}, f_{\mathrm{v'}}$ via element-wise multiplication, which ensures a richer multimodal representation both for the embedding generated from the real and generated image as defined in Eqs. (11) and (12).

$$F = \mathrm{BilinearFusion}(f_{\mathrm{t}}, f_{\mathrm{v}}), \tag{11}$$

$$F' = \mathrm{BilinearFusion}(f_{\mathrm{t'}}, f_{\mathrm{v'}}) \tag{12}$$

The resulting fused feature representations from real and generated images are concatenated and passed through a deep neural network designed for binary classification. This classification network consists of five fully connected layers, each utilizing a Rectified Linear Unit (ReLU) activation function. The first layer contains 1024 neurons, followed by the second layer with 512 neurons, the third with 256 neurons, the fourth with 128 neurons, and the final layer with 64 neurons. The progressive reduction in dimensionality enhances feature extraction and classification efficiency. To prevent overfitting, dropout regularization with a probability of 0.5 is applied after the second fully connected layer. The final layer consists of a single neuron with a sigmoid activation function, which outputs a probability value $p \in [0,1]$, representing the likelihood that the given input pair belongs to the similar or non-similar class. The effectiveness of the ViT and EfficientNet is illustrated in Fig. 5, where one can see that vital cues are extracted from ViT and EfficientNet heatmaps, which are used to fuse them to achieve successful classification.

VIT Heatmap EfficientNet Heatmap Combined Heatmap

Fig. 5. Illustrating the contribution of ViT and efficient models for fusing.

4 Experimental Results

To evaluate the proposed classification method, the standard datasets of scene text spotting were used. For each dataset, the proposed classification model classifies the images into similar and non-similar images. To validate the contribution and effectiveness of the key steps, we conducted ablation study experiments, classification accuracy for the classification method and to validate the classification step, the text spotting experiments before and after classification.

4.1 Dataset and Evaluation

In this study, for the classification of similar and non-similar images, we considered four datasets, namely, SVT (Wang et al., 2011), ICDAR 13 (Karatzas et al., 2013), CTW1500 (Yuliang et al., 2017), and Total-Text (Ch'ng et al., 2017) datasets. These datasets are popular for text spotting. For labeling similar and non-similar images for each dataset, the proposed work estimates cosine similarity for the labels generated using the Google Vision API for input and generated images. If the cosine similarity is greater than 80%, the image is classified as similar, and otherwise, it is classified as non-similar. To verify the classified similar and non-similar images, we manually cross-checked all the labels and corrected. The classes are named similarly and non-similar. The threshold for the similarity score is determined empirically as illustrated in Fig. 6, where the average classification rate is estimated for varying threshold values on four datasets. Figure 6 shows that as the cosine score increases, the average classification rate increases. However, the average classification rate starts decreasing after the 80th score. Therefore, 80 is the optimal value for the classification of similar and non-similar scene images.

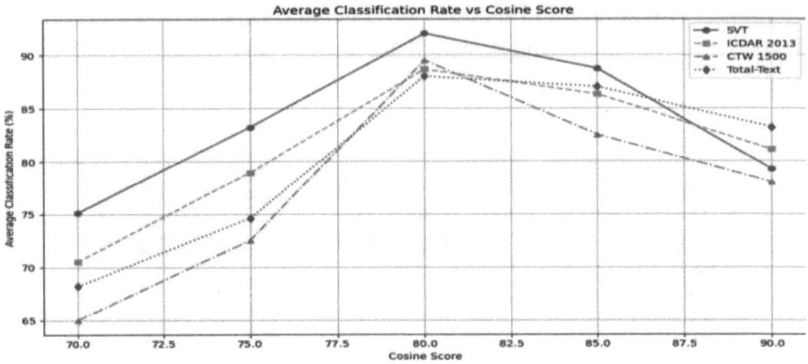

Fig. 6. Average classification rate vs cosine score. Experimentally proving an 80% threshold is an optimal value for the classification of similar and non-similar scene images.

In the SVT (Wang et al., 2011) dataset, 64 images were identified as similar, while the remaining 286 were classified as non-similar. In the ICDAR 2013 (Karatzas et al., 2013) dataset, 79 images were similar and 383 were non-similar. In the CTW1500 (Yuliang et al., 2017) dataset, 184 images were similar and 1316 were non-similar, and in the Total Text (Ch'ng et al., 2017) dataset, 92 images were similar compared to 1468 non-similar images. For measuring the performance of the classification method, we use the average classification accuracy rate, which is the mean of the classification accuracy rate of the similar and non-similar classes, which is recall. For text spotting performance, we use standard measures called recall, precision, and F-score for detection and 'None' for recognition. Here 'None' means the methods do not use lexicons for calculating the recognition rate. The state-of-the-art text spotting methods were executed on each dataset (before classification) and on each similar and non-similar class of each dataset (after classification). It is expected that the methods that use contextual information for text spotting will report better results for similar classes compared to before classification, and vice versa for non-similar classes.

Implementation Details: We implemented our multi-modal deep learning model using the Keras library. Our model is designed to compare real and synthetic images by fusing visual and textual features. The training phase varies with the model's depth and the size of the features employed. On an NVIDIA RTX A550 GPU, the training takes around 242.23 s per epoch. For example, our multi-modal model required approximately 129.53 min to train for 32 epochs using an Adam optimizer with a learning rate of 6×10^{-4} while optimizing binary cross-entropy loss. These implementation details allowed our architecture to learn complex multimodal representations while effectively controlling for overfitting. GPU usage: 12–14 GB (during diffusion + feature extraction), Inference time: 4800–5000 ms (~5 s).

4.2 Ablation Study

It is noted from the proposed methodology that for addressing the challenges of classification of similar and non-similar images, the ViT & the EfficietNet for extracting visual features, the BERT for textual feature extraction and Fully Connected (FC) layers for classification are the key steps. Therefore, to validate the effectiveness and contributions of each key step, we conducted Ablation experiments on different standard scene text datasets for calculating classification rates. The average classification rate for each experiment is listed in Table 1. When we compare the results of the combination of ViT-BERT and EfficientNet-BERT with different classifiers ((i)–(vi)), the performance of both combinations is almost similar. This shows that both combinations contribute equally to achieving the best results. However, the results of the combination of ViT and EfficientNet with BERT perform better than the individual combination, as can be seen in the experiments ((vii)–(ix)). This implies that the combination of ViT and EfficientNet with BERT is the best for the classification of similar and non-similar images. But interestingly, it is observed from the experiments on the Total-Text dataset that the combination of ViT, EfficientNet and BERT with Random Forest (viii) does not perform well compared to EfficientNet with BERT (v). The reason is that the number of images in similar and non-similar classes is imbalanced. This does not affect the proposed combination in (ix). Thus, the proposed model that fuses the merits of ViT, EfficientNet, and BERT with FC achieves the highest results. Hence, we can conclude that the key steps introduced in the proposed methodology section are all effective and contribute equally.

Similarly, the proposed method integrates image and textual features using bilinear fusion. Therefore, to show the effectiveness of the proposed bilinear fusion, the average classification rate is calculated for well-known fusion models and the results are listed in Table 2. Table 2 shows that the bilinear fusion is best for classification compared to other fusion models.

Table 1. Effectiveness of the proposed FC classifiers for classification in terms of average classification rate on four standard datasets

#	Key steps			Dataset			
	Visual Embedding	Textual Embedding	Classifier	SVT	ICDAR 2013	CTW1500	Total-Text
(i)	ViT	BERT	SVM	76.4	70.2	72.2	77.0
(ii)	ViT	BERT	Random Forest	77.5	72.4	72.7	79.5
(iii)	ViT	BERT	FC Layers	80.4	84.2	78.1	77.6
(iv)	EfficientNet	BERT	SVM	75.4	74.3	79.4	71.2
(v)	EfficientNet	BERT	Random Forest	77.2	79.4	81.3	84.2
(vi)	EfficientNet	BERT	FC Layers	78.4	75.7	83.7	80.6
(vii)	ViT + EfficientNet	BERT	SVM	85.6	82.4	84.3	81.2

(continued)

Table 1. *(continued)*

#	Key steps			Dataset			
	Visual Embedding	Textual Embedding	Classifier	SVT	ICDAR 2013	CTW1500	Total-Text
(viii)	ViT + EfficientNet	BERT	Random Forest	87.4	84.6	87.8	85.3
(ix)	**ViT + EfficientNet**	**BERT**	**FC Layers**	**92.2**	**88.65**	**89.5**	**88.00**

Table 2. Effectiveness of the proposed fusion for classification in terms of average classification rate on four standard datasets

Fusion	SVT	ICDAR 2013	CTW1500	Total-Text
ViLBERT	90.4	86.7	87.5	86.5
CLIP	91.6	87.9	88.6	87.3
Bilinear	**92.2**	**88.65**	**89.5**	**88.00**

4.3 Experiments for Classification of Similar and Non-Similar Images

Qualitative results of the proposed method for the classification of similar and non-similar images are shown in Fig. 7(a)-(b), where we can see the proposed method finds similar images and non-similar images for each dataset. The images generated are almost similar to the input images for similar classes, while different for non-similar classes. It is evident from the cosine similarity of the scores of similar and non-similar images. For similar images, the similarity score is higher than 80% while for non-similar images, the score is less than 50%.

Therefore, one can infer that the proposed classification step is accurate and promising. The quantitative results of the proposed method and existing methods on four datasets are reported in Table 3, where it is noted that our model is the best in terms of classification rate for both similar and non-similar classes compared to the existing methods. It is also noted from Table 3 that our method reports consistent results across datasets. This indicates that the proposed approach is reliable and stable for the classification of similar and non-similar images.

However, the existing methods, namely, Thapa et al. (2024), Bounab et al. (2024), and Li et al. (2020), are developed for personality traits image assessment using a multimodal approach. Since the existing model used a multi-modal approach for the classification of different personality trait images, as the proposed multimodal concept, we use the same approach for a comparative study for the classification of similar and non-similar images. Kwenda et al. (2023) and Houby et al. (2024) were developed for forest and X-ray image classification, respectively. We use these methods to compare with the proposed method to show that forest and X-ray image classification methods are not adequate for classification. Although the existing models use deep learning techniques for the classification of images, the methods do not perform well for the classification of similar and non-similar images. The key reason for the poor results is that the features extracted in their methods are not robust to the classification of similar and non-similar scene images. On the other hand, the key steps of the proposed work are effective in addressing the challenges of scene text images. Our method is the best for the classification of similar and non-similar images.

Table 3. A classification rate (CR) of the proposed and existing methods for similar and non-similar classes on four standard scene text datasets

Methods	SVT		ICDAR 2013		CTW1500		Total-Text	
	Similar	Non-Similar	Similar	Non-Similar	Similar	Non-Similar	Similar	Non-Similar
Thapa et al. (2024)	70.5	72.4	74.0	76.4	73.5	74.1	72.4	76.2
Bounab et al. (2024)	74.2	77.9	73.5	74.1	73.0	79.4	73.8	79.1
Li et al. (2020)	82.4	82.1	79.6	81.4	78.1	82.7	72.9	78.2
Kwenda et al. (2023)	78.5	79.4	76.2	77.1	73.9	75.0	72.1	73.0
Houby et al. (2024)	85.2	86.0	83.8	84.4	82.5	83.2	81.3	82.0
Ours	**91.0**	**93.4**	**87.8**	**89.5**	**86.5**	**92.5**	**86.6**	**89.8**

4.4 Text Spotting Performance Before and After Classification

To validate the effectiveness of the proposed classification, we conducted text-spotting experiments before and after the classification. The detection and recognition rates are calculated for the whole dataset, which is called before classification experiments. The detection and recognition rates are calculated for similar and non-similar classes separately, which is called after-classification experiments. For this experiment, we chose the following state-of-the-art text spotting models. ABCNetv2 (Liu et al., 2021), TESTR (Zhang et al., 2022), and DATR (Purkayastha et al., 2025). Out of three, TESTR and DATR are advanced text spotting modes compared to ABCNetv2. Therefore, it is expected that TESTR and DATR are good at extracting contextual features, which include the context between text and its background information in the scene. Thus, the TESTR and the DATR should perform better for the images of similar classes compared to the images of non-similar classes and before classification.

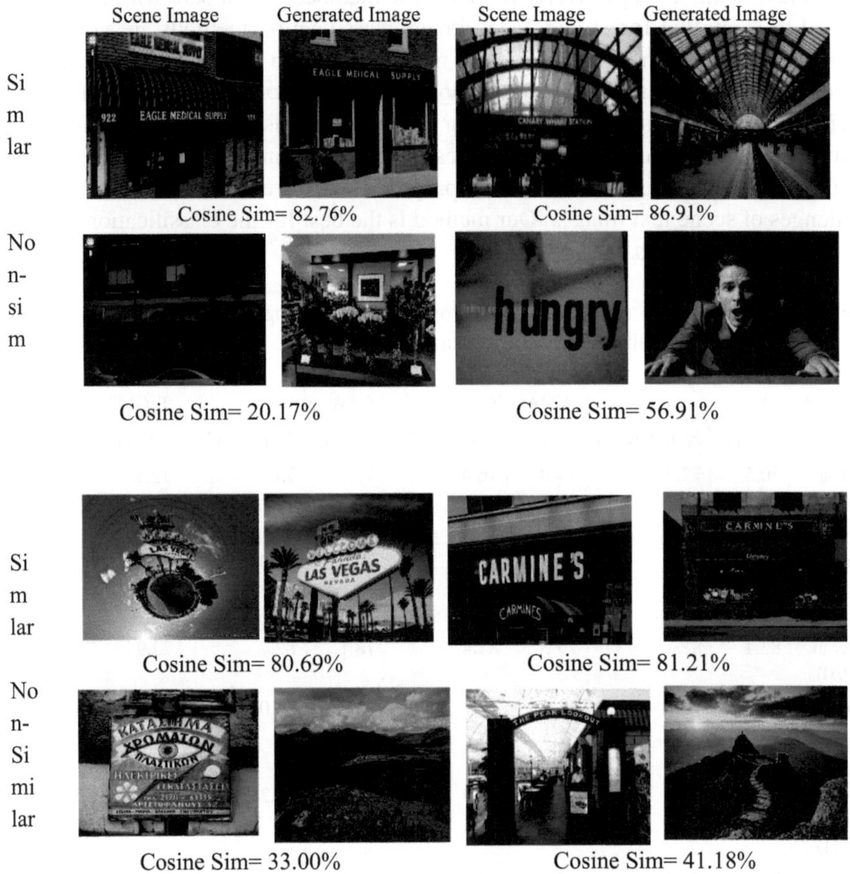

(b) Sample images chosen from CTW1500 and Total-Text were classified successfully

Fig. 7. Examples of classification of similar and non-similar images.

Qualitative results of text spotting for the images of similar and non-similar classes

Fig. 8. Sample text spotting results for the images of similar and non-similar classes. DeepSolo text spotting model is used here.

Qualitative results of the text spotting (DATR) methods for similar and non-similar images are shown in Fig. 8, where one can see the text spotting method performs well for similar images, while it does not perform well for non-similar images. For non-similar images, the DATR model misses a few text instances in the images. This is evident that the advanced methods perform better when there is a correlation between text and non-text regions in the natural scene text images. When the images miss the correlation between scene text and non-text regions, the methods may not be effective.

The same conclusions can be drawn from the quantitative results reported for different datasets in Tables 4, Table 5 and Table 6. The results in Table 4 are obtained before classification. Tables 5 and 6 present the results after the classification of different datasets. It can be noted from Tables 5 and 6 that the performance of the text-spotting methods is better for similar classes compared to non-similar classes in terms of detection and recognition. Even if we compare the performance of the methods of similar classes with the performance of classification, the performance is higher for similar classes. This is true for the detection and recognition of almost all datasets. Thus, one can infer that the text-spotting methods work better if there is a correlation between text and non-text regions in scene images. This observation motivates us to exploit the relationship between scene text and non-text regions instead of throwing non-text regions to improve text spotting performance further. In the same way, to improve the text spotting performance for the images where there is no correlation between scene text and non-text regions, one should explore the models that do not depend on non-text regions in the images. Hence, the classification of similar and non-similar images paves a new direction for the text detection and recognition research community to improve performance in adverse situations.

Table 4. Text Spotting performance of the state-of-the-art methods on different standard datasets (before classification)

Methods	SVT				ICDAR 2013				CTW1500				Total-Text			
	Detection			Recognition	Detection			Recognition	Detection			Recognition	Detection			Recognition
	R	P	F	None	R	P	F	None	R	P	F	None	R	P	F	None
ABCNet v2, (Liu, 2021)	85.3	89.0	87.10	82.4	82.3	86.50	84.30	85.7	83.8	85.6	84.7	57.5	84.1	90.2	87.0	70.4
TESTR (Zhang, 2022)	79.5	87.1	86.9	87.2	82.7	87.9	84.8	86.2	83.1	89.7	86.3	53.3	81.4	93.9	86.9	83.3
DATR (Purkayastha, 2025)	87.4	89.5	88.2	93.7	91.6	93.4	92.7	97.2	85.8	91.7	88.1	64.5	82.3	93.7	87.1	79.4

Table 5. Text Spotting performance of the state-of-the-art methods on different SVT and ICDAR 2013datasets (after classification)

Methods	SVT								ICDAR 2013							
	Similar				Non-Similar				Similar				Non-Similar			
	Detection			Recognition	Detection			Recognition	Detection			Recognition	Detection			Recognition
	R	P	F	None	R	P	F	None	R	P	F	None	R	P	F	None
ABCNet v2, (Liu, 2021)	79.1	87.1	83.1	85.4	85.3	89.0	87.1	81.3	82.5	87.6	85.0	90.5	82.3	85.1	83.70	84.1
TESTR (Zhang, 2022)	82.1	88.6.	85.3	89.2	79.3	83.4	81.3	82.2	83.4	87.7	85.7	88.4	82.7	86.3	84.40	85.1
DATR (Purkayastha, 2025)	88.4	92.1	90.5	96.1	85.2	90.3	87.5	92.6	92.5	95.2	94.1	98.4	89.7	91.1	90.5	94.4

Table 6. Text Spotting performance of the state-of-the-art methods on CTW1500 and Total-Text datasets (after classification)

Methods	CTW1500								Total-Text							
	Similar				Non-Similar				Similar				Non-Similar			
	Detection			Recognition	Detection			Recognition	Detection			Recognition	Detection			Recognition
	R	P	F	None	R	P	F	None	R	P	F	None	R	P	F	None
ABCNet v2, (Liu, 2021)	83.8	84.7	82.3	60.3	82.1	84.3	83.4	53.1	92.5	87.1	85.4	74.2	83.8	84.7	82.3	67.2
TESTR (Zhang, 2022)	80.7	87.6	85.7	58.3	83.1	86.3	81.4	50.9	87.4	94.6	86.9	85.4	84.1	90.2	87.0	82.4
DATR (Purkayastha, 2025)	87.5	95.4	89.5	68.40	86.4	89.5	87.4	62.1	84.1	95.5	89.1	82.1	81.4	93.8	86.1	77.4

4.5 Limitations

Sometimes, when the proposed generator does not match the input images, the classification rate reduces as shown in Fig. 9(a), where the generated image is different from the input image, including text. However, it is misclassified as a similar class. Similarly, for the sample shown in Fig. 9(b), the generated image and the text have a strong correlation, but it is misclassified as a non-similar class because the generated image does not match the input image.

The following are the possible reasons for the above misclassification. Generating the correct image using text depends on the generator; at the same time, the generator depends on the text prompt. The gap between the two different modalities. If this gap is huge, the classification does not work well. This is common when the text does not correlate with the background. For instance, when the text and non-text regions provide different meanings. These are the limitations of the proposed method. This can be solved by exploring CLIP and LLM models, as these models are robust enough to extract the correct meaning of the text and non-text regions without any bias. If the diffusion model integrates with the CLIP, there are chances of reducing the bias and overcoming the above limitations. This is beyond the proposed work, and therefore, this can be considered in the future.

(a) Similar class with cosine scores 81.2

(b) Non-similar class with cosine score 42.4

Fig. 9. Failure case of the proposed classification approach

5 Conclusion and Future Work

In this work, we have presented a novel idea for the classification of similar and non-similar images to improve text detection and recognition (text spotting) performance. Unlike existing methods that use characteristics of text for text spotting, the present work uses the correlation between text and non-text regions for improving text spotting performance. For the input images, the state-of-the-art text spotting model has been used for segmenting text from the background. The segmented text is used as a prompt for generating images using a diffusion generative model (Fooocus). For the generated images and input images, the proposed work combines ViT and EfficientNet for extracting visual features. For the same input and generated images, the labels are obtained using Google Vision API. The texts are supplied to BERT for extracting features. The bilinear fusion approach has been introduced to combine visual and textual features to classify similar and non-similar images, such that appropriate text-spotting methods can be chosen.

To evaluate the proposed classification method, we conducted ablation experiments, text spotting before and after classification. The effectiveness of the classification is validated by testing text-spotting methods on similar and non-similar classes. As discussed in the limitation section, when the generated image does not match input image, the classification performance degrades. Similarly, the gap between the two modalities may cause poor results for the classification. If we explore the combination of diffusion models and language models, we can find a solution to the above limitations. This could be our future work.

References

Baek, Y., Lee, B., Han, D., Yun, S., Lee, H.: Character region awareness for text detection. In: Proceedings of the IEEE/CVF Conference on Computer Vision and Pattern Recognition, pp. 9365–9374 (2019)

Baltrušaitis, T., Ahuja, C., Morency, L.P.: Multimodal machine learning: a survey and taxonomy. IEEE Trans. Pattern Anal. Mach. Intell. **41**(2), 423–443 (2018)

Barua, A., Ahmed, M.U., Begum, S.: A systematic literature review on multimodal machine learning: applications, challenges, gaps and future directions. IEEE Access **11**, 14804–14831 (2023)

Bounab, Y., Oussalah, M., Arhab, N., Bekhouche, S.: Towards job screening and personality traits estimation from video transcriptions. Expert Syst. Appl. **238**, 122016 (2024)

Chen, J., Huang, Y., Lv, T., Cui, L., Chen, Q., Wei, F.: Textdiffuser: diffusion models as text painters. Adv. Neural. Inf. Process. Syst. **36**, 9353–9387 (2023)

Ch'ng, C.K., Chan, C.S.: Total-text: a comprehensive dataset for scene text detection and recognition. In: 2017 14th IAPR International Conference on Document Analysis and Recognition (ICDAR), vol. 1, pp. 935–942. IEEE (2017)

Choi, Y., Choi, M., Kim, M., Ha, J.W., Kim, S., Choo, J.: StarGAN: unified generative adversarial networks for multi-domain image-to-image translation. In: Proceedings of the IEEE Conference on Computer Vision and Pattern Recognition, pp. 8789–8797 (2018)

Devlin, J., Chang, M.W., Lee, K., Toutanova, K.: BERT: pre-training of deep bi-directional transformers for language understanding. In: Proceedings of the 2019 Conference of the North American Chapter of the Association for Computational Linguistics: Human Language Technologies, vol. 1 (long and short papers), pp. 4171–4186 (2019)

Dosovitskiy, A., et al.: An image is worth 16x16 words: transformers for image recognition at scale. arXiv preprint arXiv:2010.11929 (2020)

El Houby, E.M.: Covid-19 detection from chest x-ray images using transfer learning. Sci. Rep. **14**(1), 11639 (2024)

Huang, X., Liu, M.Y., Belongie, S., Kautz, J.: Multimodal unsupervised image-to-image translation. In: Proceedings of the European Conference on Computer Vision (ECCV), pp. 172–189 (2018)

Isola, P., Zhu, J.Y., Zhou, T., Efros, A.A.: Image-to-image translation with conditional adversarial networks. In: Proceedings of the IEEE Conference on Computer Vision and Pattern Recognition, pp. 1125–1134 (2017)

Karatzas, D., et al.: ICDAR 2013 robust reading competition. In: 2013 12th International Conference on Document Analysis and Recognition, pp. 1484–1493. IEEE (2013)

Kiela, D., Bhooshan, S., Firooz, H., Testuggine, D.: Supervised multimodal bi-transformers for classifying images and text. CoRR abs/1909.02950 (2019). http://arxiv.org/abs/1909.02950

Kwenda, C., Gwetu, M.V., Fonou-Dombeu, J.V.: Forest image classification based on deep learning and XGBoost algorithm. In: International Conference on Computational Science, pp. 217–229. Springer (2023)

Li, Y., et al.: CR- Net: a deep classification-regression network for multimodal apparent personality analysis. Int. J. Comput. Vision **128**, 2763–2780 (2020)

Li, S., Tang, H.: Multimodal alignment and fusion: a survey. arXiv preprint arXiv:2411.17040 (2024)

Liu, Y., et al.: ABCNet V2: adaptive bezier-curve network for real-time end-to-end text spotting. IEEE Trans. Pattern Anal. Mach. Intell. **44**(11), 8048–8064 (2021)

Purkayastha, K., Sarkar, S., Palaiahnakote, S., Pal, U., Ghosal, P.: DATR: domain agnostic text recognizer. In: International Conference on Pattern Recognition, pp. 220–235. Springer (2025)

Rombach, R., Blattmann, A., Lorenz, D., Esser, P., Ommer, B.: High-resolution image synthesis with latent diffusion models. In: Proceedings of the IEEE/CVF Conference on Computer Vision and Pattern Recognition, pp. 10684–10695 (2022)

Tan, M., Le, Q.: EfficientNet: rethinking model scaling for convolutional neural networks. In: International Conference on Machine Learning, pp. 6105–6114. PMLR (2019)

Thammarak, K., Kongkla, P., Sirisathitkul, Y., Intakosum, S.: Comparative analysis of tesseract and google cloud vision for that vehicle registration certificate. Int. J. Electr. Comput. Eng. **12**(2), 1849–1858 (2022)

Thapa, L., Pandey, A., Gupta, D., Deep, A., Garg, R.: A framework for personality prediction for e-recruitment using machine learning algorithms. In: 2024 14th International Conference on Cloud Computing, Data Science Engineering (Confluence), pp. 1–5 (2024)

Wang, K., Babenko, B., Belongie, S.: End-to-end scene text recognition. In: 2011 International Conference on Computer Vision, pp. 1457–1464. IEEE (2011)

Wang, W., et al.: Shape robust text detection with progressive scale expansion network. In: Proceedings of the IEEE/CVF Conference on Computer Vision and Pattern Recognition, pp. 9336–9345 (2019)

Ye, M., et al.: DeepSolo: let trans former decoder with explicit points solo for text spotting. In: Proceedings of the IEEE/CVF Conference on Computer Vision and Pattern Recognition, pp. 19348–19357 (2023)

Yuliang, L., Lianwen, J., Shuaitao, Z., Sheng, Z.: Detecting curve text in the wild: new dataset and new solution. arXiv preprint arXiv:1712.02170 (2017)

Zhang, L., Chen, X., Wang, Y., Lu, Y., Qiao, Y.: Brush your text: synthesize any scene text on images via diffusion model. In: Proceedings of the AAAI Conference on Artificial Intelligence, vol. 38, pp. 7215–7223 (2024)

Zhang, L.: Fooocus [Computer software] (2024). https://github.com/lllyasviel/Fooocus

Zhang, X., Huang, Y., Zou, Q., Pei, Y., Zhang, R., Wang, S.: A hybrid convolutional neural network for sketch recognition. Pattern Recogn. Lett. **130**, 73–82 (2020)

Zhang, X., Su, Y., Tripathi, S., Tu, Z.: Text spotting transformers. In: Proceedings of the IEEE/CVF Conference on Computer Vision and Pattern Recognition, pp. 9519–9528 (2022)

Zhou, X., et al.: East: an efficient and accurate scene text detector. In: Proceedings of the IEEE Conference on Computer Vision and Pattern Recognition, pp. 5551–5560 (2017)

Zhu, J.Y., Park, T., Isola, P., Efros, A.A.: Unpaired image-to-image translation using cycle-consistent adversarial networks. In: Proceedings of the IEEE International Conference on Computer Vision, pp. 2223–2232 (2017a)

Zhu, J.Y., et al.: Toward multimodal image-to-image translation. Adv. Neural Inf. Process. Syst. **30** (2017b)

Enhancing Document VQA Models via Retrieval-Augmented Generation

Eric López(✉), Artemis Llabrés, and Ernest Valveny

Computer Vision Center, Universitat Autònoma de Barcelona, Barcelona, Spain
Eric.LopezCe@autonoma.cat, {allabres,ernest}@cvc.uab.cat

Abstract. Document Visual Question Answering (Document VQA) must cope with documents that span dozens of pages, yet leading systems still concatenate every page or rely on very large vision-language models, both of which are memory-hungry. Retrieval-Augmented Generation (RAG) offers an attractive alternative, first retrieving a concise set of relevant segments before generating answers from this selected evidence. In this paper, we systematically evaluate the impact of incorporating RAG into Document VQA through different retrieval variants—text-based retrieval using OCR tokens and purely visual retrieval without OCR—across multiple models and benchmarks. Evaluated on the multi-page datasets MP-DocVQA, DUDE, and InfographicVQA, the text-centric variant improves the "concatenate-all-pages" baseline by up to +22.5 ANLS, while the visual variant achieves +5.0 ANLS improvement without requiring any text extraction. An ablation confirms that retrieval and reranking components drive most of the gain, whereas the layout-guided chunking strategy—proposed in several recent works to leverage page structure—fails to help on these datasets. Our experiments demonstrate that careful evidence selection consistently boosts accuracy across multiple model sizes and multi-page benchmarks, underscoring its practical value for real-world Document VQA.

Keywords: Document Visual Question Answering · Multi-Page Scenario · Multi-Modal Scenario · Retrieval-Augmented Generation

1 Introduction

Document VQA aims to enable models to answer questions about the contents of documents, requiring multimodal reasoning over textual, spatial, and visual elements. While early research mainly addressed single-page scenarios through popular benchmarks such as DocVQA, the majority of real-world documents—including manuals, scientific papers, and technical reports—extend across multiple pages. Addressing this gap, MP-DocVQA was recently proposed as a multi-page extension of the original DocVQA dataset, posing natural language questions over sequences of up to twenty pages, accompanied by OCR-extracted text and page images.

Project page: https://pikurrot.github.io/RAG-DocVQA/.

L. Jin et al. (Eds.): ICDAR 2025 Workshops, LNCS 16225, pp. 92–107, 2026.
https://doi.org/10.1007/978-3-032-09368-4_6

Existing state-of-the-art multi-page models, such as Hi-VT5, leverage hierarchical transformer architectures to summarize individual pages before synthesizing the final answer. However, their performance deteriorates as the document length grows, since cross-page reasoning significantly increases complexity. While Large Visual Language Models (LVLMs) such as Qwen2.5-VL can handle long-context inputs, they require substantial computational resources and GPU memory, making deployment costly and inefficient.

A promising alternative is Retrieval-Augmented Generation (RAG), which has recently demonstrated effectiveness in open-domain question answering by first retrieving compact subsets of relevant evidence, thus drastically reducing the inference cost and noise introduced by irrelevant content. In the context of multi-page Document VQA, a RAG pipeline involves segmenting documents into concise textual chunks or visual patches, retrieving the most pertinent segments based on the question, and subsequently generating answers conditioned solely on these retrieved segments.

In this paper, we hypothesize that augmenting existing small and medium-sized Document VQA models with RAG can significantly mitigate the challenges posed by multi-page scenarios. Specifically, we argue that this approach yields performance improvements in both accuracy and computational efficiency across various models and datasets. To validate this hypothesis, our contributions are as follows:

- We propose and implement textual RAG variants for two representative Vision Language Models: VT5 and Qwen2.5-VL-7B-Instruct, along with a visual counterpart for Pix2Struct, an OCR-free vision transformer model.
- Through comprehensive experiments on MP-DocVQA, DUDE, and InfographicVQA, we demonstrate consistent improvements in question-answering accuracy, ANLS scores, and retrieval metrics, thereby validating our hypothesis.

2 Related Work

2.1 Datasets and Task Evolution

Document Visual Question Answering was initially formalized with the DocVQA dataset [20], comprising 50,000 questions posed over more than 12,000 industry-related single-page document images, where answers are explicitly present in the OCR-extracted text. InfographicVQA [19] expanded the task visually, introducing infographic-rich images accompanied by questions that occasionally require numerical or categorical reasoning beyond mere text extraction. Subsequently, DocCVQA [22] introduced a retrieval-oriented scenario, where questions are answered by retrieving relevant evidence from a collection of 14,362 single-page documents. Recognizing the limitations of single-page scenarios, MP-DocVQA [23] extended the original DocVQA dataset to multi-page contexts by augmenting each document-question pair with preceding and following pages, forming sequences of up to 20 pages. Concurrently, DUDE [15] provided diverse multi-page documents from various domains, featuring questions that require multi-step reasoning and both extractive and abstractive answers. Additionally, the recent MMLongBench-Doc dataset [18] features 1,091 questions over 135 long PDF documents with rich layouts, where 33% of the questions require cross-page reasoning and 22.5% are unanswerable, which tests model robustness.

2.2 From OCR-Based to OCR-Free Approaches

Initial Document VQA models largely relied on OCR-extracted text, adapting pre-trained language models such as BERT [8] with a classification head to predict the start and end indices where the answer appeared in the text, and T5 [21] for generative extractive question answering from OCR tokens. LayoutLM [28] introduced spatial understanding by augmenting textual embeddings from BERT with 2-dimensional positional information derived from OCR bounding boxes. LayoutLMv2 [27] then integrated visual embeddings from the document image and added relative positional biases to the self-attention mechanism. LayoutLMv3 [12] further refined this by encoding images as patch-level embeddings. To further leverage visual modalities, DocFormer [1] combined textual, visual, and spatial features via a novel multi-modal self-attention mechanism, using a ResNet [10] backbone for image features. Later, DocFormerv2 [2] replaced the ResNet CNN with a linear projection of image patches and adopted a T5-based encoder-decoder architecture. Addressing multi-page reasoning, Hi-VT5 [23] introduced a hierarchical transformer where the encoder summarizes each page independently, and the decoder attends to these summaries to generate the answer. GRAM proposed inserting additional document-level transformer layers between page-level encodings, using document tokens to propagate information across pages, achieving state-of-the-art performance when combined with DocFormerv2. More recently, OCR-free models emerged, eliminating explicit text extraction. Pix2Struct [16] proposed a fully end-to-end approach without explicit OCR, pretraining a vision transformer encoder to transcribe masked web screenshots into simplified HTML. Nevertheless, handling multi-page scenarios remained challenging due to growing computational complexity.

2.3 Generic Vision-Language Models

Recently, generic vision-language models (VLMs), such as Qwen2.5-VL [3], have started being explored for Document VQA tasks. These models can directly process combined visual and textual inputs thanks to extensive pre-training on large multimodal datasets. However, a major issue with these models is their high computational cost and GPU memory usage, making practical deployment difficult. To mitigate such limitations, retrieval-based strategies like ColPali [9] have been adopted, which extends the PaliGemma [4] vision-language model with a ColBERT-style [14] late interaction mechanism, generating visual page embeddings optimized for retrieval tasks, thus improving evidence selection prior to answer generation.

3 Methodology

We adapt Retrieval-Augmented Generation (RAG) to multi-page Document VQA through a three-stage pipeline comprising indexing, retrieval, and generation. In particular, we propose two complementary variants: a textual RAG, applied to two language-based models—a smaller encoderdecoder (VT5) and a larger visionlanguage model (Qwen2.5-VL-7B)—and a visual RAG, where retrieval and generation operate directly on image inputs without explicit OCR, using Pix2Struct as our baseline model (Fig. 1).

3.1 Textual RAG

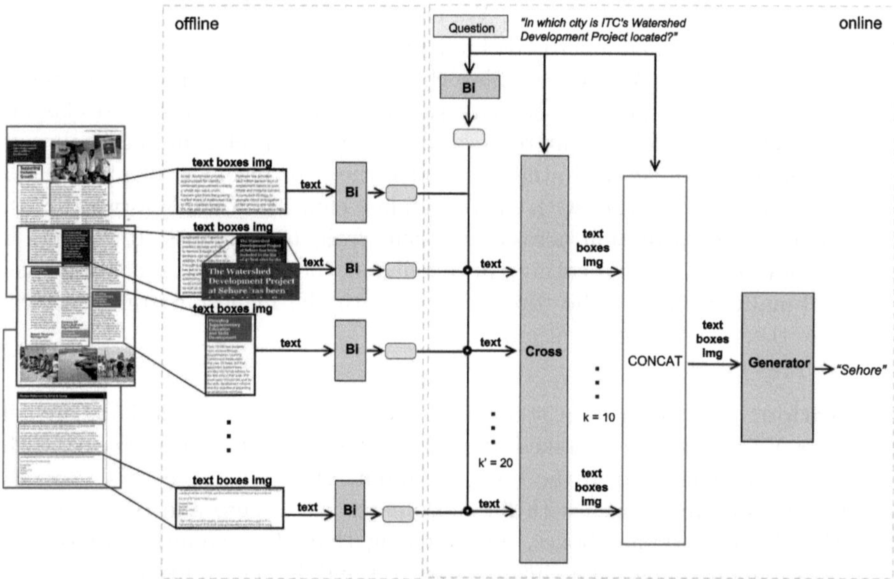

Fig. 1. Overview of the **Textual RAG** pipeline. **Offline (pink panel):** A multi-page document is segmented into chunks consisting of OCR (text and boxes) and the image crop of the chunk. A Bi-Encoder converts the chunk text into a dense embedding for later retrieval. **Online (green panel):** the user question is encoded in the same way, and cosine-similarity is used to select the top-k' chunks, whose text is then passed to a Cross-Encoder (Reranker) for a more refined filtering and ranking. The full information of the highest-ranked k chunks is then concatenated and fed to the Generator to produce an answer. (Color figure online)

Indexing. We first segment the OCR-extracted token sequences of the multi-page documents into textual chunks. The chunking process is controlled by three parameters: chunk size L (tokens per chunk), overlap O (shared tokens between neighbors), and chunk-size tolerance τ, which allows the last chunk on a page to expand to $(1+\tau) \times L$ tokens, so that small remainders are merged rather than isolated. Each chunk text is then encoded offline into a fixed-dimensional vector embedding using a bi-encoder architecture. Specifically, we use *bge-en-small-v1.5* [26], which has proven to be the most effective and efficient. We fine-tune this embedding model using contrastive learning on dataset-specific querychunk pairs, optimizing a Multiple-Negatives Ranking Loss (details in Sect. 4). Finally, we store each chunk's tokens, bounding boxes, image crop (extracted from the chunk bounding box), and embedding for fast retrieval.

Retrieval. At inference time, we encode the input query (q) using the same embedding model with shared weights, obtaining an embedding vector of the same dimensionality as the chunks. We compute the cosine similarity between the query embedding and all stored chunk embeddings, retrieving an initial candidate set of the top k' most similar chunks. To further enhance precision, we employ an additional cross-encoder reranker—a transformer-based classification model trained specifically to discriminate between relevant and irrelevant querychunk pairs. Unlike the initial bi-encoder, the cross-encoder reranker jointly encodes each query together with each candidate chunk in a single forward pass, leveraging full self-attention across all tokens from both inputs. It outputs a refined relevance score indicating the likelihood that the chunk contains the answer. Finally, we select the top k chunks with the highest reranked scores, which are then used for generation.

Generation. The generation step differs according to the underlying model used. For VT5, the k top-ranked chunks' text tokens are concatenated with the question tokens to form the semantic input. Question tokens are assigned placeholder bounding boxes with all coordinates set to zero and concatenated with the chunks' original boxes, guaranteeing that the semantic and spatial sequences remain aligned. The image crops of the different chunks are arranged into a compact adaptive grid, minimizing the total area and thereby optimizing input efficiency. As in Hi-VT5, the three input modalities—semantic, spatial, and visual—are first embedded separately. The semantic and spatial information of each token is fused by summing their corresponding embeddings:

$$\mathcal{E}_i = E_O(O_i) + E_x(x_0^i) + E_y(y_0^i) + E_x(x_1^i) + E_y(y_1^i) \tag{1}$$

where $E_O(O_i)$ is the semantic embedding of OCR token O_i, produced by a T5 language backbone, and E_x, E_y are learned embeddings for the bounding box coordinates.

The image crops grid is encoded into an embedding V by a Document Image Transformer (DIT), specifically *dit-base-finetuned-rvlcdip*. We then concatenate all the question embeddings $\mathcal{E}^q$, the text embeddings $\mathcal{E}^o$, and the visual patch embeddings V to form the final input sequence $[\mathcal{E}^q; \mathcal{E}^o; V]$ which is fed to the

VT5 decoder to generate the answer. This generation process for VT5 is illustrated in Fig. 3 (top-left).

For Qwen2.5-VL-7B, the retrieved chunks' texts are concatenated with the question and a prompt (*"Directly provide only a short answer to the question."*) to form a single textual input sequence. Each chunk's corresponding image crop is embedded independently by Qwen's vision encoder and integrated within the textual input through special placeholders. The combined multimodal embeddings are then processed by Qwen's decoder, autoregressively generating the final answer. The generation procedure for Qwen2.5-VL-7B is depicted in Fig. 3 (top-right) (Fig. 2).

3.2 Visual RAG

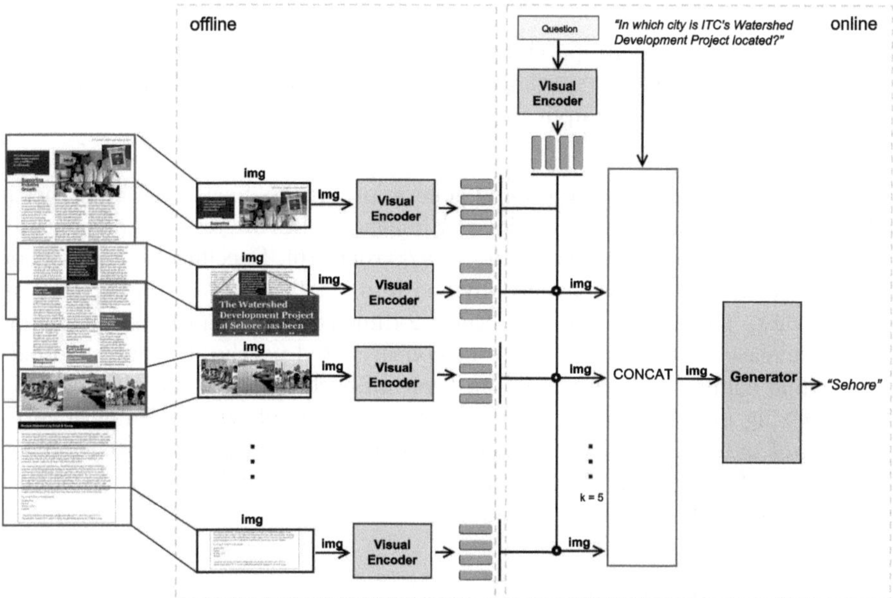

Fig. 2. Overview of the **Visual RAG** pipeline. **Offline (pink panel):** A multi-page document is segmented into image patches, each of which is passed to a Visual Encoder to produce multi-vector embeddings for later retrieval. **Online (green panel):** the user question is rendered as an image, encoded in the exact same way and matched to the patch embeddings through late interaction. The top-k image patches are concatenated (see Fig. 3 (bottom)) and passed to the generator. (Color figure online)

Indexing. Document images are horizontally segmented into overlapping visual patches. Each patch has a fixed vertical pixel size (patch size P), and overlaps by half of this height with adjacent patches to ensure no crucial information is lost across patch boundaries. Each visual patch is encoded independently into multi-vector embeddings E_p of shape $T_{\text{img}} \times 768$ using a Visual Encoder, where T_{img} is the number of image tokens. Specifically, we use the Pix2Struct vision transformer encoder, which works with an image sequence length of $T_{\text{img}} = 2048$.

Retrieval. For retrieval, the textual query is rendered as an image on a white background and embedded into multi-vector representations E_q of shape $T_{\text{img}} \times 768$ using the same Pix2Struct encoder as used for the document patches. We perform late interaction retrieval following ColBERT [14], which computes token-level similarity $S_{q,p}$ between the query q and each candidate patch p, allowing fine-grained matching:

$$S_{q,p} = \sum_{i \in [|E_q|]} \max_{j \in [|E_p|]} E_{q_i} \cdot E_{p_j} \tag{2}$$

The top-k most relevant patches are selected based on their similarity scores. If retrieved patches overlap spatially, overlapping regions are merged into single unified patches, removing duplicate visual information.

Generation. The concatenation and generation process for RAG-Pix2Struct is shown in Fig. 3 (bottom). The retrieved k image patches are resized and concatenated to form the input sequence for Pix2Struct [16]. Specifically, each image patch is scaled so it can be evenly divided into small, non-overlapping 16×16 mini-patches. These mini-patches receive 2-dimensional positional indices: row indices increase continuously across the concatenated patches (from top to bottom), while column indices restart at 1 within each patch (from left to right). This indexing preserves each patch's internal 2-D layout while forming a vertical stack of patches. Finally, the resulting sequence of T_{img} mini-patches is fed directly into Pix2Struct, which encodes them using its built-in 2-D positional encoder to generate the final answer.

4 Experimental Setup

4.1 Evaluation Metrics

We employ metrics for two complementary aspects of the task: (i) question answering, and (ii) retrieval quality of relevant document content.

(i) Question Answering Metrics. To evaluate the final output of the generator, we use standard accuracy and ANLS (Average Normalized Levenshtein Similarity), which are commonly used in Document VQA benchmarks to measure the correctness and textual similarity of predicted answers with respect to ground truth.

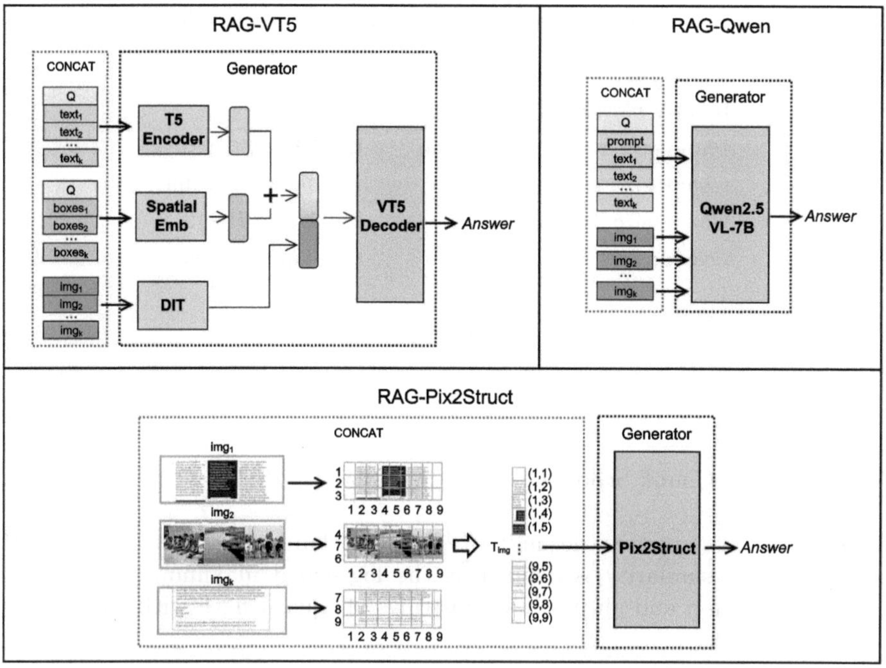

Fig. 3. Different concatenation and generation strategies for the proposed RAG methods. For RAG-VT5, the information of different modalities is separately concatenated and fed to a specialized encoder, before aggregating and passing it to the decoder. For RAG-Qwen, the text of the chunks is concatenated along with the question and an instruction prompt, and is passed together with the separate chunk image crops to the model. For RAG-Pix2Struct, only the image crops are processed, by tiling them into mini-patches, assigning a positional coordinate to each one and feeding them to the model.

(ii) Retrieval Quality Metrics. To assess the effectiveness of our retrieval pipeline, we use two additional metrics:

• *Page-level retrieval.* We adapt the retrieval precision metric, originally used by existing multi-page models to evaluate their page-retrieval modules, to measure whether any of the retrieved chunk comes from the correct page. Specifically, we define Retrieval Precision@k as:

$$\text{Retrieval Precision@}k \;=\; \mathbf{1}\Big[\;\exists\, j \in \{1,\ldots,k\} \;:\; p_{\text{gt}} = p_j^{\text{pred}}\Big] \tag{3}$$

where p_{gt} is the ground-truth answer page and p_j^{pred} is the page associated with the j-th retrieved chunk ($j = 1, \ldots, k$).

• *Chunk-level semantic matching.* We introduce Chunk Score@k as a smooth measure of how similar the ground-truth answer is to any part of the retrieved text, even if not an exact match:

Table 1. Key hyper-parameters used for the Textual RAG models (left) and the visual RAG (right).

RAG-VT5 / RAG-Qwen

Parameter	Value
Chunk size L	60
Chunk size Tolerance τ	0.2
Overlap O	10
k'	20
k	10

RAG-Pix2Struct

Parameter	Value
Patch size P	512
Overlap O_{pix}	256
k	5

$$\text{Chunk Score@}k = \log_2\left(1 + \max_{1\leq j\leq k} \text{sim}(c_j, a)\right) \tag{4}$$

where $\text{sim}(c_j, a)$ computes the similarity between chunk c_j and the ground-truth answer a. This similarity is defined as the log-scaled maximum inverse-edit-distance between a and all possible substrings of equal length within c_j.

4.2 Implementation Details

The specific hyperparameter values shown in Table 1 are the ones that yield better results in our experimentation. For the bi-encoder, we used bge-en-small-v1.5 [26] to encode both the query and the document chunks, and for the cross-encoder we used bge-reranker-v2-m3 [6,17] to rerank and filter the retrieved chunks.

To enable a fair multi-page comparison between our RAG variants and their single-page baselines, we extend each baseline with a lightweight multi-page strategy. For VT5, we adopt a "concatenate" approach analogous to the T5-concat method in [23], merging the OCR tokens—and, in our case, the page screenshots—of all pages into a single input context. For Pix2Struct, concatenating complete page images would over-shrink text, so we instead process each page independently, generate an answer per page, and retain the answer with the highest confidence; we refer to this variant as Pix2Struct-MaxConf.

4.3 Training

The bi-encoder embedding model is fine-tuned using contrastive learning. For this, we create a synthetic dataset of anchor-positive pairs, where each anchor is a query and the corresponding positive is a chunk that previously yielded good retrieval performance. Specifically, we first process each document VQA dataset through our RAG pipeline with the original embedding model. For each retrieved set of k chunks, we individually pass each chunk to the generator, obtaining k separate answers. These answers are compared with the ground truth, and the chunk producing the highest ANLS is selected for that sample. Querychunk pairs that achieve an ANLS greater than a threshold t, which we empirically set to 0.8, form the positive training set. The bi-encoder is subsequently fine-tuned using a Multiple Negatives Ranking Loss [11], where for a given positive pair (query q_1, chunk c_1), the other chunks $c_2, ..., c_b$ in the batch serve as negatives.

Due to the sparse, discontinuous, and contextually diverse nature of retrieval-augmented inputs, we further fine-tune the generator models on each target VQA dataset, keeping the bi-encoder embedding weights frozen. For VT5, we perform full fine-tuning of all layers—including the language backbone and the spatial and visual embedding modules—using the AdamW optimizer with a linear learning-rate scheduler, an initial learning rate of 2e−4, 1000 warm-up steps, and 4 epochs. For Qwen2.5-VL-7B-Instruct, we employ parameter-efficient Low-Rank Adaptation (LoRA), applying it specifically to the query and value projection matrices with parameters $\alpha = 16$, rank $= 8$ and dropout $= 0.05$. Training for both the embedding model and VT5 was performed on an NVIDIA TITAN Xp GPU (12 GB). The Qwen LoRA fine-tuning was conducted on an NVIDIA L40S GPU (46 GB). For Pix2Struct, we use versions already fine-tuned on each target dataset, so no further training was required.

5 Results

5.1 Quantitative Results

Table 2 compares four publicly available models of similar parameter scale, the single-page baselines adapted to multi-page input (Pix2Struct-baseline, VT5-baseline, Qwen2.5-VL-baseline), and their corresponding RAG versions. RAG consistently boosts the textual models: RAG-VT5 gains +13.2, +7.3, and +10.6 ANLS on MP-DocVQA, DUDE, and InfographicVQA, respectively, while RAG-Qwen improves by +22.5, +4.0, and +2.0 on the same datasets. For the visual model, RAG-Pix2Struct yields moderate gains on MP-DocVQA (+5.0 ANLS) and InfographicVQA (+3.3 ANLS) and a slight drop on DUDE (-0.8 ANLS). We attribute this to DUDE's higher reasoning complexity—scenarios where processing each full-page image (as in the baseline) can retain crucial global context that retrieval-based pruning may omit. Apart from this outlier, the gains are consistent across models and datasets, confirming the general effectiveness of retrieval-augmented generation for multi-page Document VQA. Unlike Pix2Struct-SA-retrieval, which recomputes full self-attention between the question and page features at inference time, our RAG-Pix2Struct encodes all patches offline and uses a lightweight late-interaction step. While the performance is not as high, this cuts GPU cost and latency, making real-world application more feasible.

5.2 Qualitative Results

Figure 4 illustrates qualitative examples comparing the performance of Textual and Visual RAG on two representative samples. In Example 1, we present a scenario in which a question is posed to a 17-page document. Textual retrieval

Table 2. Comparison of published multi-page Document VQA systems, our multi-page baselines, and their retrieval-augmented variants.

Method	Params (M)	ANLS (%)		
		MP-DocVQA	DUDE	InfoVQA
Published SOTA				
T5-concat [23]	223	50.5	38.7	—
Pix2Struct-SA-retrieval [13]	273	62.0	—	—
Hi-VT5 [23]	316	61.8	35.7	—
GRAM [5]	281	**73.9**	**46.2**	—
Multi-page baselines				
Pix2Struct-baseline	282	49.1	18.9	30.2
VT5-baseline	223	50.0	36.0	21.1
Qwen2.5-VL-baseline	7 000	51.2	35.9	61.6
Retrieval-augmented variants				
RAG-Pix2Struct (ours)	282	54.1	18.1	33.5
RAG-VT5 (ours)†	223 + 601	63.1	43.3	31.7
RAG-Qwen2.5-VL (ours)†	7 000 + 601	73.7	39.9	**63.6**

† Textual RAG totals include an additional 601M parameters (33M bi-encoder + 568M cross-encoder) used in the retrieval stack.

selects the top relevant text chunks, concatenates them, and passes them to the generator. Similarly, visual retrieval identifies the top image patches. Here, both models generate a correct answer, as the retrieved evidence clearly contains the necessary information.

In contrast, Example 2 depicts a case where, despite accurate retrieval, the generated answers are incorrect. For textual retrieval, the error arises because the retrieved chunks provide ambiguous context; specifically, two different retrieved tables could individually answer the question, but only one is appropriate (the first retrieved chunk corresponds to non-smokers). For visual retrieval, the model fails to identify the correct column within the retrieved image patches, as crucial contextual information—the column headers distinguishing between smoker categories—is missing.

6 Ablation Study

We conduct an ablation study to assess the contribution of individual components in our RAG pipeline, using the RAG-VT5 variant on the MP-DocVQA validation set. Alongside standard performance metrics (ANLS and accuracy), we also report retrieval-specific metrics—retrieval precision and chunk score—as defined in Sect. 4. The analysis is divided into two parts: (i) the impact of retrieval enhancements (reranking and embedding fine-tuning), and (ii) the effect of layout-guided chunking.

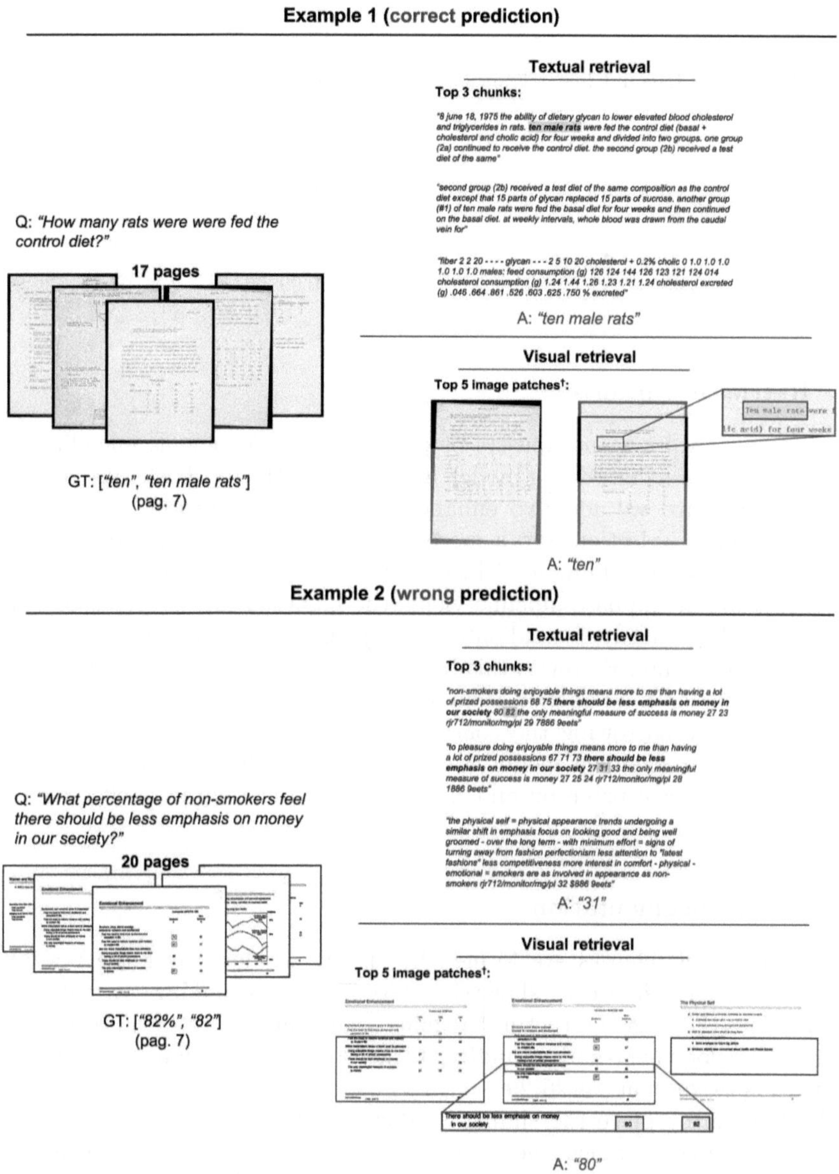

Fig. 4. Qualitative examples of Question Answering on long-context multi-page samples of MP-DocVQA. Textual RAG uses VT5 as generator, while Visual RAG uses Pix2Struct. Example 1 shows a correct retrieval and correct model generation, while Example 2 shows a correct retrieval but a wrong generation. † While 5 image patches are retrieved, some may be overlapping so they are merged into a bigger one.

Table 3. Ablation of **RAG-VT5 on MP-DocVQA validation set**: impact of adding reranking, embedder fine-tuning, layout segmentation, spatial clustering, and layout loss.

Rerank	Train Emb.	Layout	Cluster	Train Layout	ANLS	Acc	Ret. Prec.@k	Chunk Score@k
					58.23	50.74	**97.20**	**98.18**
✓					61.06	53.57	95.39	97.78
✓	✓				**61.46**	**53.82**	96.20	97.97
✓	✓	✓			57.92	49.01	95.41	95.55
✓	✓	✓	✓		58.45	49.54	96.26	95.37
✓	✓	✓	✓	✓	59.14	49.99	96.26	95.37

6.1 Retrieval Variants

The base retrieval system consists of chunking the document, encoding each chunk with a frozen embedding model, and selecting the top-k' chunks via cosine similarity. To this, we add two enhancements: a cross-encoder reranker that refines the top-k' selection into a more accurate top-k subset, and contrastive fine-tuning of the embedding model to improve the quality of initial retrieval. As shown in Table 3, the base pipeline yields the highest retrieval precision, likely because it selects a larger candidate set (k), increasing the chance of including the correct chunk. However, this broader context introduces more irrelevant content, which can hurt answer accuracy. Adding the reranker improves ANLS and accuracy by narrowing the context to the most relevant chunks. Further fine-tuning the embedding model does not significantly raise ANLS but leads to improved retrieval precision and chunk score, indicating more effective early filtering.

6.2 Layout Segmentation

While our default chunks are formed by sliding windows over the raw OCR tokens, several recent works show that layout-aware chunking—in which segments follow the page's structural regions—can improve retrieval accuracy or downstream document-understanding tasks such as question answering and information extraction [24,25,29].

We therefore replace the sliding-window chunks with regions produced by a Document Image Transformer (DIT). Specifically, we use *cmarkea/dit-base-layout-detection* [7], a DIT that outputs 11 fine-grained labels per page. Small boxes or boxes overlapping more than 50% with larger ones are considered noise and discarded. Labels are simplified to four categories: *title*, *text*, *figure*, and *table*. Centroids of these boxes form an undirected graph with edge weights as inverse Euclidean distances. Spectral clustering is then applied, selecting the number of clusters per page by maximizing the silhouette coefficient. Clusters are merged into rectangular layout chunks.

During training, each OCR token receives a layout embedding scaled by a factor of 10 to match the semantic and spatial embeddings and summed with them. Additionally, a linear head attached to every encoder output predicts each token's layout label, and the resulting cross-entropy loss is added to the main VQA loss.

Table 3 shows that layout-based chunks hurt retrieval precision and, despite a modest recovery after clustering and the auxiliary loss, never surpass the simpler window strategy. We attribute this to over-segmentation of pages, which fragments the context and prevents the generator to answer with the enough information.

7 Conclusion

In this work, we revisited multi-page Document VQA through the lens of Retrieval-Augmented Generation. We adapted RAG pipelines—text-based for VT5 and Qwen2.5-VL, and purely visual for Pix2Struct—to efficiently select and aggregate evidence from long documents. Across three multi-page benchmarks, our approach consistently improved performance over strong baselines, achieving up to +22 ANLS gains while avoiding the memory demands of processing full document sequences. Through ablation studies, we found that reranking and contrastive embedder fine-tuning are key to these gains, whereas layout-guided chunking—despite success in other document tasks—did not yield benefits on our benchmarks. These results demonstrate that careful evidence selection enables smaller models to effectively tackle real-world, multi-page document VQA.

Acknowledgments. This research has been supported by the Consolidated Research Group 2021 SGR 01559 from the Research and University Department of the Catalan Government, and by project PID2023-146426NB-100 funded by MCIU/AEI/10.13039/501100011033 and FEDER, UE.

This work has also been funded by the European Lighthouse on Safe and Secure AI (ELSA) from the European Union's Horizon Europe programme under grant agreement No 101070617.

With the support of the FI SDUR predoctoral grant program from the Department of Research and Universities of the Generalitat de Catalunya and co-financing by the European Social Fund Plus (2024FISDU_00095).

References

1. Appalaraju, S., Jasani, B., Kota, B.U., Xie, Y., Manmatha, R.: DocFormer: end-to-end transformer for document understanding. In: Proceedings of the IEEE/CVF International Conference on Computer Vision (ICCV), pp. 993–1003 (2021)
2. Appalaraju, S., Tang, P., Dong, Q., Sankaran, N., Zhou, Y., Manmatha, R.: DocformerV2: local features for document understanding. In: Proceedings of the AAAI Conference on Artificial Intelligence, vol. 38, no. 2, pp. 709–718 (2024). https://doi.org/10.1609/aaai.v38i2.27828, https://ojs.aaai.org/index.php/AAAI/article/view/27828

3. Bai, S., et al.: Qwen2.5-VL technical report. CoRR abs/2502.13923 (2025). arXiv:2502.13923
4. Beyer, L., et al.: PaliGemma: a versatile 3B VLM for transfer. CoRR abs/2407.07726 (2024). arXiv:2407.07726
5. Blau, T., et al.: GRAM: Global reasoning for multi-page VQA. In: Proceedings of the IEEE/CVF Conference on Computer Vision and Pattern Recognition (CVPR), pp. 15598–15607 (2024). https://doi.org/10.1109/CVPR52733.2024.01477
6. Chen, J., Xiao, S., Zhang, P., Luo, K., Lian, D., Liu, Z.: BGE M3-embedding: multi-lingual, multi-functionality, multi-granularity text embeddings through self-knowledge distillation (2024)
7. Delestre, C.: (2024). https://huggingface.co/cmarkea/dit-base-layout-detection
8. Devlin, J., Chang, M., Lee, K., Toutanova, K.: BERT: pre-training of deep bidirectional transformers for language understanding. In: Proceedings of NAACL-HLT (Vol. 1: Long and Short Papers), pp. 4171–4186 (2019)
9. Faysse, M., et al.: COLPALI: efficient document retrieval with vision language models. In: Proceedings of the International Conference on Learning Representations (ICLR) (2025)
10. He, K., Zhang, X., Ren, S., Sun, J.: Deep residual learning for image recognition. In: Proceedings of the IEEE Conference on Computer Vision and Pattern Recognition (CVPR), pp. 770–778 (2016). https://doi.org/10.1109/CVPR.2016.90
11. Henderson, M., et al.: Efficient natural language response suggestion for smart reply. CoRR abs/1705.00652 (2017). arXiv:1705.00652
12. Huang, Y., Lv, T., Cui, L., Lu, Y., Wei, F.: LayoutLMV3: pre-training for document AI with unified text and image masking. In: Proceedings of the 30th ACM International Conference on Multimedia (ACM-MM) (2022)
13. Kang, L., Tito, R., Valveny, E., Karatzas, D.: Multi-page document visual question answering using self-attention scoring mechanism. In: Proceedings of the 17th International Conference on Document Analysis and Recognition (ICDAR) (2024)
14. Khattab, O., Zaharia, M.: ColBERT: efficient and effective passage search via contextualized late interaction over BERT. In: Proceedings of the 43rd Int. ACM SIGIR Conference on Research and Development in Information Retrieval (SIGIR), pp. 39–48 (2020). https://doi.org/10.1145/3397271.3401075
15. Landeghem, J.V., et al.: Document Understanding Dataset and Evaluation (DUDE). In: Proceedings of the IEEE/CVF International Conference on Computer Vision (ICCV), pp. 19528–19540 (2023)
16. Lee, K., et al.: Pix2struct: screenshot parsing as pretraining for visual language understanding. In: Proceedings of the 40th International Conference on Machine Learning (ICML) (2023)
17. Li, C., Liu, Z., Xiao, S., Shao, Y.: Making large language models a better foundation for dense retrieval (2023)
18. Ma, Y., et al.: MMLONGBENCH-doc: benchmarking long-context document understanding with visualizations. In: Globerson, A., Mackey, L., Belgrave, D., Fan, A., Paquet, U., Tomczak, J., Zhang, C. (eds.) Advances in Neural Information Processing Systems, vol. 37, pp. 95963–96010. Curran Associates, Inc. (2024)
19. Mathew, M., Bagal, V., Tito, R., Karatzas, D., Valveny, E., Jawahar, C.V.: InfographicVQA. In: Proceedings of the IEEE/CVF Winter Conference on Applications of Computer Vision (WACV), pp. 1697–1706 (2022)
20. Mathew, M., Karatzas, D., Jawahar, C.V.: DOCVQA: a dataset for VQA on document images. In: Proceedings of the IEEE/CVF Winter Conference on Applications of Computer Vision (WACV), pp. 2199–2208 (2021). https://doi.org/10.1109/WACV48630.2021.00225

21. Raffel, C., et al.: Exploring the limits of transfer learning with a unified text-to-text transformer. J. Mach. Learn. Res. **21**(140), 1–67 (2020)
22. Tito, R., Karatzas, D., Valveny, E.: Document collection visual question answering. In: Proceedings of the 16th International Conference on Document Analysis and Recognition (ICDAR). LNCS, vol. 12822, pp. 778–792. Springer, Cham (2021). https://doi.org/10.1007/978-3-030-86331-9_50
23. Tito, R., Karatzas, D., Valveny, E.: Hierarchical multimodal transformers for multi-page DOCVQA. Pattern Recogn. **144**, 109834 (2023). https://doi.org/10.1016/j.patcog.2023.109834
24. Verma, P.: S2 chunking: a hybrid framework for document segmentation through integrated spatial and semantic analysis. CoRR abs/2501.05485 (2025). arXiv:2501.05485
25. Wang, D., et al.: DOCLLM: a layout-aware generative language model for multimodal document understanding. CoRR abs/2401.00908 (2023). arXiv:2401.00908
26. Xiao, S., Liu, Z., Zhang, P., Muennighoff, N.: C-pack: packaged resources to advance general Chinese embedding (2023)
27. Xu, Y., et al.: LayoutLMV2: multi-modal pre-training for visually-rich document understanding. In: Proceedings of ACL/IJCNLP (Volume 1: Long Papers), pp. 2579–2591 (2021)
28. Xu, Y., Li, M., Cui, L., Huang, S., Wei, F., Zhou, M.: LayoutLM: pre-training of text and layout for document image understanding. In: Proceedings of the 26th ACM SIGKDD International Conference on Knowledge Discovery & Data Mining (KDD), pp. 1192–1200 (2020). https://doi.org/10.1145/3394486.3403172
29. Zhao, Z., Kang, H., Wang, B., He, C.: DocLayout-YOLO: enhancing document layout analysis through diverse synthetic data and global-to-local adaptive perception. CoRR arXiv:2410.12628 (2024)

A New Multimodal Cross-Domain Network for Classification of Challenging Scene Images

Shashwat Sarkar[1], Kunal Purkayastha[1], Shivakumara Palaiahnakote[2](✉), Umapada Pal[1], Muhammad Hammad Saleem[2], and Palash Ghosal[3]

[1] Computer Vision and Pattern Recognition Unit, Indian Statistical Institute, Kolkata, India
umapada@isical.ac.in
[2] School of Science, Engineering and Environment, University of Salford, Manchester, UK
{s.palaiahnakote,m.h.saleem}@salford.ac.uk
[3] Department of Information Technology, Sikkim Manipal Institute of Technology, Sikkim Manipal University, Sikkim, India

Abstract. Classification of natural-scene images affected by adverse effects into their respective challenge type is a challenging problem. The state-of-the-art usually focuses on developing a model that extracts invariant features to address the above challenges. However, this direction is ineffective and inefficient for solving industry-oriented problems. This is because industry-related problems are specific and have a limited dataset and complexity. Therefore, this work introduces a new multi-modal approach called data-centric, which focuses on the classification of images of challenges (complexities), namely, quality degradation, illumination effects, and contrast issues. Once the complexity is known, the existing method can be used to achieve the expected results without proposing a new model. Our work proposes a dual ResNet50 for extracting spatial-domain and discrete wavelet-domain based features, which outputs image features. To extract the features that assess the quality of the images, the proposed work introduces a dual-CLIP-based text encoder by considering class labels as input. Therefore, the proposed model leverages image and textual features for the classification. The approach integrates image features in both spatial and wavelet domains, along with the textual context of the image quality, such as defect class labels. To evaluate the proposed method, the classification rate is estimated on our dataset and compared with the state-of-the-art methods.

Keywords: Defect classification · Hybrid image-text labeling · Quality image estimation · CLIP · Noisy images · Data-centric approach

1 Introduction

In real-time applications that classify challenging scene images from self-driving cars to person re-identification and sports analytics we often face images suffering from quality degradation, uneven illumination, and low contrast caused by open environmental factors. The quality of the images degrades due to blur, ringing, and sensor noise introduced by camera motion, poor optics, or lossy compression. Illumination inconsistency

L. Jin et al. (Eds.): ICDAR 2025 Workshops, LNCS 16225, pp. 108–123, 2026.
https://doi.org/10.1007/978-3-032-09368-4_7

reflects spatial-temporal luminance drift from non-uniform lighting, shadows, or lens flare and Contrast issues result due to limited dynamic range or atmospheric scattering compressing the histogram or collapsing edge gradients.

As a result, attaining high challenge-classification accuracy in such adverse conditions is complex and remains an open problem for the document-analysis community. Despite substantial advances in deep-learning-based classification architectures, quality degradation, illumination inconsistency, and contrast deficiency still create a critical bottleneck in real-world deployments. To address this challenge, researchers typically attempt to develop a model robust to these issues, which is known as the model-centric approach. However, this approach is neither feasible, efficient, nor cost-effective for industry-oriented applications, which often do not require a completely new robust model but rather a modification of existing models to suit the complexity of the dataset (challenges). Therefore, this study proposes a data-centric approach to identify images suffering from different challenges, as Fig. 1(a) shows quality degradation, Fig. 1(b) depicts illumination inconsistencies in the image, and Fig. 1(c) has contrast issues.

(a) Quality (b) Illumination Effect (c) Contrast

Fig. 1. Images with Quality degradation, Illumination Effect and Contrast issues

To address these issues, we propose a multimodal deep-learning model to classify challenging images. The key basis for proposing the model is that each challenge acts as a form of noise, degrading the image quality at different levels and resulting in unique quality degradations for each class. This observation motivated us to integrate spatial and discrete-domain feature extraction using a dual ResNet-50 architecture alongside semantic descriptors processed through parallel CLIP-based text encoders. This multimodal representation enables the model to capture both visual and contextual cues using two modes of information i.e. image and text, related to degradation factors, enhancing its ability to classify images affected by different challenges. Leveraging a cross-domain network by utilizing features from spatial and frequency domains enables the model with better understanding of the core underlying challenge of an image. Notably, while the spatial features excel at preserving object boundaries and text structures, the discrete wavelet-domain features are sensitive to text blur and irregular stroke spacing. The extracted features are projected into a shared 256-dimensional latent space before undergoing multimodal fusion through self-attention mechanisms, which allows the model to adaptively weight features based on their relevance. A final classification head then processes the fused representation using a multi-layer perception (MLP) to predict the actual class label.

The key contributions of the proposed work are as follows. (i) Introducing the data-centric approach, which is an alternative to the model-centric approach to classify the images of different challenges. (ii) Exploring dual ResNet50 and CLIP text encoder for capturing spatial and discrete wavelet-domain based features. (iii) Fusing the images and textual features for classifying the images of different challenges successfully.

The organization of the paper is as follows. The methods related to data-centric and model-centric are reviewed in Sect. 2. Section 3 presents a complete architecture of the proposed classification model. An experimental analysis to validate the proposed method in terms of classification rate is discussed in Sect. 4. Section 5 concludes with a summary of the proposed method based on performance.

2 Related Work

This section reviews the methods of different categories, such as the methods invariant to different complexities and the methods that detect images affected by different distortions.

2.1 Methods Invariant to Different Complexities

Although recent advances have significantly improved the robustness of scene-image analysis, most studies remain detection-centric and omit a preliminary, systematic classification of the underlying degradations. Zhang et al. [1] introduced the Kernel Proposal Network, which generates and iteratively grows text kernels to cope with complex layouts. Wang et al. [2] proposed a fuzzy semantics approach that integrates fuzzy theory with deep learning to handle ambiguities in text shapes and orientations. Cai et al. [3] developed a dynamic convolution method that adapts convolutional kernels to local geometry.

Liao et al. [4] presented a real-time scene text detection-centric system combining differentiable binarization and adaptive scale fusion. Chen et al. [5] introduced FAST, a minimalist kernel-representation scheme that greatly accelerates arbitrary-shaped text localisation, yet it still presumes a single detector can cope with every form of degradation without first identifying what kind of challenge is present. Duan et al. [6] extended this detection-centric trend with ODM, an image–text alignment pre-training strategy that bolsters robustness in complex backgrounds, but likewise operates without an upstream challenge classifier. Qin et al. [7] focused on instance representation learning, addressing challenges like false positives and instance adhesion. Zhang et al. [8] proposed a disentangled-representation framework that isolates intrinsic attributes of scene-text images, thereby advancing downstream tasks such as text analysis, removal, and editing. In summary, these methods are innovative but may not be cost-effective for industry-oriented applications, which often prioritize efficiency over novelty. Developing a model that works across all types of distortions remains a significant challenge.

2.2 Methods for Detecting or Classifying Noise, Low Resolution, and Poor-Quality Images

Several works have attempted to mitigate degradation by directly enhancing text quality or boosting recognition accuracy. Uchigasaki et al. [9] developed a scene text image quality assessment model to evaluate text readability in compressed images, guiding compression algorithms to maintain text quality. Yu et al. [10] proposed a textual reasoning and multiscale cross-convolution network for scene text image super-resolution, effectively enhancing the visual quality of low-resolution text images. Ma et al. [11] introduced TATT, a text attention network that focuses on character regions to super-resolve text images with spatial distortion robustness. Additionally, Ye et al. [12] introduced TextSSR, a diffusion-based data-synthesis pipeline that generates high-quality synthetic images to strengthen recognition models.

Although the above methods used elegant deep-learning approaches to detect distortions in images, none of them considered the images of different challenges as a single challenge to solve the problem. Each method focuses on specific complexities or distortions. This highlights the difficulty in developing a generalized model that works for different challenges.

3 Proposed Methodology

As mentioned in the previous section, there is a need for developing a data-centric approach to classify the images of different challenges. This work considers three classes that represent different complexities, namely, (i) quality degradation (poor quality), (ii) illumination effects, and (iii) contrast issues. It is also noted that all three classes represent kinds of noise and distortion, and hence, the images affected by these challenges suffer from degradation. This observation motivated us to extract the features that are sensitive to different degradations. It is also true that the spatial domain features are good for capturing object boundaries, edges, and structure of objects, while discrete wavelet domain features are good for capturing blur and irregular stroke information in the images. These observations are the main basis for introducing a dual ResNet50 network for extracting spatial, discrete wavelet domain-based features from the input images. As mentioned earlier, assessing quality is important for the classification of images of different challenges. Inspired by the success of extracting semantic features from images and text information using CLIP, we propose dual CLIP for fusing visual and textual features (class labels) to assess the quality of the images. Finally, the fused features are used for classification. The block diagram of the proposed work can be seen in Fig. 2.

3.1 Image Feature Extraction: Hierarchical Representation Learning from Spatial and Discrete Wavelet Domains

The first core unit of the model is the Image Feature Extraction Unit, which processes the input image and its discrete wavelet domain counterpart in parallel to capture spatial and discrete wavelet-based features, as shown in Fig. 2. A dual-branch ResNet-50-based convolutional network design is employed, where one branch processes spatial-domain

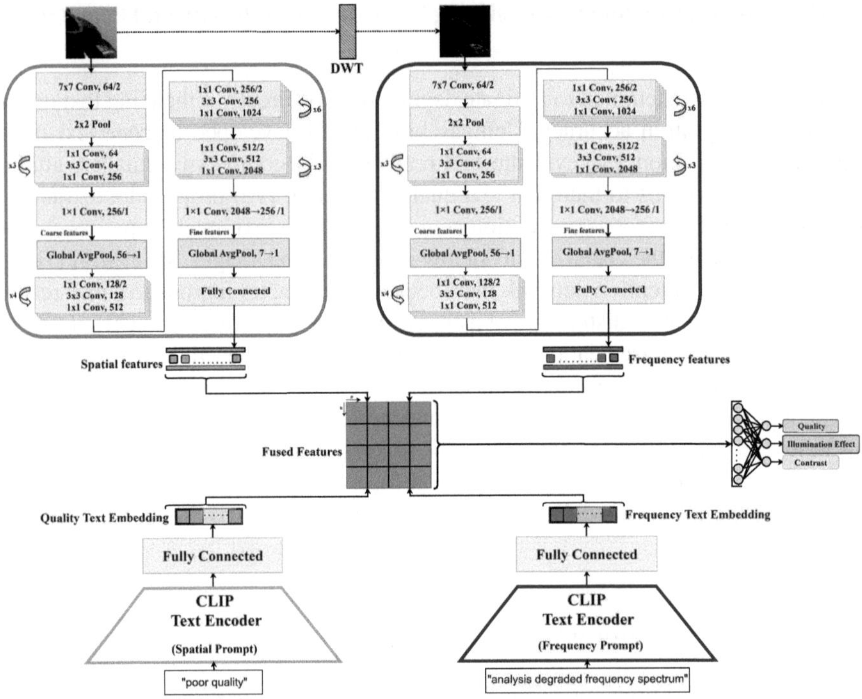

Fig. 2. Proposed Model architecture (best visible at 50% zoom)

representations while the other extracts discrete wavelet-based spectral information. This separation is essential because spatial features excel at preserving object boundaries and text structures, whereas frequency features capture high-frequency distortions and degradation artifacts that directly impact text readability.

For an input image $\mathcal{I} \in R^{H \times W \times 3}$, the spatial processing pathway follows a ResNet-50 architecture, where each convolutional layer performs localized feature extraction and abstraction. The transformation at each layer l is mathematically formulated as defined in Eq. (1).

$$z_{\text{spatial},l} = \sigma\left(W_l^{(s)} * z_{\text{spatial},l-1} + b_l^{(s)}\right) \tag{1}$$

Where $W_1^{(s)} \in R^{k \times k \times c}$ represents the convolutional filter weights, $b_l^{(s)}$ is the bias term, and $\sigma(\cdot)$ is the ReLU non-linearity, which ensures stable gradient propagation and preserves sparse activations. The hierarchical feature extraction within the spatial pathway is further improved using residual connections, mathematically expressed as in Eq. (2).

$$z_{\text{spatial},1} = \sigma\left(W_1^{(s)} * z_{\text{spatial},1-1} + b_1^{(s)}\right) + z_{\text{spatial},1-1} \tag{2}$$

This is particularly beneficial for low-contrast text regions, where small pixel variations may otherwise be lost due to activation saturation in deep networks. To complement spatial feature extraction as shown in Fig. 3(a), the second branch processes the wavelet-domain representation of the image as shown in Fig. 3(b), which provides localized frequency information about underlying degradation patterns. Unlike Fourier Transform, which analyzes frequency content globally, the Discrete Wavelet Transform (DWT) decomposes the image into multiple frequency sub-bands while maintaining spatial localization. This property makes it ideal for analyzing text-based degradation effects such as blurring, noise, and compression artifacts.

The wavelet decomposition of an image I using a 2D DWT is mathematically defined as in Eq. (3).

$$W(I) = \{I_{LL}, I_{LH}, I_{HL}, I_{HH}\} = \{I * \psi_{LL},\ I * \psi_{LH},\ I * \psi_{HL},\ I * \psi_{HH}\} \quad (3)$$

Where: $I_{LL} = I * \psi_{LL}$ represents the approximation coefficients, capturing the low-frequency (coarse)structure. $I_{LH} = I * \psi_{LH}$ captures horizontal details (edges and text contours). $I_{HL} = I * \psi_{HL}$ captures vertical details (text edges and structures). $I_{HH} = I * \psi_{HH}$ extracts high-frequency noise and fine text structures.

Here, $\psi_{LL}, \psi_{LH}, \psi_{HL}, \psi_{HH}$ are the corresponding wavelet basis functions (filters), and $*$ denotes the convolution operation.

Since the raw wavelet coefficients span different scales and magnitudes, a normalization step is applied to enhance contrast across sub-bands as defined in Eq. (4).

$$W(I) = \log(1 + |W(I)|) \quad (4)$$

This transformed representation is then processed through a parallel ResNet-50 network, maintaining an identical hierarchical feature extraction structure to the spatial pathway, as shown in Eq. (5).

$$z_{wavelet,1} = \sigma\left(W_1^{(w)} * W(I) + b_1^{(w)}\right) \quad (5)$$

Coarse-level wavelet features ($z_{wavelet\text{-}coarse}$) capture low-frequency degradation effects, such as illumination inconsistencies and compression artifacts. **Fine-level wavelet features** ($z_{wavelet\text{-}fine}$) encode high-frequency distortions, such as blurring and noise.

To ensure effective multimodal fusion, both spatial and wavelet features are projected into a shared feature space using two 1×1 convolutional layers, ensuring dimensional consistency across modalities. The use of these projection layers serves multiple roles. **Dimensionality Reduction**: ResNet-50 outputs high-dimensional feature maps (e.g., 2048 channels in deeper layers), which require compression for efficient processing. **Feature Space Alignment**: Since spatial and wavelet features are extracted in separate domains, they must be projected into a common latent space to enable seamless fusion. **Multiscale Learning**: By extracting both coarse and fine representations, the model captures local and global text degradation effects, enhancing robustness across multiple imaging conditions. The features extracted from both the spatial domain and the discrete wavelet domain are combined using feature concatenation. We chose concatenation as it allows the model to preserve the distinct characteristics of each modality while enabling the classifier to learn the best combination during training. This has been shown to work better for classifier performance, as we can observe from Table 3 in the Ablation section.

3.2 Text Feature Unit: Context-Aware Encoding for Scene-Level Quality Estimation in Challenge Classification

Scene-level quality degradation often stems from complex image distortions that are difficult to characterize using purely visual features. Contrast variations and illumination inconsistencies do not always manifest as strong pixel-based artifacts but instead influence the way text appears to a detector. Although convolutional and spectral features can quantify spatial and frequency distortions, they provide little semantic insight into how those degradations undermine reliable challenge classification. Our approach addresses this by introducing dual CLIP-based text encoders, which provide scene-aware contextual embeddings to enhance model robustness across different image degradation scenarios. By incorporating scene-aware semantic features, the network gains an explicit understanding of how text-related failures occur, strengthening its ability to classify domain-specific degradation patterns.

Tokenization and Contextual Encoding of Scene Descriptions. The model incorporates two textual descriptors per image: the Scene Quality Descriptor (T_{scene}), which categorizes the primary degradation factor (e.g., "poor contrast" or "uneven illumination"), and the Frequency-Based Descriptor (T_{freq}), which describes degradation in spectral space (e.g., "analysis: low contrast frequency pattern"). These descriptors are mapped to textual embeddings using a CLIP-based text encoder, ensuring that degradation types are captured in a structured, contextual manner. The CLIP tokenizer first converts each descriptor into subword units, allowing the model to process phrases effectively, even when domain-specific terminology is involved.

After tokenization, the sequence is passed through CLIP's transformer-based text encoder, generating a dense feature representation. Unlike traditional embeddings, this approach captures relationships between words, ensuring that text-based degradation descriptions align with visual features extracted from the image. The resulting embeddings undergo a linear transformation to match the dimensionality of the image feature space, ensuring that both modalities can be effectively combined for classification. This transformation is necessary as CLIP's text encoder outputs a high-dimensional feature vector that must be projected into the same space as the spatial and frequency features before fusion.

Self-attention for Multi-scale Semantic Representation. CLIP's encoder leverages multi-head self-attention to capture dependencies between words, enabling the model to dynamically weigh important tokens. This is especially beneficial when processing descriptors related to image quality and illumination effect, as different aspects require varying degrees of emphasis. For instance, the term "low quality" becomes particularly relevant in situations where compression artifacts or blurriness compromise clarity, whereas "uneven illumination" is more significant when inconsistent lighting obscures critical details. The attention mechanism ensures that such quality and illumination-related terms are assigned appropriate importance within the textual representation.

Once self-attention is applied, the output passes through a feedforward layer that refines the learned features before multimodal fusion. This step enhances the text encoding by reinforcing semantic relationships while filtering out redundant information. As a result, the textual embeddings provide a compact, context-aware representation of degradation descriptions, which, when fused with spatial and discrete wavelet-based image features, enhances classification accuracy by introducing additional degradation-specific constraints.

Dual CLIP Encoders for Scene and Frequency-Aware Text Representations. To handle scene-level and frequency-based degradation independently, the model employs two parallel CLIP encoders. The first encoder processes the quality descriptor (T_{scene}) as shown in Fig. 3(c), which captures broad degradation effects like "poor contrast" or "poor quality." The second encoder processes the frequency descriptor (T_{freq}) as shown in Fig. 3(d), which encodes spectral domain characteristics such as "analysis: degraded frequency spectrum" or "analysis: irregular frequency domain signal." This dual-encoder configuration models spatial and spectral degradation patterns independently while remaining applicable to a wide range of challenge-classification scenarios.

Both textual embeddings are projected into a 256-dimensional space to align with image-based representations before fusion. This transformation allows the model to directly compare spatial, frequency, and textual information in a unified manner, ensuring that all modalities contribute equally to classification. Without this projection step, inconsistencies between feature distributions would hinder effective multimodal learning. By fusing scene and frequency-aware text embeddings, the model becomes more robust to diverse degradations, thereby increasing classification accuracy in adverse imaging conditions.

3.3 Classification Unit: Multimodal Feature Fusion with Self-attention for Classification

To integrate spatial, frequency, and textual embeddings, we first stack their representations as defined in Eq. (6).

$$Z = \text{Stack}\left(z_{\text{spatial-fine}}, z_{\text{dwt - fine}}, z_{\text{quality-text}}, z_{\text{freq - text}}\right) \tag{6}$$

Let $Z \in R^{n \times d}$ be the stacked features, where n is the number of feature vectors (one per modality) and d is their dimensionality. We then apply a multi-head self-attention mechanism as shown in Eq. (7).

$$Z_{\text{attn}} = \text{Concat}(\text{head}_1, \ldots, \text{head}_k)W^O \tag{7}$$

where each attention head head_i is given by Eq. (8).

$$\text{head}_i = \text{Attention}\left(ZW_i^Q, ZW_i^K, ZW_i^V\right) = \text{Softmax}\left(\frac{ZW_i^Q\left(ZW_i^K\right)^\top}{\sqrt{d_k}}\right)\left(ZW_i^V\right) \tag{8}$$

Here, $W_i^Q, W_i^K, W_i^V \in R^{d \times d_k}$ are the projection matrices for queries, keys, and values, respectively, and $W^O \in R^{(h \cdot d_v) \times d}$ is the learnable matrix that projects the concatenated heads back to the model dimension (is the number of heads, is the dimension of each head's value vectors).

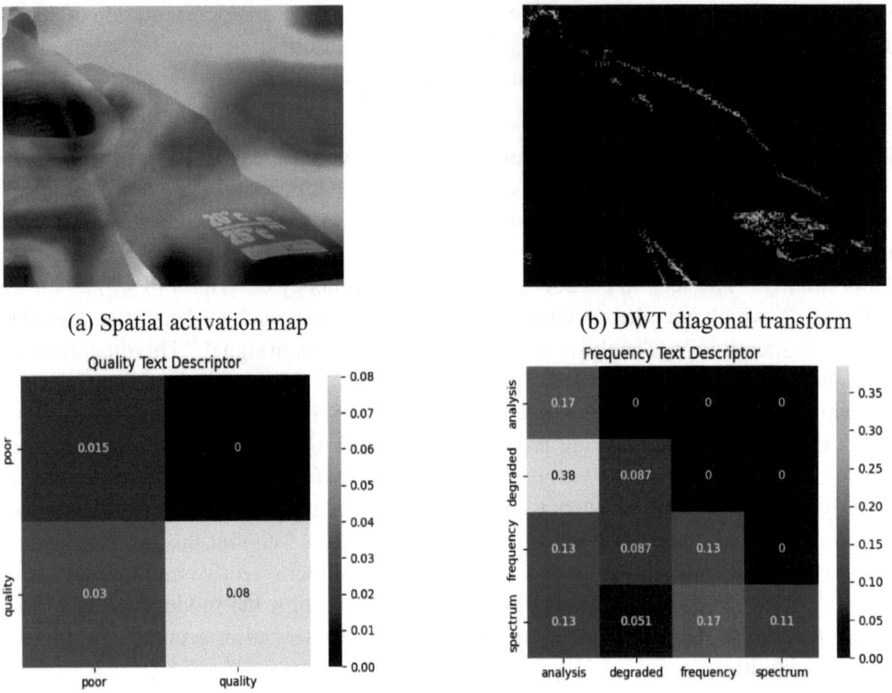

(a) Spatial activation map

(b) DWT diagonal transform

(c) Quality Text Descriptor

(d) Frequency Text Descriptor

Fig. 3. Intermediate layer representations in the proposed approach.

In our approach, we stack the features because this approach preserves the unique, modality-specific information inherent in the spatial, frequency, and textual embeddings (see Fig. 4). By aligning them along a new axis, we create a unified representation that allows the multi-head self-attention mechanism to effectively learn and leverage the interdependencies among the modalities. This stacking ensures that no critical detail from any modality is lost, enabling the attention module to dynamically emphasize the most relevant features for robust classification of challenging scene images. Therefore, the modality-specific features (spatial, frequency, and two textual embeddings) are stacked into a 4×256 tensor and passed to a multi-head self-attention module. Attention weights are computed via a scaled dot-product between learned query and key projections, capturing pairwise relevance across modalities. These weights are optimized during training to minimize classification loss, allowing the model to adaptively assign higher weights to more informative modalities (e.g., frequency under poor illumination). Thus, the attention mechanism learns to focus across modalities in a task-specific, input-dependent manner.

Finally, we flatten Z_{attn} and feed it into a multi-layer perceptron (MLP) for classification, producing the final probability distribution over three challenging classes as defined in Eq. (9).

$$\hat{y} = \mathrm{Softmax}(W_o \cdot \mathrm{Flatten}(Z_{attn})) \tag{9}$$

Where $W_o \in R^{C \times (h \cdot d_v \cdot \alpha)}$ (with C being the number of classes and the flattened spatial dimension, if applicable). Thus, W^O is specific to merging the multi-head outputs, while W_o is dedicated to generating the final class logits in the MLP's output layer.

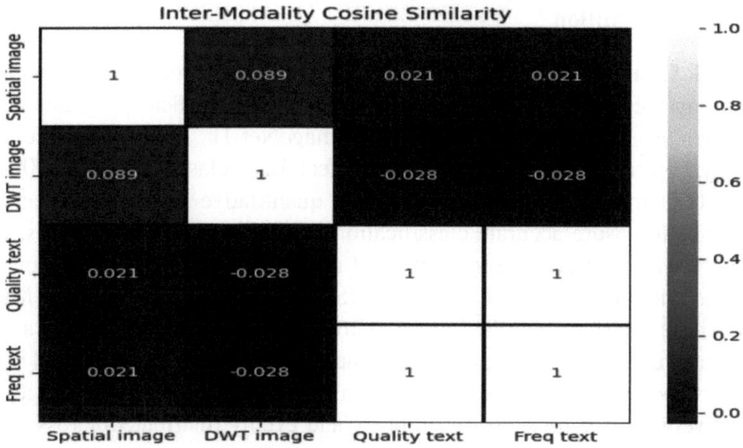

Fig. 4. Inter-Modality Cosine Similarity Representation

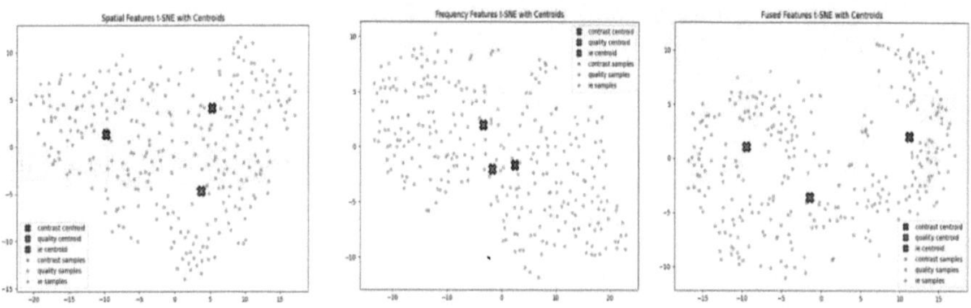

(a) Spatial Features (b) Frequency Features (c) Fused Features

Fig. 5. t-SNE Maps for each feature extraction branch

The effectiveness of the key component of the proposed method is illustrated in Fig. 5, where it is noted that spatial and frequency features contribute to classifying the images of different complexities, but not as accurately compared to fused features. The fusion features integrate the vital information of spatial and frequency to achieve successful results.

4 Experimental Results

To evaluate the proposed method, we use the standard datasets (Underwater images, Person and vehicle re-identification, drone images, and datasets of natural scene images) of different categories to choose the images that suffer from different challenges. Since it is not so easy to collect such challenging images, each class of different challenges has a small number of images. Therefore, the proposed work uses Imagenet-1k for training and testing, and we use our constructed dataset.

4.1 Dataset Description

Imagenet-1k Challenging Scene Image Classification Subset: To construct a dataset for pretraining the proposed multimodal model for the classification of images of different challenges, we curated a subset from ImageNet-1k, comprising three classes: poor quality, contrast issues and illumination effect. Each class contains 10,000 images, totaling 30,000. Image selection was based on quantitative image quality assessment (IQA) metrics to ensure accurate classification of challenging scene images. Initially, ImageNet-1k was preprocessed to filter valid images, followed by the computation of IQA metrics such as Signal-to-Noise Ratio (SNR), Shannon Entropy and Michelson Contrast. These metrics quantified noise, entropy, contrast, frequency characteristics, and edge structures. The worst-performing metric per image was identified via GPU-accelerated processing, determining the dominant challenge for each image. Images were iteratively refined for class separability and evenly distributed across categories (see Fig. 6 for different categories).

(a) Quality (b) Illumination Effect (c) Contrast

Fig. 6. Samples of images per class from the ImageNet-1k subset.

Proposed Multimodal Dataset for classifying Challenging Scene Images: In addition to the ImageNet-1k subset, we constructed a specialized dataset focusing on various real-world scenarios encompassing different degradation factors, which we have used for fine-tuning our approach. This dataset spans multiple domains, including underwater text [13], text on glass surfaces, and license plate text [13] from datasets curated at ISI Kolkata along with natural scene text from the datasets CTW1500 [14], ICDAR2015 [15] and Totaltext [16]. The dataset was designed to capture instances where state-of-the-art text detection-centric models fail to accurately localize text due to degradation factors. To curate this dataset, we first ran inference using state-of-the-art text spotting models, DNTextSpotter [17] and DeepSolo [18] on images collected from the aforementioned domains. Instances where the model failed to detect text instances correctly were identified and extracted. These failure cases were then manually observed into three challenging categories (as shown in Fig. 7) based on the most dominant degradation factor affecting text visibility: poor quality, contrast issues, and illumination effect. The dataset comprises a total of 1041 images.

Unlike traditional text detection benchmarks, which focus on correctly localized text, our dataset specifically targets failure cases, enabling models to learn degradation-aware representations. By integrating both controlled large-scale datasets (ImageNet-1k subset) and real-world failure scenarios, our approach ensures that the classification model is robust across both synthetic and practical settings.

(a) Quality (b) Illumination Effect (c) Contrast

Fig. 7. Samples of images per class from our Proposed Multimodal Dataset for classifying Challenging Scene Images

Implementation Details: The network was pre-trained on ImageNet1k and then fine-tuned on our proposed dataset using the Adam optimizer, with an initial learning rate of 2×10^{-4} and a weight decay of 1×10^{-5}. For pre-training and fine-tuning, the images in the dataset were resized to 224×224 before passing them into the network. A StepLR scheduler was applied, reducing the learning rate by a factor of 0.1 every 5 epochs to stabilize convergence. Training was conducted for 50 epochs on an A1000 GPU, using a batch size of 64.

4.2 Ablation Studies

As discussed in the proposed methodology section, the spatial and frequency features are effective for the classification of images with different complexities. To show the contribution of spatial and frequency features, we calculated the accuracy on our dataset as reported in Table 1. We validate the domain of features that can be fused for effective results in Table 1. As evident from the report results in Table 1, we observe that spatial features are effective in capturing local structure and contrast, while frequency-domain features offer robustness to global degradations like blur and illumination. Hybrid fusion leverages the strengths of both. Our experiments confirm that combining both domains leads to improved classification accuracy.

The encoder is indeed the main component of any deep learning model for feature extraction. We validate the effectiveness of different text encoders in extracting meaningful semantic information for challenging image classification. In the proposed method, we used dual CLIP encoders for feature extraction. BERT is indeed a well-known model

Table 1. Comparison study on the domain of features for fusion.

Feature Fusion	Classification Accuracy (%)
Spatial + Spatial	65.37
Frequency + Frequency	61.46
Spatial + Frequency	**70.24**

for encoding textual features. Therefore, the BERT is considered a baseline model for ablation experiments. Experiments are conducted for different combinations as listed in the encoder section of Table 2. Similarly, as discussed in the proposed method, the DWT is used for feature extraction. To show that DWT is better than other frequency domains, we conducted experiments using FFT, Polar, and DCT, and the results are reported in the image-feature section of Table 2. In Table 2, the CLIP encoder-CLIP encoder and the Spatial and DWT images output almost the same results. Overall, the combination of spatial and DWT images is the best one for classification compared to other pair combinations.

Table 2. Comparison study on different encoders and image stream combinations.

Quality - Frequency	Classification Accuracy (%)
BERT encoder - BERT encoder	68.58
BERT encoder - CLIP encoder	68.24
CLIP encoder - BERT encoder	67.91
CLIP encoder - CLIP encoder	**70.24**
Spatial Image + FFT Image	61.46
Spatial Image + DCT Image	65.37
Spatial Image + Polar Image	69.27
Spatial Image + DWT Image	**70.24**

The results reported in Table 2 show that for the encoder stream, the pair of CLIPs is the best one compared to other combinations. It is also noted from Table 2 that the combination of spatial + DWT image features is the best one compared to other frequency domains. This is because DWT can extract rich local information compared to the polar and FFT domains. The fusion step is also important. However, feature concatenation is the best one compared to other combinations. Since spatial and frequency features contribute equally to classification, simple concatenation is sufficient for achieving the best results, as shown in Table 3, where the concentration is the best compared to other fusion approaches. Weighted fusion and feature splicing may lose vital cues due to operations and weight determination.

Table 3. Comparison study on different feature fusion techniques.

Feature Combination Method	Classification Accuracy (%)
Weighted Fusion	69.27
Feature Splicing	68.29
Feature Concatenation	**70.24**

4.3 Experiments on Classification

The accuracy of the proposed method is determined for evaluating the classification on our CTW1500 datasets. As we could not find a publicly available dataset that directly aligns with our specific classification objective, we used the public CTW1500 dataset and introduced class-specific degradations: Gaussian noise and blur for quality, illumination masks with gamma shifts for IE, and dynamic range suppression for contrast to compare our proposed method to other existing methods as stated in Table 4. The results are shown in Table 4, where the proposed classification is better than other classification methods on both datasets. This shows that the proposed model is capable of classifying complex, challenging images. However, the existing methods were not developed for handling images of different complexities, and one can expect poor performance compared to the proposed method.

Table 4. Comparison of classification accuracy across different methods

Methods	Our dataset	CTW1500 dataset
ResNet-50 [19]	66.83	78.00
ConvNet [20]	67.32	80.67
ViT [21]	68.78	73.00
XCiT [22]	67.33	68.34
GCViT [23]	69.67	80.67
Our Method	**70.24**	**83.33**

5 Conclusion and Future Work

This paper introduced multimodal deep learning for the classification of images of different challenges, namely, poor quality, contrast and illumination effect. By leveraging spatial and discrete wavelet-domain feature extraction through a dual ResNet-50 architecture and integrating semantic embeddings from CLIP-based text encoders, the proposed approach effectively classifies the images of different challenges. The incorporation of self-attention for multimodal fusion further enhances the model's robustness, enabling reliable generalization across distinct degradation categories. Experimental results demonstrate the effectiveness of our method in classifying the images of different complexities.

While the proposed model achieves promising results, certain challenges remain intact. One key limitation is its reliance on a single predominant fault classification per image, which may not fully capture cases where multiple degradation factors coexist. In future work, we aim to extend the model towards a multi-label classification framework along with further research into the robustness of the class labels with respect to variations in text description. Additionally, incorporating an adaptive weighting mechanism enables the ability to emphasize the most critical degradation factors dynamically. Another avenue for improvement is exploring a fuzzy logic-based deep learning model for improving the classification rate.

References

1. Zhang, S.X., et al.: Kernel proposal network for arbitrary shape text detection. IEEE Trans. Neural Netw. Learn. Syst. **33**(6), 2526–2538 (2022)
2. Wang, F., Xu, X., Chen, Y., Wu, F., Bai, X.: Fuzzy semantics for arbitrary-shaped scene text detection. IEEE Trans. Image Process. **32**, 1–12 (2023)
3. Cai, Y., Liu, Y., Shen, C., Chen, Q.: Arbitrarily shaped scene text detection with dynamic convolution. Pattern Recogn. **127**, 108608 (2022)
4. Liao, M., Zou, Z., Wan, Z., Yao, C., Bai, X.: Real-time scene text detection with differentiable binarization and adaptive scale fusion. arXiv preprint arXiv:2207.12382 (2022)
5. Chen, Z., Wang, W., Xie, E., Yang, Z., Lu, T., Luo, P.: FAST: searching for a faster arbitrarily-shaped text detector with minimalist kernel representation. arXiv preprint arXiv:2111.02394 (2021)
6. Duan, H., et al.: ODM: a text-image further alignment pre-training approach for scene text detection. In: Proceedings of the IEEE/CVF Conference on Computer Vision and Pattern Recognition (CVPR) (2024)
7. Qin, X., et al.: Towards robust real-time scene text detection: from semantic to instance representation learning. arXiv preprint arXiv:2308.07202 (2023)
8. Zhang, B., Xie, H., Gao, Z., Wang, Y.: Choose what you need: disentangled representation learning for scene text recognition, removal and editing. arXiv preprint arXiv:2405.04377 (2024)
9. Uchigasaki, S., Miyazaki, T., Omachi, S.: Deep image compression using scene text quality assessment. Pattern Recogn. **142**, 109696 (2023)
10. Yu, L., Li, X., Yu, Q., Li, G., Jin, D., Qi, M.: Scene text image super-resolution via textual reasoning and multiscale cross-convolution. Appl. Intell. **54**, 1997–2008 (2024)
11. Ma, J., Liang, Z., Zhang, L.: A text attention network for spatial deformation robust scene text image super-resolution. arXiv preprint arXiv:2203.09388 (2022)
12. Ye, X., Du, Y., Tao, Y., Chen, Z.: TextSSR: diffusion-based data synthesis for scene text recognition. arXiv preprint arXiv:2412.01137 (2024)
13. Purkayastha, K., Sarkar, S., Palaiahnakote, S., Pal, U., Ghosal, P.: DATR: domain agnostic text recognizer. In: Antonacopoulos, A., Chaudhuri, S., Chellappa, R., Liu, C.L., Bhattacharya, S., Pal, U. (eds.) ICPR 2024. LNCS, vol. 15317, Springer, Cham (2025). https://doi.org/10.1007/978-3-031-78447-7_15
14. Liu, Y., Jin, L., Zhang, S., Zhang, S.: Curved scene text detection via transverse and longitudinal sequence connection. Pattern Recognit. **90**, 337–345 (2019). arXiv preprint arXiv:1712.02170 (2017)

15. Karatzas, D., et al.: ICDAR 2015 competition on robust reading. In: Proceedings of the 13th International Conference on Document Analysis and Recognition (ICDAR), pp. 1156–1160 (2015)
16. Chng, C.K., Chan, C.S.: Total-Text: a comprehensive dataset for scene text detection and recognition. In: Proceedings of the 14th IAPR International Conference on Document Analysis and Recognition (ICDAR). arXiv preprint arXiv:1710.10400 (2017)
17. Xie, Y., et al.: DNTextSpotter: arbitrary-shaped scene text spotting via improved denoising training. arXiv preprint arXiv:2408.00355 (2024)
18. Ye, M., et al.: DeepSolo: let transformer decoder with explicit points solo for text spotting. Proceedings of the IEEE/CVF Conference on Computer Vision and Pattern Recognition (CVPR) (2023). arXiv preprint arXiv:2211.10772 (2022)
19. He, K., Zhang, X., Ren, S., Sun, J.: Deep residual learning for image recognition arXiv preprint arXiv:1512.03385 (2015)
20. Liu, Z., Mao, H., Wu, C., Feichtenhofer, C., Darrell, T., Xie, S.: A ConvNet for the 2020s. arXiv preprint arXiv:2201.03545 (2022)
21. DosoViTskiy, A., et al.: An image is worth 16x16 words: transformers for image recognition at scale. arXiv preprint arXiv:2010.11929 (2021)
22. Ali, M., Touvron, H., Synnaeve, G., Jégou, H., Cord, M.: XCiT: cross-covariance image transformers. In: Proceedings of the IEEE/CVF International Conference on Computer Vision (ICCV) (2021)
23. Hatamizadeh, A., Yin, H., Heinrich, G., Kautz, J., Molchanov, P.: Global context vision transformers. arXiv preprint arXiv:2206.09959 (2023)

TextBite: A Historical Czech Document Dataset for Logical Page Segmentation

Martin Kostelník(✉), Karel Beneš, and Michal Hradiš

Faculty of Information Technology, Brno University of Technology, Brno, Czech Republic
{ikostelnik,ibenes,ihradis}@fit.vut.cz

Abstract. Logical page segmentation is an important step in document analysis, enabling better semantic representations, information retrieval, and text understanding. Previous approaches define logical segmentation either through text or geometric objects, relying on OCR or precise geometry. To avoid the need for OCR, we define the task purely as segmentation in the image domain. Furthermore, to ensure the evaluation remains unaffected by geometrical variations that do not impact text segmentation, we propose to use only foreground text pixels in the evaluation metric and disregard all background pixels. To support research in logical document segmentation, we introduce TextBite, a dataset of historical Czech documents spanning the 18th to 20th centuries, featuring diverse layouts from newspapers, dictionaries, and handwritten records. The dataset comprises 8,449 page images with 78,863 annotated segments of logically and thematically coherent text. We propose a set of baseline methods combining text region detection and relation prediction. The dataset, baselines and evaluation framework can be accessed at https://github.com/DCGM/textbite-dataset.

Keywords: Dataset · Czech Historical Documents · Page Segmentation · Document Intelligence · Document Layout Analysis

1 Introduction

Logical page segmentation is a process of dividing a document page into smaller, meaningful, semantically coherent units. It is a crucial task in the analysis of documents, enabling the extraction of meaningful content from scanned or digitized texts. For example, such segmentation allows the creation of finer text embeddings, which can in turn improve search and retrieval results. Consequently, the work of teachers, librarians, or researchers is simplified by enabling them to find documents related to their topics of interest [22,31].

Traditional approaches to logical page segmentation problem typically fall into two categories: (1) text-based segmentation, where boundaries are detected in a sequential text stream [33], and (2) object-detection-based methods [38,39], which detect paragraphs of text and subsequently connect them together to form

L. Jin et al. (Eds.): ICDAR 2025 Workshops, LNCS 16225, pp. 124–140, 2026.
https://doi.org/10.1007/978-3-032-09368-4_8

the final segmentation. However, these methods are inherently bound to evaluation schemes, which both suffer from key limitations. Text-based approaches depend on Optical Character Recognition (OCR), tightly coupling evaluation performance with OCR accuracy. They also require a consistent reading order, which OCR may not reliably provide. On the other hand, geometric object-detection-based methods may overly penalize minor, inconsequential deviations in the shapes or sizes of annotated objects.

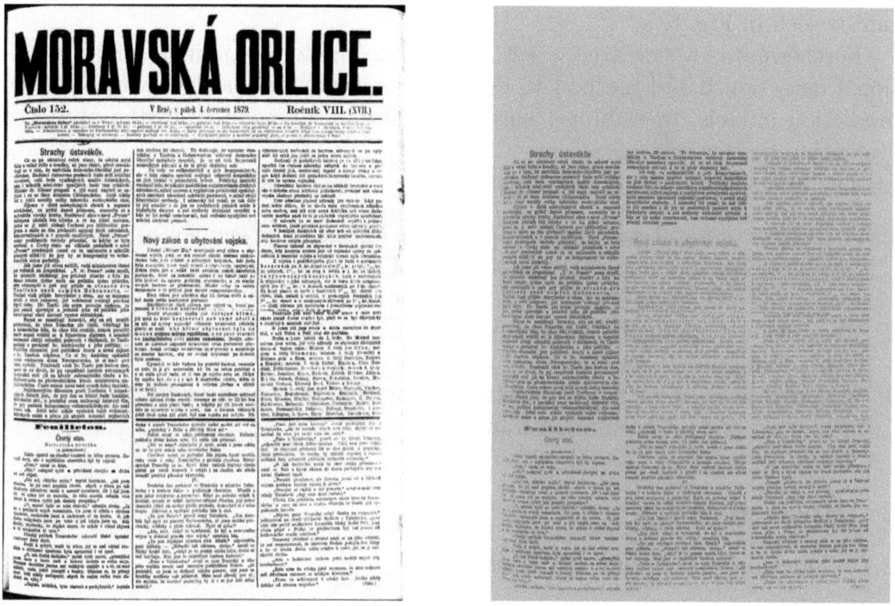

(a) Input page (b) Logical segmentation

Fig. 1. Example of logical segmentation of a newspaper page as proposed in TextBite. Each segment is denoted in a different color. Only the colored pixels are considered in the evaluation. Page metadata, such as the title or edition are not included in the segmentation, as they are not a thematically coherent segments. However, our evaluation scheme does not penalize marking these parts of the page as additional segments.

In this work, we propose evaluating logical page segmentation as a clustering problem. Rather than performing evaluation at the text level or relying on detection-based metrics, we treat segments as clusters of pixels within the document image. Additionally, we argue that background pixels carry no semantic information relevant to the represented text and thus should be excluded from evaluation. Foreground pixels, which correspond directly to individual letters, can be reliably identified using relatively simple segmentation methods, such as thresholding or binarization. By considering only the pixels forming individual letters, our approach removes dependency on OCR accuracy, eliminates variations arising from different detection methods, and provides a unifying evaluation metric applicable to all types of logical page segmentation approaches.

To support further research in this direction, we introduce a new dataset for logical page segmentation, based on historical Czech documents called TextBite. Our dataset comprises of 8,449 manually annotated pages with varying layouts and containing both printed and handwritten pages. This publicly available dataset can be used to develop and benchmark document segmentation systems. An example of a page from the TextBite dataset together with its segmentation can be seen in Fig. 1.

Furthermore, we created a set of baseline models, ranging from a traditional detection-based method to detection methods augmented with relation-prediction models, to establish baseline results and demonstrate the proposed evaluation framework.

The main contributions of this paper can be summarized as follows:

1. We propose a novel evaluation approach to logical page segmentation based on pixel clustering, making it independent of OCR and insensitive to irrelevant geometric variations of detected text regions.
2. We provide a hand-annotated dataset for logical page segmentation and other tasks, containing 8,449 pages.
3. We provide a set of baseline solutions for this task along with their implementation and an evaluation framework.

2 Related Work

2.1 Document Layout Datasets

Many datasets exist for layout analysis of both modern and historical documents; however, to our knowledge, no existing benchmarks employ an evaluation strategy similar to ours. Typically, benchmarks for these datasets frame evaluation either as a detection problem or as a relation-prediction problem and assess performance accordingly.

Some datasets, such as PubLayNet [43], DocLayNet [25] and M6Doc [4] frame layout analysis as a detection task, providing annotations for document elements like text, tables, and figures. Similarly, DocBank [18] offers large-scale OCR and detection annotations, making it a common benchmark for layout-based pretraining. Other datasets, such as CompHRDoc [39] and its superset HRDoc [23], go beyond detection by modeling hierarchical relationships, where elements are not only identified, but also classified and connected within a document's structure. These datasets introduce parent-child relations between document components, making them particularly useful for tasks involving nested and multi-column structures, such as historical newspapers. Datasets like ROOR [42] focus on reading order prediction, evaluating how well models can reconstruct the logical sequence of content, which is crucial for multi-column or complex layouts.

For historical documents, datasets such as Europeana Newspapers [5], and TexBiG [35] introduce additional challenges related to the varying styles of printing, ornamental decorations, and manuscript-specific layouts. HBA 1.0 [24] provides pixel-level annotations for historical books, marking textual content and elements such as initials and illustrations. The U-DIADS-Bib dataset [44] focuses on pixel-precise document layout analysis using segmentation maps and is composed of historical handwritten manuscripts.

Despite the diversity of these benchmarks, none explicitly evaluate document segmentation as a pixel clustering task. Most existing approaches rely on bounding box detection, region classification, or relation modeling.

2.2 Document Layout Analysis and Segmentation Methods

One way to achieve a logical segmentation is to first extract the textual content and then apply text segmentation methods. Older methods like TextTiling [10] or GraphSeg [7] use algorithmic techniques to detect boundaries in a continuous block of text. A more recent approach, proposed by Somasundaran et al. [33] utilizes two hierarchically connected transformer modules in a two learning objective setting. However, pure text-based approaches suffer in complex layouts by being dependent on reading order and OCR results.

To overcome these shortcomings, segmentation at the page level is necessary, as it enables effective integration of visual and spatial features. Early methods relied on convolutional neural networks for detecting layout elements [8,13,21,29]. More recent approaches, such as DLAFormer [38], integrate object detection, information retrieval and reading order prediction within an end-to-end model, improving both the detection accuracy and relation prediction. Alternatively, Wang et al. [39] introduced a three-stage framework that constructs a hierarchical document structure, facilitating information retrieval from segmented pages.

Pre-trained transformer models are also widely employed in this domain [6, 27,36]. LayoutLM and its successors [11,40,41] integrate spatial embeddings with text tokens. Document Image Transformer (DiT) [17] introduces a self-supervised training approach using large-scale unlabeled images. Similarly, DocFormer [1] is trained in an unsupervised manner, using text, vision, and spatial information within a multi-modal attention framework. LiLT [37] takes a language-independent approach, allowing pre-training in one language with subsequent fine-tuning in another.

Lastly, several pixel level segmentation methods have also been developed. Sang et al. [30] proposed a pixel-wise spatial attention module designed to capture long distance pixel dependencies and they employed a conditional random field to model contextual information. DocSegTr [2], on the other hand, treats the task as instance-level segmentation, proposing an end-to-end transformer model enhanced with a twin attention module.

3 TextBite Dataset

The TextBite dataset consists of scanned historical Czech documents from various sources with diverse layouts. It includes simpler layouts, such as book pages and dictionaries, as well as more complex multi-column formats from newspapers, periodicals, and other printed materials. Additionally, part of the dataset contains handwritten documents, primarily records from schools and public organizations, introducing extra segmentation challenges due to their more loosely structured layouts. Examples of pages from the dataset are visualized in Fig. 2. In total, the dataset contains 8,449 annotated pages, from which 7,346 pages are printed and 1,103 are handwritten. The pages contain a total of 78,863 segments. To ensure consistency in the evaluation, we split the dataset into a development and test subset. The test subset contains 964 pages, of which 185 are handwritten. The annotations are provided in an extended `COCO` format [20].

In the annotations, each segment is represented by a set of axis aligned bounding boxes, which are connected by directed relationships, representing reading order. To include these relationships in the `COCO` format, a new top-level key `relations` is added. Each relation entry specifies a source and a target bounding box. Subsequently, this style of annotations implicitly creates an intra-segment reading order.

In addition to the layout annotations, we provide a textual representation of the pages produced by Optical Character Recognition (OCR) tool `PERO-OCR` [14–16]. These come in the form of XML files in the `PAGE-XML` [26] format, which includes an enclosing polygon for each individual textline along with the transcriptions and their confidences. Lastly, we provide the OCR results in the `ALTO` [19] format, which includes polygons for individual words in the page image.

Apart from the proposed logical page segmentation, the dataset can be used in multiple research tasks, such as reading order prediction, title detection and association or document layout analysis in general.

3.1 Data Acquisition

Several Czech libraries are currently digitizing historical documents to preserve national heritage, making most of these documents publicly accessible for personal and research use. Librarians actively support this effort by organizing, sorting, and filtering the documents, creating collaboration opportunities between libraries and public institutions, such as universities.

As a result, these document collections became the primary data source for the TextBite dataset. The printed document pages were sampled through multiple iterations. Initially, around 4,000 pages spanning various document types and historical periods were manually selected from a single library. Then, approximately 1,000 newspaper title pages were selected and further filtered by annotators. The remaining printed pages were randomly sampled from several libraries without specific criteria regarding document types but were carefully filtered to ensure suitability for segmentation tasks.

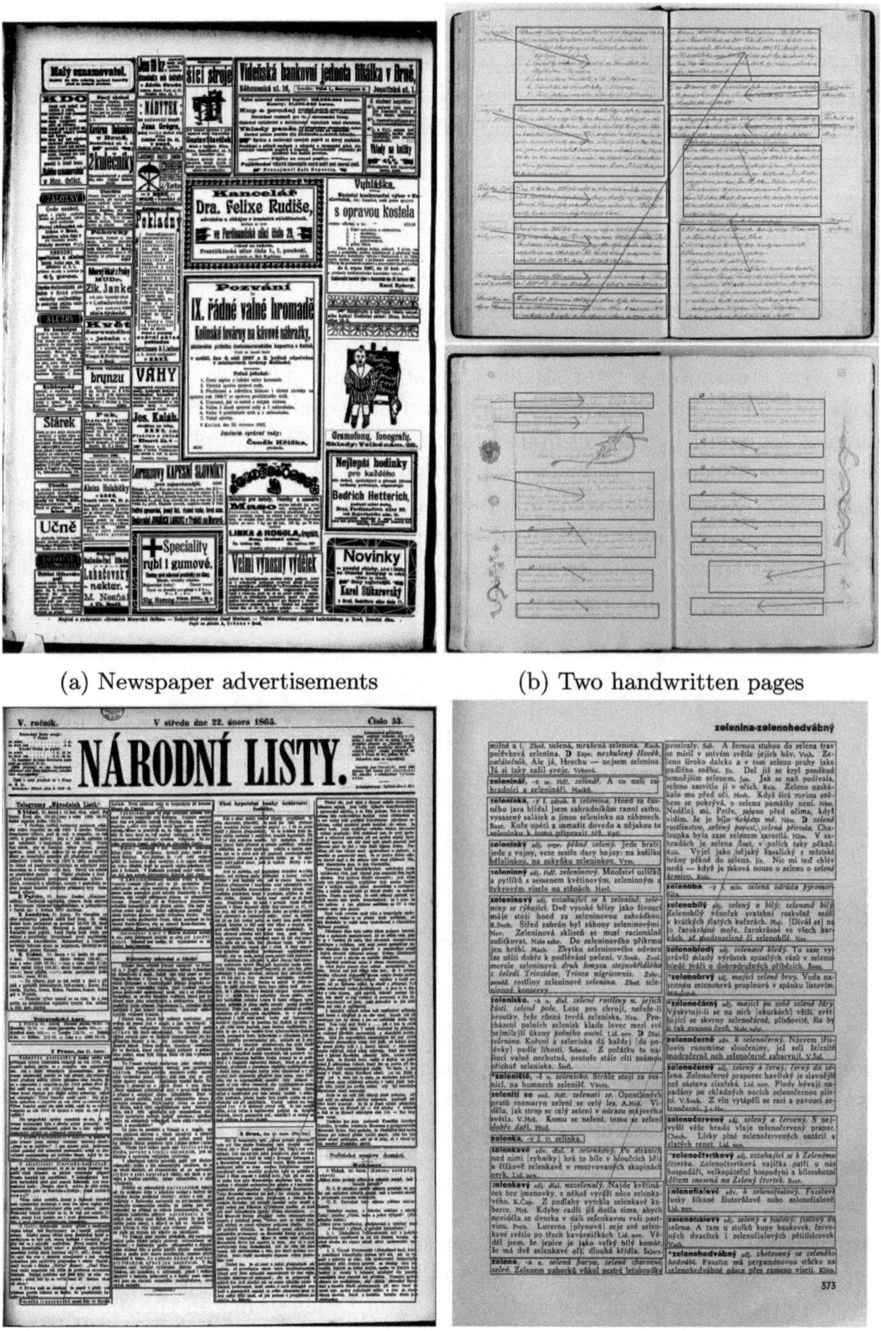

(a) Newspaper advertisements

(b) Two handwritten pages

(c) Newspaper title page

(d) Dictionary page

Fig. 2. Examples of annotated pages in TextBite dataset with various layouts.

We specifically selected only pages written in the Czech language, although occasional traces of other languages appear. German is the most prevalent secondary language, while languages such as French occur only occasionally—for example, as dictionary keys in translation dictionaries. Filtering also excluded pages that were damaged, rotated, overly blurry, or otherwise unsuitable for segmentation.

3.2 Data Annotation

Annotation Structure. Each page contains two types of annotations: bounding boxes and relations. Bounding boxes enclose a single-column text region and are categorized as text region, title, or page number. Page number annotations were created at a later stage as an additional sub-task, but in the eyes of logical segmentation, they are excluded from the evaluation. Relations establish coherent segments by linking bounding boxes, forming an implicit reading order. Titles are connected to their corresponding text blocks unless fully enclosed in other region (see Fig. 3). This happens in dictionaries and newspapers, where the title is included in the text and usually begins with it (named paragraphs). In such case, the title annotation has no effect on the evaluation. Note that the page numbers have no relations.

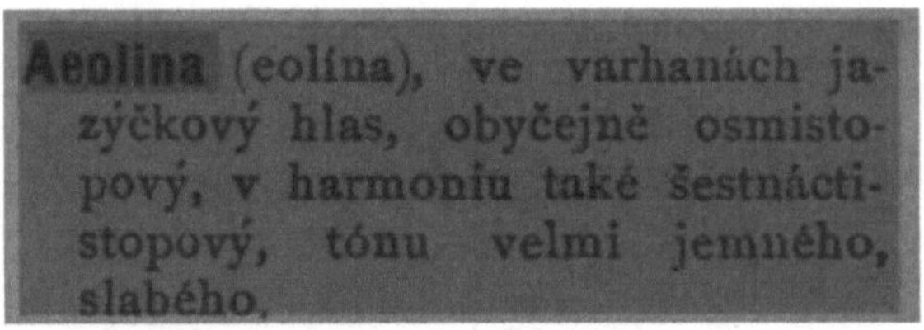

Fig. 3. Cropped section of an annotated image showcasing an enclosed title region (green) in a text region (blue). (Color figure online)

We did not annotate any page metadata, as it does not contribute to the core informational content of the page. Elements such as headers, footers, and other details of the publication were excluded to keep the focus on meaningful text. For example, in newspapers, details such as the publication date, newspaper name, or edition were not annotated.

Annotation Process. The annotation process was conducted using the open-source data labeling tool `Label Studio` [34]. After setting up the labeling interface, we prepared a set of instructions to ensure consistency in the annotations. Additional, more detailed guidelines and feedback were given to annotators during initial training. More than ten people participated in the annotation process, including university students, librarians, and researchers. Most of the annotation work was carried out by students and librarians. Researchers served mostly as supervisors. After the annotations were completed, researchers manually checked and corrected the test portion of the dataset, whereas students reviewed the development portion.

3.3 Dataset Characteristics

The TextBite dataset presents a diverse range of historical Czech documents with varying complexity in their layouts. Table 1 provides an overview of the number of annotated regions in different categories for both printed and handwritten documents. Notably, printed documents usually contain a higher number of text regions and titles, while handwritten documents contain relatively fewer, but often more complicated, text segments. Figure 4 presents an overview of the dataset's characteristics, including the number of textlines, image dimensions, segment count per image, and region count per image.

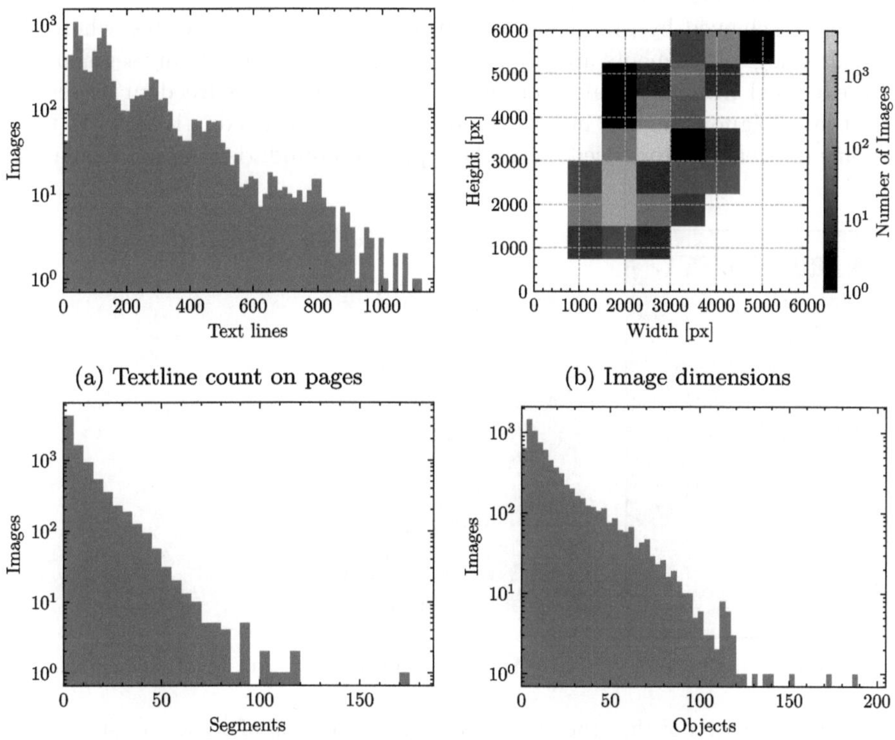

(a) Textline count on pages

(b) Image dimensions

(c) Segment count on pages

(d) Object count on pages

Fig. 4. Data characteristics of the TextBite dataset.

Table 1. Number of annotated regions of their respective class in both printed and handwritten pages.

Category	Printed	Handwritten
Title	46,140	2,648
Text	76,469	3,989
Page number	5,248	965

3.4 Evaluation

The TextBite dataset is annotated with linked bounding boxes that define the logical segments. While the annotations were created with reasonable precision, the exact placement of box boundaries is irrelevant to the evaluation. Instead, the evaluation process focuses on textual content within the annotated regions rather than on their precise spatial boundaries.

This is achieved by producing a per-pixel mask, which describes where the individual logical segments are (Fig. 5). Pixels not belonging to any segment are then excluded from the evaluation, giving system designers freedom in dealing with the ambiguous regions in-between text regions and making it easier to train systems on weakly supervised data where precise boundaries are not available.

(a) Human annotation (b) Annotation mask (c) Pixel segmentation

Fig. 5. Construction of the ground truth pixel segmentation of a page. First, we turn the human annotation in the form of connected regions 5 ainto their masked representation 5 b. This is then intersected with textlines obtained from OCR and thresholded, which results in the final pixel segmentation 5 c.

We produce the pixel annotations by narrowing the output from annotators in two ways: First, we intersect the manually annotated regions with textlines as detected by the PERO-OCR. This way, we constrain the pixel annotations so that larger areas of empty space do not contribute to the definitions of logical segments, which happens e.g. when the text is not justified. Second, we use thresholding to further limit the extent of the pixel annotation. Specifically, we use adaptive thresholding, with the threshold value estimated dynamically from a patch of 301×301 pixels. Using such fine segmentation, we (a) further prune areas without text and (b) allow direct evaluation of systems operating at the pixel level.

Finally, we evaluate the segmentation output using the Rand index [28]. Since the evaluation focuses only on text regions, we compute the Rand index over pixels within the ground truth pixel segmentation. For each pair of pixels, we mark if they are in the same segment in the ground truth and in the system output. The Rand index is then the proportion of pixel pairs where these assignments agree, with its value ranging from 0 (no agreement) to 1 (perfect agreement). If a pixel is assigned to a segment in the ground truth but is missing in the system output, it is treated as a misassignment, reducing the score. In practice, when 10 % of the considered pixels are misassigned, the Rand index is typically around 0.83, with the exact value depending on the specific misassignments.

Since pixels marked as background in the ground truth are not taken into consideration, segmentation systems are free to mark page numbers, page-level titles etc. as additional segments without hurting their Rand index. On the other hand, if the system output fails to label a part of a ground truth segment, the corresponding pixels are treated as assigned to an extra cluster, decreasing the Rand index.

While the exact pixel annotation arises from this automatic process and is thus not gold-label in a strict sense, the possible imperfections do not jeopardize the evaluation scheme: Since the pixel annotations are derived from manually verified human annotations that accurately localize text, they do not introduce incorrect clusters. In the case when the textline detection or thresholding act too aggressively and omit a portion of the text, the result is just a decreased weight of the specific segment, which still allows for a fair comparison between different systems.

4 Baseline Experiments

To establish a reference point for future research and facilitate further development, we have implemented a set of baseline methods. These methods serve as benchmarks, allowing for an objective evaluation of performance and providing insights into the challenges posed by the task. By assessing the effectiveness of these baselines, we aim to create a foundation for future improvements and encourage the development of more advanced approaches.

The development portion of the dataset was split into a training and validation set. The validation set contains 1,000 images, of which 200 are handwritten. The remaining 6,400 images were used to train the models. To estimate the uncertainty of evaluation results, we computed 95% confidence intervals using bootstrapping.

4.1 YOLO Detection Model

The first baseline method, which also serves as the basis for the other methods, detects regions of text using the YOLOv11 [13] object detector finetuned to the training set. This model identifies bounding boxes on a page, which represent distinct segments. To ensure consistency, any detected bounding box that is fully enclosed within a larger bounding box of is filtered out.

To train the detector, we used the publicly available Ultralytics [12] implementation. We experimented with three model variants and different image resolutions, with the results summarized in Table 2. Ultimately, we selected the small variant with a resolution of 1,200 pixels. All models were trained with a batch size of 16, a learning rate of $2e^{-4}$, and the Adam optimizer. Training was terminated after 20 epochs of no improvement on the validation set. The best-performing model, evaluated on the test data, achieved a Rand index of 83.9 % with a 95% confidence interval of $[82.6, 85.1]$.

This pure detection approach has a significant limitation. The model treats all predicted regions as independent segments, whereas, in reality, a segment may span multiple columns, include an associated title positioned above the main text, or it may consist of multiple bounding boxes. To address this, we introduce a second stage into the pipeline designed to merge detected bounding boxes into coherent segments.

Table 2. Training results of YOLOv11 object detection models in three sizes with five different resolutions. The best value in each category is highlighted.

Metric	mAP50					mAP50-95				
Resolution	640	800	1000	1200	1400	640	800	1000	1200	1400
YOLOv11n	86.4	89.1	90.7	91.3	91.5	65.4	68.2	69.9	70.9	70.8
YOLOv11s	88.0	90.0	91.2	**92.0**	–	67.9	69.9	71.4	**72.2**	–
YOLOv11m	90.0	91.4	–	–	–	70.3	71.4	–	–	–

4.2 Graph Neural Network

The first approach to detection merging is based on a graph neural network trained as a binary edge classifier. Each edge in the graph represents a potential connection between two detected regions, with the label indicating whether they should be merged.

Graphs are constructed as complete undirected graphs, where each node represents a detected region. During annotation, relationships were only defined between consecutive regions rather than all regions within a segment. To ensure that all detections belonging to a single segment are merged, we assigned positive labels to all edges within the same segment. Edge classification is performed by computing the cosine similarity between the updated node representations and comparing it to a classification threshold. The overall model architecture is illustrated in Fig. 6.

The input features include 29 geometric node features and 12 edge features, which combine both geometric and text features. Nodes contain only geometric features representing the position of the region both in absolute and relative values. Geometric edge features capture spatial relationships, including absolute and relative distances between the centers of the region and the closest edges. Text-based features are created by computing a cosine similarity between text embeddings of the two connected regions. The text embeddings are obtained using the CZERT [32] masked language model.

Since both absolute and relative features are used, their scales vary significantly. To address this, we apply Z-score normalization, using mean and standard deviation values computed from the training set to normalize the data.

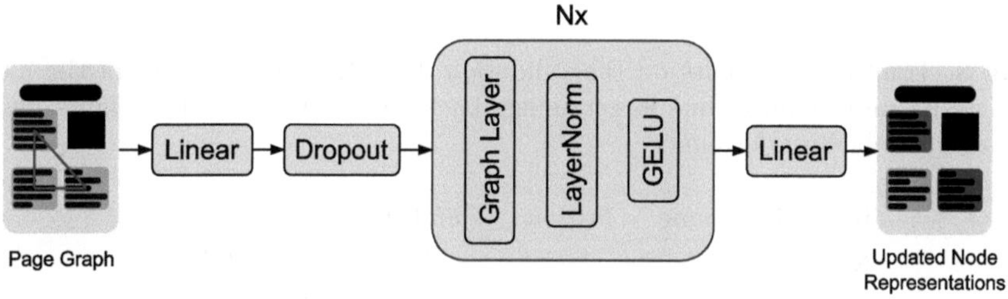

Fig. 6. Graph neural network structure.

We experimented with various model configurations. The best-performing model was trained with the Residual Gated Graph Convolution operator [3], a batch size of 16, dimensionality of 512, 3 graph layers and AdamW optimizer with $1e^{-3}$ learning rate. On the test data, the model achieved a Rand index of 92.5 % with a 95% confidence interval $[91.6, 93.3]$.

4.3 Transformer

The second region merging method is based on a transformer architecture with the task set up as an exclusive relation prediction. The input queries $Q = \{q_1, q_2, ..., q_n\}$ are the detected bounding boxes represented by 2D sinusoidal positional embedding vectors. Then, relation scores are calculated using the model output feature vectors $O = \{o_1, o_2, ..., o_N\}$ as:

$$s_{i,j} = FC_a^D(o_i) \cdot FC_b^D(o_j); i \in \{1..N\}, j \in \{1..N\},$$

where FC^D is a fully-connected layer with dimensionality D. The highest scoring query q_j is then chosen as the relation for query q_i. The relations are used to form chains of regions, which are then grouped together to create the final segments.

In its basic form, the model consists of a transformer encoder. We also experimented with the usage of images as additional information for the model. In this case, a transformer decoder with parallel decoding is used. The image is first resized to the resolution of 1024×1024 and then processed by an image backbone and the resulting feature vectors are used in the cross-attention layer of the decoder. The results of these experiments can be seen in Table 3. We chose ResNet-18 [9] and ResNet-50 as the backbones and we also experimented with freezing their weights during training. Adding the visual information increases the performance, but no significant difference can be observed between the ResNet variants and frozen weights. Due to these results, we chose the model with the frozen ResNet-18 image backbone as the final model.

The models were trained with 6 layers, hidden dimension of 128, 2 attention heads, batch size of 4, learning rate of $2e^{-4}$ and the inputs were padded to 80 tokens. The cross-entropy loss function was used to train the model, acting on the relation scores. On the test data, the model achieved a Rand index of 86.9 with a 95% confidence interval $[85.9, 88.0]$.

Table 3. Transformer results on the validation data. Including images in the model improves its performance, but larger image backbone or freezing its weights has no effect on the model performance.

Image Backbone	Frozen	Rand Index	CI@0.95
None	–	86.5	[85.4, 87.5]
ResNet-18	No	88.3	[87.3, 89.2]
ResNet-18	Yes	88.5	[87.5, 89.5]
ResNet-50	No	87.5	[86.5, 88.5]
ResNet-50	Yes	88.9	[88.0, 89.8]

4.4 Results Discussion

The results in Table 4 show that the pure YOLOv11 detection model achieves a reasonable baseline performance, but its inability to detect more complicated segments limits its potential. Incorporating a Graph Neural Network significantly improves the results, achieving the highest performance with a 95% confidence interval of $[91.6, 93.3]$. The transformer-based approach, while outperforming the standalone YOLOv11 model, does not reach the performance of the GNN. Additionally, using image features in the transformer improved performance, indicating that visual context can be beneficial.

5 Conclusions

In this paper, we introduced the TextBite dataset for logical page segmentation and document layout analysis. It consists of 8,449 manually annotated

Table 4. Final results of all the proposed methods on the test portion of the TextBite dataset.

Method	Rand Index	CI@0.95
YOLOv11	83.9	[82.6, 85.1]
GNN	92.5	[91.6, 93.3]
Transformer	86.9	[85.9, 88.0]

pages from historical Czech documents, encompassing a diverse range of layouts from periodicals, books, and handwritten manuscripts. The dataset includes annotations in an extended `COCO` format along with transcriptions generated by `PERO-OCR`. It is publicly available online[1], providing a valuable resource for research in logical document segmentation and layout analysis.

To address the limitations of existing segmentation evaluation methods, we proposed a novel approach that formulates logical page segmentation as a pixel clustering problem. By representing segments as clusters of foreground pixels corresponding to individual letters, our approach eliminates the dependency on OCR accuracy and mitigates the impact of geometric variations in detected text regions. This unifying evaluation framework enables fair comparisons across different logical page segmentation methods.

To establish baselines, we implemented and evaluated three methods: a traditional object detector, region merging with a graph neural network, and region merging with a transformer model. The best-performing method—YOLOv11 combined with a graph neural network—achieved a Rand index of 92.5%. These results highlight the potential of our evaluation framework and dataset to advance research in document segmentation and layout analysis.

As future work, we envision the TextBite dataset serving as a foundation for developing new methods in logical page segmentation. Moreover, the dataset's diversity across historical document types provides a challenging testbed for models designed to generalize across different visual and structural document styles.

Acknowledgments. This work has been supported by the Ministry of Culture Czech Republic in NAKI III project semANT - Semantic Document Exploration (DH23P03OVV060).

Disclosure of Interests. The authors have no competing interests to declare that are relevant to the content of this article.

References

1. Appalaraju, S., Jasani, B., Kota, B.U., Xie, Y., Manmatha, R.: DocFormer: end-to-end transformer for document understanding. In: Proceedings of the IEEE/CVF International Conference on Computer Vision, pp. 993–1003 (2021)

[1] The dataset is available at: https://github.com/DCGM/textbite-dataset.

2. Biswas, S., Banerjee, A., Lladós, J., Pal, U.: DocSegTr: an instance-level end-to-end document image segmentation transformer. arXiv preprint arXiv:2201.11438 (2022)
3. Bresson, X., Laurent, T.: Residual gated graph convnets. arXiv preprint arXiv:1711.07553 (2017)
4. Cheng, H., et al.: M6Doc: a large-scale multi-format, multi-type, multi-layout, multi-language, multi-annotation category dataset for modern document layout analysis. In: Proceedings of the IEEE/CVF Conference on Computer Vision and Pattern Recognition, pp. 15138–15147 (2023)
5. Clausner, C., Papadopoulos, C., Pletschacher, S., Antonacopoulos, A.: The ENP image and ground truth dataset of historical newspapers. In: 2015 13th International Conference on Document Analysis and Recognition (ICDAR), pp. 931–935 (2015). https://doi.org/10.1109/ICDAR.2015.7333898
6. Da, C., Luo, C., Zheng, Q., Yao, C.: Vision grid transformer for document layout analysis. In: Proceedings of the IEEE/CVF International Conference on Computer Vision, pp. 19462–19472 (2023)
7. Glavaš, G., Nanni, F., Ponzetto, S.P.: Unsupervised text segmentation using semantic relatedness graphs. In: Gardent, C., Bernardi, R., Titov, I. (eds.) Proceedings of the Fifth Joint Conference on Lexical and Computational Semantics, pp. 125–130. Association for Computational Linguistics, Berlin, Germany (2016). https://doi.org/10.18653/v1/S16-2016, https://aclanthology.org/S16-2016/
8. He, K., Gkioxari, G., Dollár, P., Girshick, R.: Mask R-CNN. In: 2017 IEEE International Conference on Computer Vision (ICCV), pp. 2980–2988 (2017).https://doi.org/10.1109/ICCV.2017.322
9. He, K., Zhang, X., Ren, S., Sun, J.: Deep residual learning for image recognition. In: Proceedings of the IEEE Conference on Computer Vision and Pattern Recognition, pp. 770–778 (2016)
10. Hearst, M.A.: TextTiling: segmenting text into multi-paragraph subtopic passages. Comput. Linguist. **23**(1), 33–64 (1997)
11. Huang, Y., Lv, T., Cui, L., Lu, Y., Wei, F.: LayoutlMv3: pre-training for document AI with unified text and image masking. In: Proceedings of the 30th ACM International Conference on Multimedia, pp. 4083–4091 (2022)
12. Jocher, G., Qiu, J., Chaurasia, A.: Ultralytics YOLO (2023). https://github.com/ultralytics/ultralytics
13. Khanam, R., Hussain, M.: YOLOv11: An overview of the key architectural enhancements. arXiv preprint arXiv:2410.17725 (2024)
14. Kišš, M., Beneš, K., Hradiš, M.: AT-ST: self-training adaptation strategy for OCR in domains with limited transcriptions. In: International Conference on Document Analysis and Recognition, pp. 463–477. Springer (2021)
15. Kodym, O., Hradiš, M.: Page layout analysis system for unconstrained historic documents. In: Document Analysis and Recognition–ICDAR 2021: 16th International Conference, Lausanne, Switzerland, September 5–10, 2021, Proceedings, Part II 16, pp. 492–506. Springer (2021)
16. Kohút, J., Hradiš, M.: TS-Net: OCR trained to switch between text transcription styles. In: Lladós, J., Lopresti, D., Uchida, S. (eds.) ICDAR 2021. LNCS, vol. 12824, pp. 478–493. Springer, Cham (2021). https://doi.org/10.1007/978-3-030-86337-1_32
17. Li, J., Xu, Y., Lv, T., Cui, L., Zhang, C., Wei, F.: DiT: self-supervised pre-training for document image transformer. In: Proceedings of the 30th ACM International Conference on Multimedia, pp. 3530–3539 (2022)

18. Li, M., et al.: DocBank: a benchmark dataset for document layout analysis. In: Proceedings of the 28th International Conference on Computational Linguistics, pp. 949–960 (2020)
19. Library of Congress: ALTO (Analyzed Layout and Text Object) XML Schema (2024). https://www.loc.gov/standards/alto/. Accessed 03 Mar 2025
20. Lin, T.-Y., et al.: Microsoft COCO: common objects in context. In: Fleet, D., Pajdla, T., Schiele, B., Tuytelaars, T. (eds.) ECCV 2014. LNCS, vol. 8693, pp. 740–755. Springer, Cham (2014). https://doi.org/10.1007/978-3-319-10602-1_48
21. Liu, W., et al.: SSD: single shot multibox detector. In: Leibe, B., Matas, J., Sebe, N., Welling, M. (eds.) ECCV 2016. LNCS, vol. 9905, pp. 21–37. Springer, Cham (2016). https://doi.org/10.1007/978-3-319-46448-0_2
22. Liu, Z., Simon, C.E., Caspani, F.: Passage segmentation of documents for extractive question answering. arXiv preprint arXiv:2501.09940 (2025)
23. Ma, J., et al.: HRDoc: dataset and baseline method toward hierarchical reconstruction of document structures. In: Proceedings of the AAAI Conference on Artificial Intelligence. vol. 37, pp. 1870–1877 (2023)
24. Mehri, M., Héroux, P., Mullot, R., Moreux, J.P., Coüasnon, B., Barrett, B.: HBA 1.0: a pixel-based annotated dataset for historical book analysis. In: Proceedings of the 4th International Workshop on Historical Document Imaging and Processing, pp. 107–112. HIP '17, Association for Computing Machinery, New York, NY, USA (2017). https://doi.org/10.1145/3151509.3151528, https://doi.org/10.1145/3151509.3151528
25. Pfitzmann, B., Auer, C., Dolfi, M., Nassar, A.S., Staar, P.: DocLayNet: a large human-annotated dataset for document-layout segmentation. In: Proceedings of the 28th ACM SIGKDD Conference on Knowledge Discovery and Data Mining, pp. 3743–3751 (2022)
26. Pletschacher, S., Antonacopoulos, A.: The page (page analysis and ground-truth elements) format framework. In: 2010 20th International Conference on Pattern Recognition, pp. 257–260 (2010). https://doi.org/10.1109/ICPR.2010.72
27. Powalski, R., Borchmann, Ł, Jurkiewicz, D., Dwojak, T., Pietruszka, M., Pałka, G.: Going full-TILT boogie on document understanding with text-image-layout transformer. In: Lladós, J., Lopresti, D., Uchida, S. (eds.) ICDAR 2021. LNCS, vol. 12822, pp. 732–747. Springer, Cham (2021). https://doi.org/10.1007/978-3-030-86331-9_47
28. Rand, W.M.: Objective criteria for the evaluation of clustering methods. J. Am. Stat. Assoc. **66**(336), 846–850 (1971). http://www.jstor.org/stable/2284239
29. Ren, S., He, K., Girshick, R., Sun, J.: Faster R-CNN: towards real-time object detection with region proposal networks. Adv. Neural Inf. Process. Syst. **28** (2015)
30. Sang, Y., Zeng, Y., Liu, R., Yang, F., Yao, Z., Pan, Y.: Exploiting spatial attention and contextual information for document image segmentation. In: Gama, J., Li, T., Yu, Y., Chen, E., Zheng, Y., Teng, F. (eds.) Advances in Knowledge Discovery and Data Mining, pp. 261–274. Springer International Publishing, Cham (2022)
31. Shtekh, G., Kazakova, P., Nikitinsky, N., Skachkov, N.: Applying topic segmentation to document-level information retrieval. In: Proceedings of the 14th Central and Eastern European Software Engineering Conference Russia. CEE-SECR '18, Association for Computing Machinery, New York, NY, USA (2018).https://doi.org/10.1145/3290621.3290630
32. Sido, J., et al.: Czert–Czech BERT-like model for language representation. In: Proceedings of the International Conference on Recent Advances in Natural Language Processing (RANLP 2021), pp. 1326–1338 (2021)

33. Somasundaran, S., et al.: Two-level transformer and auxiliary coherence modeling for improved text segmentation. In: Proceedings of the AAAI Conference on Artificial Intelligence. vol. 34, pp. 7797–7804 (2020)
34. Tkachenko, M., Malyuk, M., Holmanyuk, A., Liubimov, N.: Label Studio: Data labeling software (2020-2025). https://github.com/HumanSignal/label-studio, open source software available from https://github.com/HumanSignal/label-studio
35. Tschirschwitz, D., Klemstein, F., Stein, B., Rodehorst, V.: A dataset for analysing complex document layouts in the digital humanities and its evaluation with krippendorff's alpha. In: DAGM German Conference on Pattern Recognition, pp. 354–374. Springer (2022)
36. Tu, Y., Guo, Y., Chen, H., Tang, J.: LayoutMask: enhance text-layout interaction in multi-modal pre-training for document understanding. arXiv preprint arXiv:2305.18721 (2023)
37. Wang, J., Jin, L., Ding, K.: Lilt: A simple yet effective language-independent layout transformer for structured document understanding. arXiv preprint arXiv:2202.13669 (2022)
38. Wang, J., Hu, K., Huo, Q.: DLAFormer: an end-to-end transformer for document layout analysis. In: International Conference on Document Analysis and Recognition, pp. 40–57. Springer (2024)
39. Wang, J., Hu, K., Zhong, Z., Sun, L., Huo, Q.: Detect-order-construct: a tree construction based approach for hierarchical document structure analysis. Pattern Recogn. **156**, 110836 (2024)
40. Xu, Y., et al.: LayoutLMv2: Multi-modal pre-training for visually-rich document understanding. arXiv preprint arXiv:2012.14740 (2020)
41. Xu, Y., Li, M., Cui, L., Huang, S., Wei, F., Zhou, M.: LayoutLM: pre-training of text and layout for document image understanding. In: Proceedings of the 26th ACM SIGKDD International Conference on Knowledge Discovery & Data Mining, pp. 1192–1200 (2020)
42. Zhang, C., et al.: Modeling layout reading order as ordering relations for visually-rich document understanding. In: Proceedings of the 2024 Conference on Empirical Methods in Natural Language Processing, pp. 9658–9678 (2024)
43. Zhong, X., Tang, J., Jimeno Yepes, A.: PubLayNet: largest dataset ever for document layout analysis. In: 2019 International Conference on Document Analysis and Recognition (ICDAR), pp. 1015–1022 (2019).https://doi.org/10.1109/ICDAR.2019.00166
44. Zottin, S., De Nardin, A., Colombi, E., Piciarelli, C., Pavan, F., Foresti, G.L.: U-DIADS-Bib: a full and few-shot pixel-precise dataset for document layout analysis of ancient manuscripts. Neural Comput. Appl. (2024). https://doi.org/10.1007/s00521-023-09356-5

Few-Part-Shot Font Generation

Masaki Akiba[1(✉)], Shumpei Takezaki[1], Daichi Haraguchi[2], and Seiich Uchida[1]

[1] Kyushu University, Fukuoka, Japan
{masaki.akiba,shumpei.takezaki}@human.ait.kyushu-u.ac.jp,
uchida@ait.kyushu-u.ac.jp
[2] CyberAgent, Tokyo, Japan
haraguchi_daichi_xa@cyberagent.co.jp

Abstract. This paper proposes a novel model of few-part-shot font generation, which designs an entire font based on a set of partial design elements, i.e., partial shapes. Unlike conventional few-shot font generation, which requires entire character shapes for a couple of character classes, our approach only needs partial shapes as input. The proposed model not only improves the efficiency of font creation but also provides insights into how partial design details influence the entire structure of the individual characters.

Keywords: Part-based image generation · Font style · Few-shot font generation

1 Introduction

New fonts are created and published almost daily through the considerable efforts of typeface designers. Even for the basic uppercase letters ("A" to "Z"), they must define 26 distinct character shapes. Furthermore, each shape must be carefully refined to ensure the font's style remains unique and consistent across all letters.

Various software tools have been developed to assist in font design. For example, `FontLab`[1] and `Glyphs`[2] are widely used font editors, enabling designers to create typefaces from scratch or select a reference font and gradually refine it to achieve their desired style. These tools now offer advanced features for creating so-called *variable fonts*, in which attributes such as slant and stroke width can be controlled by parameters. While these tools reduce manual effort, designers must still make numerous fine adjustments to keep the font style unique and consistent across the entire set of characters.

Recently, generative models—such as Generative Adversarial Networks (GANs) [4] and diffusion models [6]—have been applied to font design, giving rise to a semi-automatic framework called *few-shot font generation*. It generates all characters in the alphabet using only one or a few fully designed reference

[1] https://www.fontlab.com.
[2] https://glyphsapp.com.

L. Jin et al. (Eds.): ICDAR 2025 Workshops, LNCS 16225, pp. 141–157, 2026.
https://doi.org/10.1007/978-3-032-09368-4_9

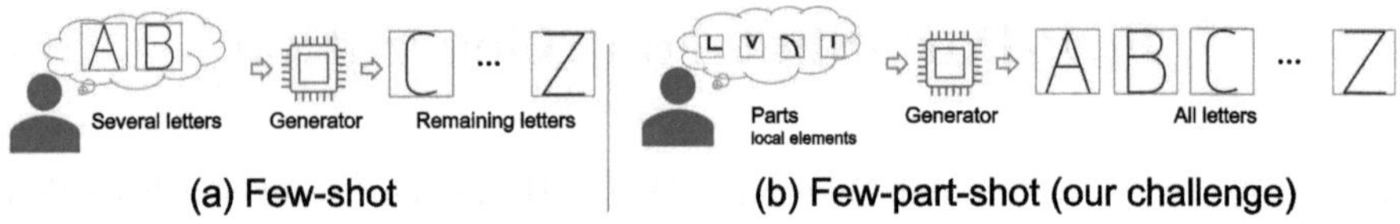

Fig. 1. (a) Conventional few-shot font generation. (b) Our challenge of few-part-shot font generation. We aim to clarify the capabilities and limitations of (b) by benchmarking it against (a).

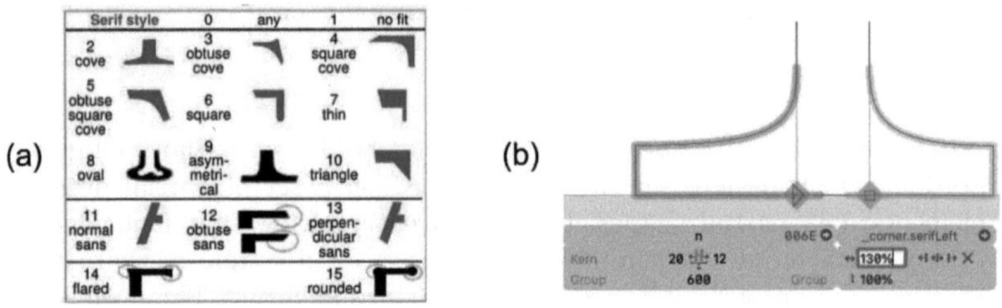

Fig. 2. (a) PANOSE Font Classification Chart (in part) (source: https://archive.org/details/Panose-chart). (b) A snapshot of the part-driven font design function in `Glyphs` (source: https://glyphsapp.com/learn/reusing-shapes-corner-components).

characters[3]. For instance, as illustrated in Fig. 1(a), by providing the character images of "A" and "B" in a certain style, the remaining letters "C" through "Z" can be generated in that same style. Naturally, this framework holds great potential to reduce the effort required in font design.

This paper challenges a new framework called few-*part*-shot font generation, which designs an entire font from a small set of user-specified parts. *Parts* are defined as representative local shapes capturing stroke thickness, terminations, corners, curves, intersections, and decorative elements. Parts play important roles in specifying font styles. The PANOSE chart[4] in Fig. 2(a) is a well-known font style classification scheme and uses part shapes. The above font-designing software tools (`FontLab` and `Glyphs`) already offer functionalities of part-oriented design, as shown in Fig. 2(b).

Figure 1(b) illustrates the idea of few-part-shot font generation. This idea stems from the observation that the same or similar parts appear in multiple characters. For instance, defining a vertical stroke end as a part facilitates the design of characters such as "I," "K," or "P," all of which share a similar vertical stroke pattern. Consequently, even a small set of well-chosen parts can effectively represent the majority of local shapes in the alphabet.

Despite the anticipated difficulties, the challenge of this new few-part-shot font generation offers three novel insights. First, from a practical perspective, it offers a new degree of flexibility in font creation. As illustrated in Fig. 1(a), the

[3] If a framework assumes one reference character, it is called one-shot font generation.
[4] https://archive.org/details/Panose-chart.

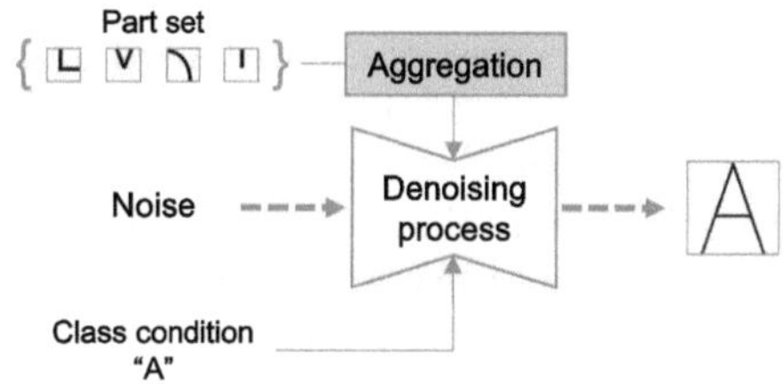

Fig. 3. Overview of the proposed model. The part set is aggregated into a single style feature vector. The diffusion model then generates the target font image conditioned on this vector and the character class.

existing framework requires at least one fully designed reference character image; however, in an extreme example where the single reference character is "O," no information about straight lines would be provided, potentially hindering the generation of other characters. By contrast, our new framework imposes no requirement to complete the design of an entire character image; users can simply supply arbitrary parts, particularly those shapes deemed critical for the desired font style.

Second, there is a scientific motivation in that this framework can reveal the extent to which a limited set of parts correlates with an entire alphabet. If a small collection of parts can reproduce most character shapes, those parts effectively function as a compressed representation of the overall shapes in the alphabet.[5] To our knowledge, no studies have addressed the question of how parts can faithfully reconstruct entire characters, suggesting that our attempt could be a new starting point for font research.

Third and finally, it is also crucial to clarify the limitations of this framework. Few-part-shot generation cannot, in principle, perfectly reconstruct certain global characteristics from only local parts. A simple illustration is a rectangle: from the four 90-degree corners of a rectangle, we cannot reconstruct the original rectangle because the corners do not fix its aspect ratio. Applying this to fonts, even if all parts come from a "condensed" font (i.e., a narrow font), there is no guarantee that the original condensed appearance can be recovered precisely. Likewise, overall features such as oblique angles may lie outside the representational reach of a small set of parts. Despite these potential shortcomings, our experiments will demonstrate that few-part-shot generation often produces surprisingly reasonable results in practice. Moreover, future work can focus on a complementary combination of the conventional entire-shape-based and our part-based generation frameworks.

In our practical implementation of few-part-shot font generation, we employ a diffusion model that is conditioned on a set of parts. This setup ensures that the generated character images reflect the stylistic cues embedded in those parts.

[5] it is well-known that parts capture character "classes," through various part-based classification techniques such as bag-of-visual-words [17,21]. In contrast, our focus here is on how well parts capture complete character "shapes.".

Figure 3 presents a simplified overview of the proposed model. To handle a variable number of user-defined parts, we use DeepSets [23], which transforms each part into a feature vector and then aggregates all such vectors into a single, unified representation. This design allows our model to integrate any number of parts without modifying the network architecture, thereby keeping the style feature vector consistent regardless of the set's size.

2 Related Work

2.1 Few-Shot Font Generation

A common approach to example-based font generation is few-shot font generation, which aims to generate complete fonts by extracting style features from a small number of example characters. In the early stages of the research, EMD [24] and AIGS-Net [3] extracted both content and style features directly from whole character images to synthesize the target font.

In recent years, many studies have shifted their focus to the generation of fonts for complex writing systems such as Chinese and Hangul, which contain component (radical) structures [9,13,14]. To address this complexity, methods that leverage component-based content and style feature extraction have been proposed. In the GAN-based approaches, LF-Font [13] and DM-Font [1] uses component labels for style extraction, and CG-GAN [9] utilizes weak component labels in the discriminator for one-shot font generation. FS-Font [16] captures fine-grained local styles from the whole image via cross-attention. Furthermore, Diff-Font [5], a diffusion-based approach, incorporates stroke-level information to enhance font generation quality. These component-based approaches have achieved high accuracy by decomposing complex characters, such as Chinese characters, into their components, such as radicals.

Thanks to the development of the diffusion model, several methods that do not rely on component information have achieved state-of-the-art performance. For example, MSD-Font [2] employs a three-stage reverse diffusion process. Font-Diffuser [22], on the other hand, achieves one-shot font generation by utilizing multi-scale feature maps while ensuring font style consistency through its Style Contrastive Refinement module, which aligns the font style features from generated font and reference font image.

This paper focuses on Latin alphabets, which do not have component structures and proposes a novel approach that extracts font style features from parts (i.e., local shapes) of characters to generate complete font images. This perspective can be seen as further refining the granularity of feature extraction compared to conventional component-based methods.

2.2 Part-Based Analysis of Fonts

The analysis of fonts based on individual parts is not a common topic in computer science but was first introduced by Ueda et al. [18]. Their study focused on not analyzing the global shape of a font but examining individual parts, aiming to identify which parts contribute to specific impressions, such as "legible," "elegant," or "comic-like." As an extension of the first study, they further investigated the association between specific parts and impressions [19]. They hypothesized that the combinations of parts, rather than individual parts alone, play an important role in shaping font impressions. To examine this hypothesis, they employed a self-attention mechanism [20] to capture dependencies between parts and analyzed their joint influence on impressions. Consequently, those trials emphasize how local parts of font are important.

3 Few-Part-Shot Font Generation

Our proposed few-part-shot font generation method has a relatively simple architecture. Specifically, it consists of two main stages: (1) font style feature extraction from the given set of parts and (2) font image generation using a diffusion model. As will be shown in the experiments, even this rather simple architecture still successfully generates font images from the limited information contained in the parts. This finding not only validates the sufficiency of our proposed method but also highlights that the overall design of a character strongly depends on its local shapes—in other words, knowing the local shapes provides a clear pathway to reconstructing the entire character.

3.1 Font Style Extraction from Parts

We now describe how to pre-train the model to extract a consistent font style representation from an arbitrary number of parts. To learn the mapping from a given set of parts to a style feature vector, we employ DeepSets [23]. Furthermore, we use contrastive learning to ensure that the style feature vector can capture subtle variations in font style while remaining invariant to character classes (e.g., "A" and "B" of the same font should share the same style feature).

Aggregating a Set of Parts Using DeepSets. DeepSets is a type of neural network model designed to handle sets with an arbitrary number of elements. As depicted in Fig. 4(a), DeepSets consists of two components: (i) a network that transforms each element of the set into a vector, and (ii) an operator that aggregates the resulting vectors. In our case, each part is given as a small image (i.e., a patch), so we use a CNN for the transformation. The aggregation is simply a standard vector addition. After the addition, the resultant vector is normalized to have a unit norm, yielding the aggregated style feature vector. This normalization is intended to reduce the dependence of the feature vector on the number of parts in the set.

Contrastive Learning. The style feature vector obtained in DeepSets is refined via contrastive learning [12], as illustrated in Fig. 4(b). Contrastive learning encourages feature vectors with the same label to be pulled closer together,

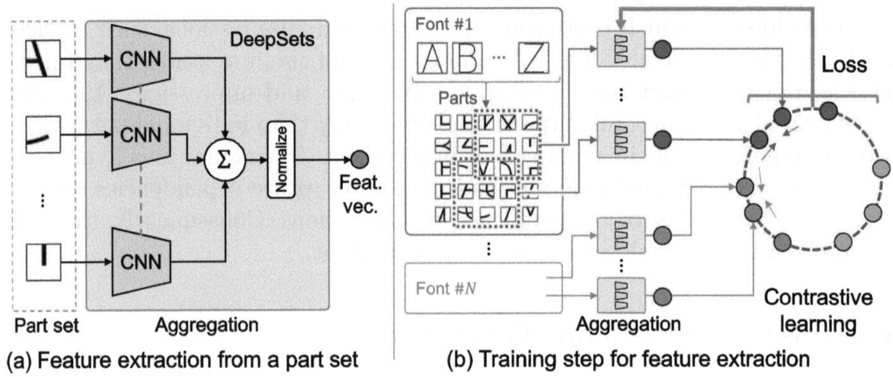

Fig. 4. Style feature extraction from a set of parts using DeepSets and contrastive learning. (a) Each part in the part set is encoded into a feature vector using a CNN. The resulting vectors are summed and then normalized to produce a single aggregated style vector. (b) The feature extractor is pretrained using contrastive learning, which pulls together part sets from the same font and pushes apart those from different fonts.

while pushing those with different labels apart. We apply this mechanism to sets of parts originating from the same font, forcing their style vectors to be similar. We also apply it to sets from different fonts, driving their style vectors apart and enabling the model to learn a stable representation of the font style. Through this procedure, the internal CNN of DeepSets is trained to produce style vectors that depend as little as possible on the specific parts chosen. Furthermore, by training with part sets of varying sizes, we ensure that the resulting style feature vector remains robust regardless of the size of the set.

3.2 Font Generation with the Style Feature from a Part Set

As shown in Fig. 3, we employ a standard conditional diffusion model to generate complete font images from the style feature vectors produced by DeepSets. Concretely, we train a U-Net that progressively transforms random noise into an entire character image (rather than just a partial one), conditioning on three key inputs: (1) the character class label (e.g., "A"—"Z"), (2) the style feature vector extracted from the parts, and (3) the diffusion process timestep, which is omitted in Fig. 3 for simplicity.

During training, we integrate both the character label (converted into an embedding) and the style feature vector (obtained from a user-provided set of parts via DeepSets) into the U-Net through cross-attention layers. This design allows the network to adapt its denoising process to the target character and the specified font style. In addition, we randomly vary the number of parts in each set, enabling the model to handle different amounts of local shape information. As a result, at inference time, the U-Net can generate high-quality font images for any character using only the chosen style parts.

4 Experimental Setup

4.1 Dataset

In this study, we used 26 uppercase letters ("A"–"Z") from the Google Fonts[6] dataset for model training and evaluation. The collected dataset consists of a total of 1,804 font families and 6,796 fonts. We divided them into three mutually exclusive font-family subsets to minimize the similarity between training and testing fonts. Specifically, 5,326 fonts from 1,443 font families were used for training. 746 fonts from 181 font families were used as the validation set for hyperparameter tuning and early stopping. The remaining 724 fonts from 180 font families are used for quantitative and qualitative evaluations. Each character in the font sets is rendered to be a 128×128-pixel bitmap image.

4.2 Preparation of Part Set

To train a neural network for few-part-shot font generation, we need a dataset containing only the essential parts of each font. It would be ideal if we have a public dataset of parts prepared by professional designers. However, no such public dataset exists at present. While it is possible to extract parts randomly from font images, this naive approach lacks objectivity and reproducibility, making it unsuitable for constructing a reliable dataset.

To address this issue, we propose an objective method that automatically extracts $K > 0$ representative parts from each font using the Scale-Invariant Feature Transform (SIFT) [11] and K-medoids clustering. This method enables consistent selection of characteristic parts within a font.

Extracting only the "meaningful" parts of a font is inherently challenging, since the importance of each part can vary depending on the designer's intention and the perception of font users. Nevertheless, we assume three key properties of font design: (1) local shapes such as serifs, stroke terminals, intersections, and corners are widely regarded as representative stylistic elements, as shown in Fig. 2; (2) designers typically do not reuse exactly identical parts multiple times; and (3) purely blank regions (i.e., background-only patches) are not considered as valid design components.

Based on these assumptions, we *simulate* the selection of meaningful parts by using SIFT to detect key points along the font strokes and generate corresponding feature descriptors, which are then clustered via K-medoids. We finally extract K parts (i.e., image patches) corresponding to the medoid key points. The part size is a hyper-parameter to be fixed before training; we use 32×32 or 64×64 for 128×128 font images This procedure offers an objective and reproducible way to extract representative parts from any given font. In our experiments, we set K at 26, resulting in 26 representative parts per font, although it does NOT aim to extract a single part from each of the 26 alphabets.

Figure 5 shows results of automatically extracting 26 representative parts from each font. Their size is 32×32 pixels. Despite the variations in overall font

[6] https://fonts.google.com.

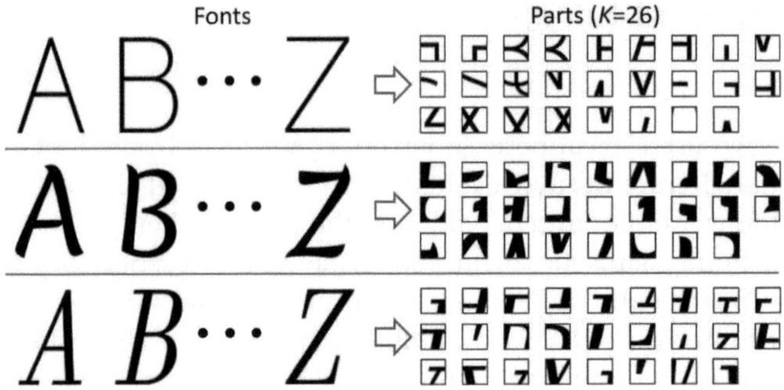

Fig. 5. Examples of three different fonts (left) and their corresponding set of $K = 26$ parts (right) as 32×32-pixel images. The extracted parts include corners, strokes, and terminations that characterize the font style.

design (e.g., stroke thickness, curvature, or decorative elements), the method successfully identifies salient local shapes that capture key stylistic features of the font. This demonstrates how the proposed pipeline is able to distill a font's characteristic elements into a manageable collection of parts.

4.3 Implementation Details

Font Style Extractor. A convolutional neural network (CNN) is employed as a feature extractor for individual parts (i.e., tiny image patches) in the set, producing a 256-dimensional output vector. During training the CNN in the DeepSets architecture of Fig. 4(a), the number of parts used is randomly varied between 1 and 8. The mini-batch size is set to 64. For contrastive learning, the number of negative samples is set to 63, and the temperature parameter is fixed at 1. The model is optimized using the Adam optimizer with a learning rate of 10^{-4}. Additionally, early stopping is applied, terminating training if the validation loss does not improve for 500 epochs.

Diffusion Model. We employ a U-Net [7] for noise estimation in the diffusion process. The diffusion model operates with 1,000 timesteps. The character classes used in training consist of 26 uppercase letters 'A'–'Z,' and each character class condition (represented as a 26-dimensional one-hot vector) is embedded into a 256-dimensional vector. While training the denoising model, the CNN in DeepSets is also updated. Again, the number of parts during training is randomly varied between 1 and 8. The mini-batch size is set to 64. The model is optimized using the Adam optimizer with a learning rate of 10^{-4}. Training is conducted for approximately 200k iterations. During evaluation, the generated results are obtained after 200 sampling steps.

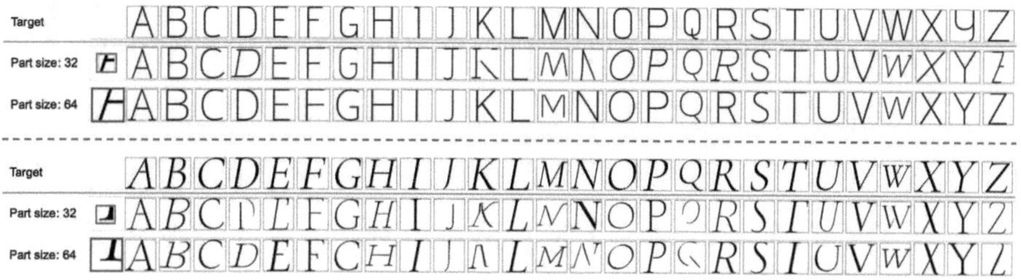

Fig. 6. Results of one-part-shot font generation. Two different part sizes (32×32 and 64×64) were used.

5 Experimental Analysis of Few-Part-Shot Font Generation

5.1 Qualitative Analysis

One-Part-Shot Generation for Observing the Basic Performance. As the simplest demonstration of our method's behavior, Fig. 6 shows the results of *one*-part-shot font generation with $K = 1$. The part used in the upper example is manually selected from near the horizontal stroke endpoint of the letter "A" in the target font, and the lower example uses a part from near another stroke endpoint of the same letter "A." Even though each part is just a small fragment, it carries information about stroke thickness and curvature (or straightness). The generated alphabets reflect these features, indicating that a single part is surprisingly rich in style-related information. Although the results do not exactly match the target fonts, one can see how effectively the local shape properties are generalized into the entire alphabet.

Effect of Part Size. Figure 6 also shows the difference when using two part sizes (32×32 and 64×64) for the same region in each font. Naturally, a larger part provides more global cues. In the upper row, the 32×32 part captures most features of the target font, yet some letters (like "D" or "R") appear in an italicized style that was not intended. Because italicization is a more global property, relying solely on a smaller local part can lead to ambiguity. In contrast, using a 64×64 part stabilizes the style, suggesting that even moderately enlarged parts can help preserve the intended global design.

In the lower row of the figure, most generated letters already include serif details even at a 32×32 part size. This implies that local decorations, such as serifs, can be captured accurately from relatively small parts. However, in cases where the target font is italic, certain letters still fail to reflect italicization at the smaller part size. When the part size is increased to 64×64, more letters correctly exhibit the italic style. Nevertheless, achieving perfect global consistency remains challenging with local parts alone, highlighting a fundamental limitation of few-part-shot font generation.

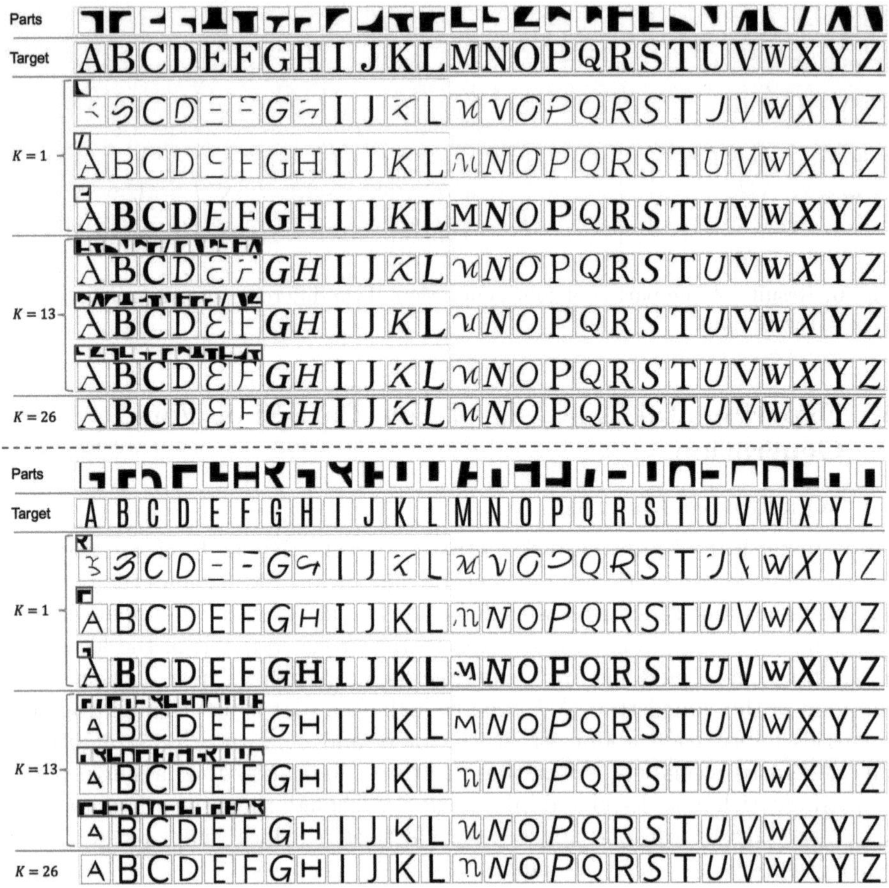

Fig. 7. Results of our few-part-shot font generation method under different part sets and the number of parts. The parts (32 × 32) used for generating the displayed result are highlighted with a red frame. For $K = 26$, all of the 26 parts shown at the top were used. (Color figure online)

Effect of Number of Parts and Their Variation. Figure 7 shows the results of our few-part-shot font generation when varying both the individual parts and the number of parts. For each of the two font examples shown, we extract 26 distinct parts from the target font. Note that the lower example in Fig. 7 features a "condensed" font with a notably narrow width, providing an interesting challenge for our method. By comparing the two examples, we can examine how well our approach generalizes to different styles with varying constraints.

One-part-shot generation ($K = 1$) often produces images that exhibit instability or inconsistency across 'A' to 'Z,' depending on which single part is chosen. Nonetheless, as demonstrated in Fig. 6, selecting a particularly informative part (such as the third one-part-shot in the upper example) can yield generated results remarkably close to the target style. This suggests that even a single part can carry substantial style information if chosen wisely.

Increasing the number of parts K raises the likelihood of a stable, coherent output. At $K = 13$, the generated characters become more consistent across all letters, as illustrated by the fully serifed style in the upper example and the consistent sans-serif style in the lower example. In the upper example, certain letters (e.g., 'D' and 'P') appear nearly indistinguishable from their counterparts in the target font. A crucial observation here is that the results become more robust to the specific choice of parts: different subsets of 13 parts often yield very similar character images. This phenomenon points to a strong correlation between local shapes and overall font structure. When $K = 26$, the generated outputs closely resemble those obtained at $K = 13$, indicating that even a larger set of parts may not entirely capture all aspects of the global style. We will revisit this point in a later quantitative evaluation.

In Sect. 1, we discussed the fundamental limitation of relying on parts to describe an entire shape—the rectangle example shows that merely knowing the four 90-degree corners does not guarantee recovery of its aspect ratio. This highlights the inherent ambiguity when reconstructing a global shape from local elements. In the lower example of Fig. 7, the generated letters fail to maintain the narrow proportions of the condensed style, suggesting that the parts alone cannot fully capture certain global traits. Nonetheless, it is notable that much of the local styling (such as serif or sans-serif features) is still successfully reproduced.

Despite this fundamental limitation, many of the generated fonts in Fig. 7 exhibit a high degree of similarity to their respective targets, underscoring the promise of few-part-shot font generation. In practical settings, a designer could augment these part-based inputs with a small number of full-character references or additional global constraints, thereby refining the overall aesthetic. Such a hybrid approach would not only help overcome the current limitations of part-only generation but could also improve the fidelity of existing full-character methods by incorporating more precise local structures. This combination of global and local shape information suggests a valuable direction for future research on font generation.

5.2 Quantitative Analysis

Evaluation Metrics

We employ three evaluation metrics: L1 distance, SSIM, and local L1 distance. The L1 distance measures the mean absolute pixel-wise difference, thereby capturing the pixel-level error between the target and generated images. SSIM evaluates image quality by considering luminance, contrast, and structural information, rather than direct pixel-wise differences.

In addition to these two metrics, we propose the local L1 distance, which focuses on the local parts of font images. Specifically, we use SIFT to detect keypoints on the target images and extract corresponding local parts. We then locate the same regions in the generated images and measure the L1 distance

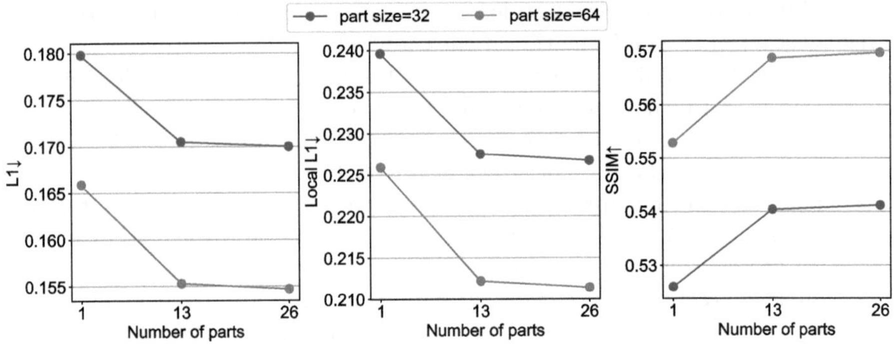

Fig. 8. Quantitative performance of few-part-shot font generation as a function of part size (32 vs. 64) and number of parts (K = 1, 13, 26). The three panels respectively show L1 distance (left), local L1 distance (middle), and SSIM (right).

between the target and generated patches. This approach allows us to assess how accurately local font structures are reproduced.

Quantitative Evaluation by Part Size and Number of Parts. Figure 8 presents the quantitative results of few-part-shot font generation when varying both the part size and the number of parts. Focusing first on part size, we observe that larger parts yield better performance in all three metrics. This result suggests that increasing the part size enriches the style information contained within a single part, thereby improving generation quality.

Regarding the number of parts, increasing K from 1 to 13 significantly improves all three metrics. However, the gain from 13 to 26 is relatively modest, indicating a form of diminishing marginal returns that often arises in set-based optimization problems. While adding more parts can capture additional local details, it does not necessarily guarantee a dramatic increase in global fidelity. This phenomenon also highlights a limitation: there appears to be an upper bound to how much additional style information can be conveyed by simply supplying more parts.[7] (From a positive viewpoint, human designers do not need to give more and more parts to improve the quality of the generated font images.)

Interestingly, when the part size is 64, using 26 parts of size 32 still cannot surpass the accuracy of a single 64×64 part in terms of L1 distance and SSIM. This finding reaffirms the importance of part size for capturing critical style cues. On the other hand, if we focus on local L1 distance, using 26 parts at a size of 32 yields performance nearly comparable to that of a single 64×64 part. Hence, a sufficient number of smaller parts can effectively represent local structures, even if they do not convey as much global information as a larger part might.

[7] This phenomenon, where additional inputs yield progressively smaller benefits, is commonly referred to as *diminishing marginal returns*. It is observed in various fields such as economics and resource allocation.

5.3 Comparative Experiments

Evaluation Metrics. In addition to the evaluation metrics described in Sect. 5.2, we use MSSIM, StyleFID, and ContentFID [10]. MSSIM is similar to SSIM and evaluates image quality by considering luminance, contrast, and structural information rather than direct pixel-wise differences. FID assesses the difference between the feature distributions of the generated and real images using the Fréchet distance. StyleFID is computed using feature vectors from a CNN-based font style classifier, while ContentFID is derived from a CNN-based character class classifier.

Comparative Methods. One major difficulty in evaluating our approach is the lack of suitable baselines. To the best of our knowledge, this is the first attempt to generate fonts directly from a set of parts, so no existing models allow for a direct comparison. Moreover, many recent few-shot font generation methods focus on Chinese characters, leveraging radical (sub-character) structures [9,13,14]. Since our current work targets the more decorative Latin alphabet, we cannot fairly compare with such radical-based approaches.

To address this challenge, we adopt a *one*-shot font generation setting for fair comparison and prepare three comparative methods:

- First, we use FontDiffuser [22] – a state-of-the-art model for one-shot font generation without assuming radical structures – to generate the Latin alphabet. FontDiffuser achieves high performance in one-shot font generation through a carefully designed training scheme, including two-stage learning and style refinement.
- Second, we also create another baseline model that employs the same diffusion backbone as ours but replaces our style feature extraction module with the encoder from FANnet [15]. This baseline is necessary because FontDiffuser and our proposed method differ in their underlying diffusion architectures, complicating a direct comparison. Kondo *et al.* [8] similarly used FANnet to extract font style features, combining it with a diffusion model for font generation. We thus regard it as a reasonable one-shot baseline, which we call **FANnet + DM**. Like FontDiffuser, this baseline processes a single full-character image rather than a small part.
- Third, we modify the proposed method to be evaluated in the same one-shot framework ($K = 1$) by setting the part size at 128×128, i.e., the size of the original character images. We denote this version as "Ours as 1-shot" in the later result.

We compare these methods with our few-part-shot method with the part size of 32×32 and $K = 13$ parts. This setup ($32 \times 32 \times 13$ total pixels) roughly equals one 128×128 image in terms of pixel count, allowing for an approximate comparison of information capacity.

Results. Table 1 presents the quantitative comparison of our method (the bottom row) against other one-shot font generation approaches (FontDiffuser, FANnet + DM, and Ours as 1-shot). Our method significantly outperforms FANnet+DM in ContentFID and StyleFID, despite relying on far less information in the one-part scenario. This finding suggests that font generation can be achieved with minimal input and still achieve performance on par with simpler one-shot methods. In contrast, FontDiffuser exhibits higher performance than our model on all metrics, likely due to its carefully designed two-step training strategy that provides superior global coverage of font styles.

Nevertheless, our approach produces results comparable to or better than FontDiffuser in certain cases, which we examine in detail in the next section. More importantly, by allowing designers to supply only local shapes as opposed to a fully designed character, our few-part-shot framework offers significant flexibility in practical font creation. Although it currently cannot surpass state-of-the-art methods that use entire characters, the fact that it can come close underscores its potential. With the option to incorporate a few complete character references or additional global constraints, our method could be extended further, closing the gap with FontDiffuser while maintaining the benefits of local part-based design. Such a hybrid strategy opens up new avenues for efficient and creative font generation workflows.

Class-Wise Comparison with FontDiffuser. Figure 9 shows a box-plot of L1 distances of FontDiffuser and ours for each character class, revealing variability in generation performance depending on the specific letter. The results for 'I' and 'J' (highlighted with a red box) show that our method performs comparably to FontDiffuser on these two characters. Qualitative analysis indicates a recurring trend, as illustrated in the top-right inset of the figure. FontDiffuser often introduces serifs when generating 'I' and produces a thicker shape for 'J,' especially in fonts where 'I' and 'J' are much narrower than other letters. In contrast, our method generates 'I' and 'J' with widths closer to those in the target fonts. Since our method leverages only local parts, it avoids certain global tendencies that might distort thin characters into a more uniform style. Moreover, 'I' and 'J' are very simple characters, composed of a single, mostly straight stroke; thus, their shapes can be reliably inferred from a small set of local design cues.

6 Conclusion, Limitation, and Future Work

In this paper, we challenged *few-part-shot font generation*, a new extension of few-shot font generation that relies solely on localized parts rather than fully designed character images. By providing only small patches that capture stroke endings, corners, or decorative elements, our approach can generate fonts that largely reflect the intended style. Through a series of qualitative and quantitative experiments, we confirmed several capabilities and limitations:

- Even a single, carefully selected part can convey enough local style cues to produce recognizable variations of an entire alphabet.

Table 1. Quantitative comparison with the existing methods for one-shot font generation. "Ours as 1-shot" is a one-shot version of the proposed method using the whole image of a single character (i.e., part size was 128×128 and $K = 1$). For the proposed method (Ours), the part size was 32×32 and $K = 13$. The column "part" shows whether one-part-shot or not. The column "#shot" shows the number of shots for generation.

	part	#s	L1↓	Local L1↓	SSIM↑	MSSIM↑	StyleFID↓	ContentFID↓
FontDiffuser		1	0.128	0.172	0.602	0.560	53.785	4.026
FANnet+DM		1	0.164	0.227	0.565	0.445	51.139	133.107
Ours as 1-shot		1	0.135	0.185	0.604	0.513	15.918	18.276
Ours ($K = 13$)	✓	13	0.171	0.228	0.540	0.426	15.296	9.060

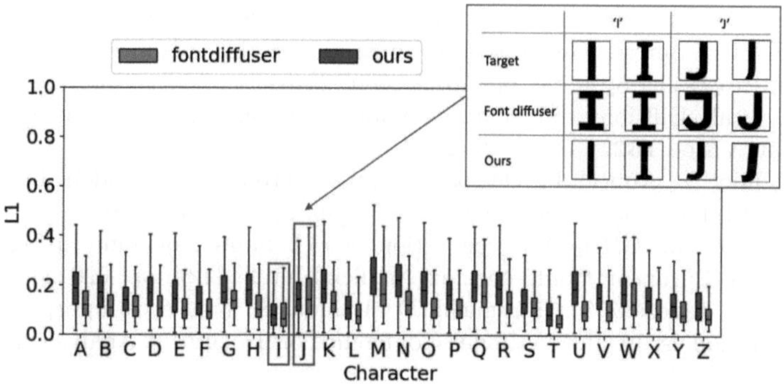

Fig. 9. Class-wise evaluation of generated images with L1 distance. For the proposed method, the part size is set at 32×32 and $K = 13$.

- Increasing either the part size or the number of parts further stabilizes and improves the quality of the generated fonts, though the benefits diminish beyond a certain point (i.e., diminishing marginal returns).
- A fundamental limitation arises in reproducing certain global characteristics (e.g., oblique angles or condensed shapes). Local parts alone cannot guarantee global consistency because they lack holistic information about overall proportions.

Despite the limitation, our few-part-shot approach offers a more flexible and efficient way to prototype fonts without requiring a fully finished reference character. Looking ahead, we plan to investigate hybrid methods that combine part-based inputs with a small number of complete-character references, hoping to capture both local and global structure more effectively. We also envision applying this method to more complex scripts (e.g., non-Latin alphabets) and exploring richer editing capabilities such as style interpolation or partial shape manipulation. By unifying local and global design cues in one framework, we aim to further bridge the gap between the creativity of human font designers and the power of modern generative models.

Acknowledgments. This work was supported by JSPS KAKENHI-JP22H00540, JST CRONOS-JPMJCS24K4, and Hoso Bunka Foundation.

References

1. Cha, J., Chun, S., Lee, G., Lee, B., Kim, S., Lee, H.: Few-shot compositional font generation with dual memory. In: European Conference on Computer Vision (ECCV) (2020)
2. Fu, B., et al.: Generate like experts: multi-stage font generation by incorporating font transfer process into diffusion models. In: Computer Vision and Pattern Recognition, pp. 6892–6901 (2024)
3. Gao, Y., Guo, Y., Lian, Z., Tang, Y., Xiao, J.: Artistic glyph image synthesis via one-stage few-shot learning. ACM Trans. Graph. **38**, 1–12 (2019)
4. Goodfellow, I., et al.: Generative adversarial nets. In: Advances in Neural Information Processing Systems, pp. 2672–2680 (2014)
5. He, H., et al.: Diff-font: diffusion model for robust one-shot font generation. Int. J. Comput. Vision **132**, 5372–5386 (2024)
6. Ho, J., Jain, A., Abbeel, P.: Denoising diffusion probabilistic models. In: Advances in Neural Information Processing Systems, pp. 6840–6851 (2020)
7. Kang, L., et al.: GRIF-DM: generation of rich impression fonts using diffusion models. In: European Conference on Artificial Intelligence, pp. 226–233 (2024)
8. Kondo, T., Takezaki, S., Haraguchi, D., Uchida, S.: Font style interpolation with diffusion models. In: International Conference on Document Analysis and Recognition, pp. 86–103 (2024)
9. Kong, Y., et al.: Look closer to supervise better: one-shot font generation via component-based discriminator. In: Conference on Computer Vision and Pattern Recognition, pp. 13482–13491 (2022)
10. Li, C., Taniguchi, Y., Lu, M., Konomi, S.: Few-shot font style transfer between different languages. In: Winter Conference on Applications of Computer Vision, pp. 433–442 (2021)
11. Lowe, D.G.: Distinctive image features from scale-invariant keypoint. Int. J. Comput. Vision **60**, 91–110 (2004)
12. Oord, A.v.d., Li, Y., Vinyals, O.: Representation Learning with Contrastive Predictive Coding. arXiv preprint arXiv:1807.03748 (2018)
13. Park, S., Chun, S., Cha, J., Lee, B., Shim, H.: Few-shot font generation with localized style representations and factorization. In: AAAI Conference on Artificial Intelligence, pp. 2393–2402 (2021)
14. Park, S., Chun, S., Cha, J., Lee, B., Shim, H.: Multiple heads are better than one: few-shot font generation with multiple localized experts. In: International Conference on Computer Vision, pp. 13900–13909 (2021)
15. Roy, P., Bhattacharya, S., Ghosh, S., Pal, U.: STEFANN: scene text editor using font adaptive neural network. In: Computer Vision and Pattern Recognition, pp. 13228–13237 (2020)
16. Tang, L., et al.: Few-shot font generation by learning fine-grained local styles. In: Proceedings of the IEEE/CVF Conference on Computer Vision and Pattern Recognition (CVPR), pp. 7895–7904 (2022)
17. Uchida, S., Liwicki, M.: Part-based recognition of handwritten characters. In: International Conference on Frontiers in Handwriting Recognition, pp. 545–550 (2010)

18. Ueda, M., Kimura, A., Uchida, S.: Which parts determine the impression of the font? In: International Conference on Document Analysis and Recognition, pp. 723–738 (2021)
19. Ueda, M., Kimura, A., Uchida, S.: Font shape-to-impression translation. In: International Workshop on Document Analysis Systems, pp. 3–17 (2022)
20. Vaswani, A., et al.: Attention is all you need. In: Advances in Neural Information Processing Systems (2017)
21. Wang, S., Uchida, S., Liwicki, M.: Part-based recognition of arbitrary fonts. In: International Conference on Document Analysis and Recognition, pp. 170–174 (2013)
22. Yang, Z., Peng, D., Kong, Y., Zhang, Y., Yao, C., Jin, L.: FontDiffuser: one-shot font generation via denoising diffusion with multi-scale content aggregation and style contrastive learning. In: AAAI Conference on Artificial Intelligence, pp. 6603–6611 (2024)
23. Zaheer, M., Kottur, S., Ravanbakhsh, S., Poczos, B., Salakhutdinov, R.R., Smola, A.J.: Deep sets. In: Advances in Neural Information Processing Systems (2017)
24. Zhang, Y., Zhang, Y., Cai, W.: Separating style and content for generalized style transfer. In: Computer Vision and Pattern Recognition, pp. 8447–8455 (2018)

Non-linear Audio-Visual Storytelling from Scanned Comics: A Character-Centric Approach

Ansh Kushwaha[1], Sandeep Khanna[1,2], Lenin Khangjrakpam[1], Chiranjoy Chattopadhyay[1](✉), and Gaurav Bhatnagar[2]

[1] FLAME University, Pune, India
{ansh.kushwaha,chiranjoy.chattopadhyay}@flame.edu.in
[2] Indian Institute of Technology Jodhpur, Jodhpur, India
{khanna.1,goravb}@iitj.ac.in

Abstract. We introduce a framework for transform static comic scans into non-linear, character-focused audiovisual narratives using automated analysis and synthesis. Our hierarchical panel and speech bubble segmentation algorithm incorporating spatial and semantic features. A character relationship graph and an emotion trajectory modeling system are proposed that facilitate dynamic narrative generation. A character-focused panel sequencing system allows nonlinear storytelling by adjusting the narrative flow to the user-chosen character. We present a system that synchronizes text animations in speech bubbles with audio for coherent audiovisual sequences. Experiments on benchmark datasets demonstrate the effectiveness of the framework in panel segmentation accuracy, character identification precision, and the quality of audiovisual narratives generated with synchronized text animations. Quantitative and qualitative analyses show significant improvements over existing approaches in handling complex layouts and maintaining narrative coherence.

Keywords: Comics · Non-Linear Storytelling · Audio-Visual · Panel · Speech Bubble · Segmentation

1 Introduction

Comics, as a multimodal medium, offer a unique field for studying automated narrative comprehension. Their combination of visual and textual elements holds the potential for new document image analysis techniques. However, the complexity of layouts, diverse artistic styles, and subtle character interactions pose challenges to current automated systems. This paper addresses the problem of transforming scanned comic book pages into dynamic, character-centric, and non-linear audiovisual narratives. Existing approaches to comic analysis often focus on isolated tasks such as panel segmentation, speech bubble detection, or character recognition. Figure 1 shows the process of constructing a non-linear

L. Jin et al. (Eds.): ICDAR 2025 Workshops, LNCS 16225, pp. 158–174, 2026.
https://doi.org/10.1007/978-3-032-09368-4_10

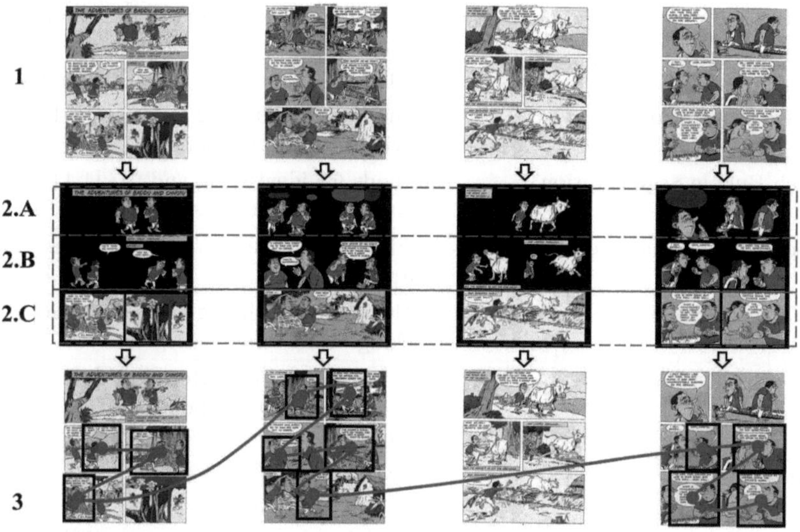

Fig. 1. Visualization of the proposed approach.

narrative centered on characters from a scanned comic page. Initially, a user-selected character is tracked across multiple pages, forming a sequence of appearances. Frames with the character's speech are selected to ensure meaningful interactions. These frames are then cropped and ordered, deviating from the original panel sequence (as in stage 3 of Fig. 1) to generate an audiovisual representation that prioritizes the character's dialogue-driven storyline.

Table 1 provides a comprehensive overview of the core algorithms that drive comic understanding and related applications. Panel and speech bubble segmentation methods have evolved from traditional image processing to deep learning approaches. Initially, active contour models and morphological operations were used for isolation [19,22]. More recent studies have integrated convolutional neural networks and graph-based methods to enhance the precision and robustness of these tasks [6,10,11]. Furthermore, research on character associations and character recognition of speech bubbles has further advanced the field by linking dialogue with characters and accurately tracking them across panels [18,23,25]. These developments not only reflect the steady progression of algorithmic techniques, but also establish the technical foundation upon which our work is built.

Knowledge-driven systems merge image cues with semantic data, improving narrative analysis [21]. Deep learning, including zero-shot learning, expands comic analysis with tasks such as speaker identification and narrative extraction [16,25]. The advances in animation and multimedia transform static comics into dynamic platforms [3,9]. These applications demonstrate state-of-the-art capabilities and highlight the potential for new interactive multimedia experiences. Based on the literature, this article makes the following key contributions:

Table 1. Summary of Categories and References in Comic Understanding

Category	Description	References
Panel Segmentation	Techniques for isolating individual panels in comics.	[10,11,14,19]
Speech Bubble Segmentation	Methods for identifying and isolating speech bubbles.	[4,11,22]
Speech Bubble-Character Association	Linking speech bubbles to characters.	[18,23,25]
Character Recognition	Methods for identifying and tracking characters.	[12,26]
Knowledge-Driven	Systems combining bottom-up and top-down information for comic analysis.	[21]
Deep Learning	Approaches for various comic understanding tasks.	[6,8,15,24]
Zero-Shot Learning	Approaches for predicting speaker names without annotated data.	[16]
Animation and Multimedia	Animating comics and generating multimedia content.	[3,9]

- A novel approach for transforming scanned comic book pages into character-centric nonlinear audiovisual narratives.
- Improved approaches for segmenting panels and speech bubble, as well as a novel speech bubble to speaker association algorithm.
- A dynamic audio-visual generation module that supports personalized and interactive storytelling, as well as text-based animation.
- A comprehensive evaluation of our system on diverse datasets.

The paper is organized as follows: Sect. 2 details our approach and algorithms. Section 3 presents experimental results, comparing them with state-of-the-art methods. Section 4 concludes and suggests future work.

2 Details of the Proposed Architecture

In this section, details of the proposed architecture as depicted in Fig. 2 are described in detail.

2.1 Panel Segmentation and Layout Analysis

The proposed approach is detailed in Algorithm 1. Given an input comic page $\mathcal{I}$, we first obtain its grayscale representation and apply a Gaussian filter with a kernel 9×9 to produce a smoothed image. Edge detection using the Canny

Algorithm 1. Panel Segmentation Algorithm

```
Input: 𝓘, the path to the comic page image.
Output: 𝓜, a binary mask representing panel regions.
Let I ∈ ℝ^{m×n} be the grayscale image obtained from 𝓘.
I' = 𝒢(I), where 𝒢 Gaussian filter, 9 × 9 kernel.
E = 𝒞(I'), where 𝒞 represents the Canny edge detection operator.
E' = 𝒟(E, S), 𝒟= dilation, S= structural element.
Extract external contours 𝒦 = {K_i} from E'.
for each contour K_i ∈ 𝒦 do
    Calculate the minimum area rectangle R_i = ℛ(K_i) enclosing K_i.
    Determine the corner points P_i = 𝒫(R_i) of R_i.
    if Area(K_i) > 0.01 · mn then          ▷ m= page height, n= page width
        Draw a filled white contour on 𝓜 for the bounding box R_i.
    else
        Draw a filled black contour on 𝓜 for the bounding box R_i.
𝓜 = ℰ(𝓜, S'), ℰ= erosion, S' = structuring element.
Return 𝓜
```

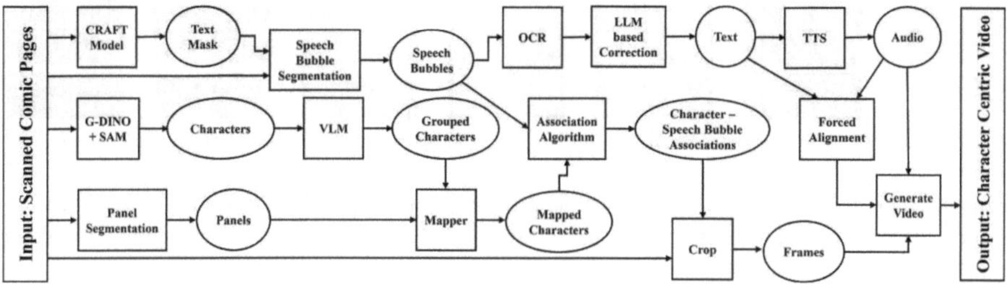

Fig. 2. A block-diagram of the proposed architecture.

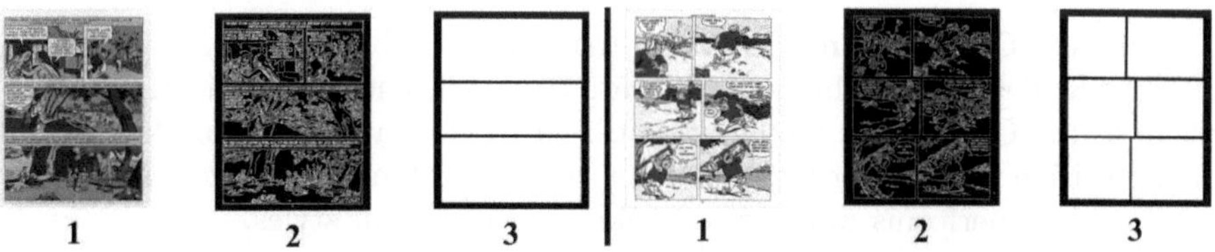

Fig. 3. Panel Segmentation: 1) Input Comic Page, 2) Edges, 3) Generated Mask.

[2] operator $\mathcal{C}$ yields an edge map E. To enhance structural continuity, we apply morphological dilation $\mathcal{D}$ with the structuring element S, resulting in E'. The contours $\mathcal{K}$ are extracted from E', and for each K_i, the minimum-area bounding rectangle R_i is calculated. The set of corner points P_i is determined, and the contours satisfying $\text{Area}(K_i) > 0.01 \cdot mn$ are retained as valid panels, forming a binary mask $\mathcal{M}$. Finally, a morphological erosion $\mathcal{E}$ with the structuring element S' refines the mask. Figure 3 shows the panel segmentation steps for a given image.

Algorithm 2. Main Speech Bubble Segmentation Algorithm

Input: $\mathcal{I}$ (path to comic page image), δ (boolean flag)
Output: $\mathcal{M}$ (binary mask representing speech bubble regions)
$T_d \leftarrow \text{TextMaskGeneration}(\mathcal{I})$
$G', \mathbb{K} \leftarrow \text{EdgeProcessing}(I, \delta)$
$\Omega \leftarrow \text{CandidateMaskGeneration}(G', T_d, \mathbb{K})$
$\mathcal{M} \leftarrow \text{FinalMaskRefinement}(\Omega, \delta)$
Return $\mathcal{M}$

2.2 Speech Bubble Segmentation

Speech bubbles in comics convey essential narrative information, but identifying them is challenging due to their varied shapes, sizes, and placements. The Algorithm 2 describes the main steps of the approach.

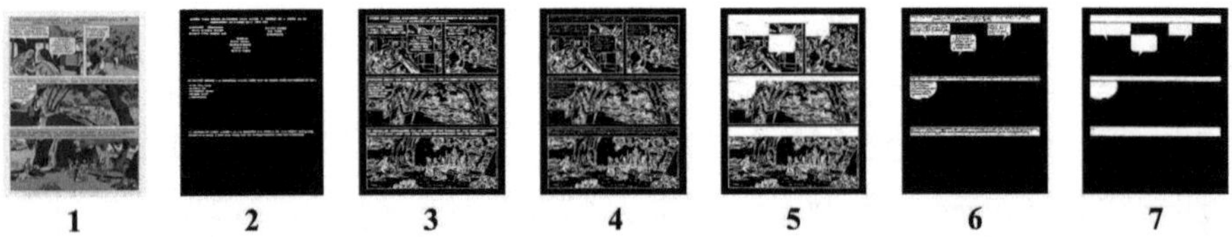

Fig. 4. Speech Bubble Segmentation: 1) Input Comic Page, 2) Text Mask, 3) Edges, 4) Edges - Text Mask Difference, 5) Flood-Fill, 6) Edge Removal, 7) Smoothen & Bubble Fill.

Text Mask Generation: Observing that speech bubbles always contain text, we use text detection techniques to locate them efficiently. As indicated in Algorithm 3, the CRAFT [1] model was utilized to generate a text mask, ensuring high precision in the identification of text regions within comic pages. This model is robust and performs well across a wide variety of font styles.

Algorithm 3. Speech Bubble Segmentation: Text Mask Generation

Input: $\mathcal{I}$ (path to comic page image)
Output: T_d (dilated text mask)
Let $I \in \mathbb{R}^{h \times w \times 3}$ be the RGB image read from $\mathcal{I}$
Define $T = \mathcal{F}(I)$ where $\mathcal{F}$ is the CRAFT text detection model
Compute $\tau = \mathcal{P}_{99}(T)$, the 99th percentile threshold of T
Define $T_b = \{(i,j) | T_{i,j} \geq \tau\}$, the binary text mask
Apply morphological dilation: $T_d = \mathcal{D}(T_b, B_1, 2)$ with structuring element B_1
Return T_d

Algorithm 4. Speech Bubble Segmentation: Edge Processing

```
Input: I_c (original image), δ (boolean flag for page-level processing)
Output: G′ (processed grayscale image), 𝕂 (set of contours)
Let I = φ(I_c) represent the grayscale version of I_c
Compute edge map E = 𝒞(I) using Canny edge detector 𝒞
Apply dilation: E_d = 𝒟(E, B_1)
Extract external contours 𝕂 = {K_1, K_2, ..., K_n} from E_d
if δ = True then
    Initialize P as zero matrix of size h × w
    ∀K_i ∈ 𝕂 where Area(K_i) > 0.01 · hw, set P = P ∪ K_i
    Apply erosion: P_e = ℰ(P, B_2)
    Define G′ = G ⊙ P_e, where ⊙ denotes element-wise multiplication
else
    G′ = G
Return G′, 𝕂
```

Edge Processing: Algorithm 4 processes an image to segment speech bubbles by converting it to grayscale, detecting edges, dilating edges, and extracting contours. If page-level processing is enabled, it further refines the image by excluding small contours and applying erosion. The processed grayscale image and the set of contours are returned as output.

Candidate Mask Generation: Algorithm 5 processes a grayscale image G', a dilated text mask T_d, and contours $\mathbb{K}$ to produce candidate masks Ω. The algorithm adjusts the size of T_d to be consistent with G', subsequently sets the relevant pixels in G' to zero, and delineates the contours in G'. For every text contour $C_j \in \mathbb{C}$ derived from T_d, the algorithm specifies a bounding box β_j and a center γ_j. Conditional on $G'_{\gamma_j} = 0$, a flood-fill operation ϕ is conducted at γ_j, subsequently integrating the result into Ω. Ultimately, the algorithm yields Ω.

Algorithm 5. Speech Bubble Segmentation: Candidate Mask Generation

```
Input: G′ (processed grayscale image), T_d (dilated text mask), 𝕂 (contours)
Output: Ω (set of candidate masks)
If necessary, resize T_d to dimensions of G′
Set G′_{i,j} = 0 ∀(i, j) ∈ T_d
∀K_i ∈ 𝕂, draw K_i on G′ with thickness 2
Extract text contours ℂ = {C_1, C_2, ..., C_m} from T_d
Initialize Ω = ∅, the set of candidate masks
for each C_j ∈ ℂ do
    Define bounding box β_j = (x_j, y_j, w_j, h_j) for C_j
    Let center γ_j = (x_j + w_j/2, y_j + h_j/2)
    if G′_{γ_j} = 0 then
        Ω = Ω ∪ {φ(G′, γ_j)}, where φ is the flood-fill operation
Return Ω
```

Algorithm 6. Speech Bubble Segmentation: Final Mask Refinement

Input: Ω (set of candidate masks), δ (boolean flag)
Output: $\mathcal{M}$ (final speech bubble mask)
Initialize $\mathcal{M} = \mathbf{0}^{h \times w}$ and $\Psi = \emptyset$
if $\Omega \neq \emptyset$ **then**
  for each $\omega_i \in \Omega$ **do**
    $\mathcal{M} = \max(\mathcal{M}, \omega_i)$
    $\Psi = \Psi \cup \{\mu(\omega_i)\}$
if $\delta =$ True **then**
  Define $\Gamma = \{(i,j) \| \|(i,j) - (\frac{h}{2}, \frac{w}{2})\| < r\}$
  for each $(i,j) \in \Gamma$ where $\mathcal{M}_{i,j} = 1$ **do**
    $\mathcal{M} = \mathcal{M} \setminus \phi(\mathcal{M}, (i,j))$
  Extract contours $\mathbb{F}$ from $\mathcal{M}$
  for each $F_l \in \mathbb{F}$ **do**
    Calculate $\alpha_l = \frac{4\pi \cdot \text{Area}(F_l)}{\text{Perimeter}(F_l)^2}$ and $\beta_l = \frac{\text{Area}(F_l)}{\text{Area}(\text{ConvexHull}(F_l))}$
    if $\alpha_l > 0.8 \lor \beta_l > 0.8$ **then**
      $\mathcal{M} = \mathcal{M} \cup F_l$
    else
      $\mathcal{M} = \mathcal{M} \setminus F_l$
$\Psi = \Psi \cup \{\mu(\mathcal{M})\}$
if $\delta = \text{True} \land \max(\Psi) \geq 0.6$ **then**
  $\mathcal{M} = \mathbf{0}^{h \times w}$
Return $\mathcal{M}$

Final Mask Refinement: The final mask refinement algorithm, denoted $\mathcal{R}$, takes a set of candidate masks Ω and a Boolean flag δ as input, producing a refined speech bubble mask $\mathcal{M}$. Initially, $\mathcal{M}$ is initialized as a zero matrix of size $h \times w$. If Ω is not empty, the algorithm iterates on each candidate mask $\omega_i \in \Omega$, updating $\mathcal{M}$ using the maximum operation with ω_i and collecting the mean values $\mu(\omega_i)$ in the set Ψ. If $\delta =$ True, an additional refinement involves analyzing pixels around the center of the image and applying geometric criteria to retain or remove regions. Finally, the algorithm checks if the maximum mean value in Ψ exceeds a threshold of 0.6; if so, it resets $\mathcal{M}$ to a zero matrix.

2.3 Text Extraction, Correction and Audio Generation

Converting segmented speech bubble content into audio narratives involves OCR, Text Correction, and TTS. This pipeline combines advanced OCR with LLM corrections for high-quality audio narratives in audio-visual media.

1. **Text Extraction:** The PaddleOCR [5] engine, denoted as $\mathcal{O}_{\text{Paddle}}$, is employed to extract textual data $\mathcal{T} = \{t_1, t_2, \ldots, t_n\}$ from the identified speech bubbles. The OCR pipeline comprises detection ($\mathcal{D}$), classification ($\mathcal{C}$), and recognition ($\mathcal{R}$) stages, mapping input images $\mathcal{I}$ to raw text outputs as $\mathcal{O}_{\text{Paddle}} : \mathcal{I} \rightarrow \mathcal{T}$.

2. **Post-processing:** Due to inherent OCR inaccuracies (e.g., spelling errors, case inconsistencies), the extracted text $\mathcal{T}$ is post-processed using a locally deployed Large Language Model (LLM), denoted as $\mathcal{L}$. The LLM performs the transformation $\mathcal{L} : \mathcal{T} \rightarrow \mathcal{T}'$, yielding grammatically corrected and contextually refined text $\mathcal{T}'$. This step also ensures an accurate handling of case-sensitive distinctions (e.g. "us" vs. "US") and improves semantic integrity.
3. **Text-to-Speech Conversion:** The refined $\mathcal{T}'$ is subsequently converted into audio signals $\mathcal{A}$ using the gTTS [7] engine, formalized as $\mathcal{TTS} : \mathcal{T}' \rightarrow \mathcal{A}$.

2.4 Character Detection and Segmentation

Character detection and segmentation in comic panels is crucial for effective audio-visual narratives. We use the grounded DINO model (G-DINO) [17], denoted $\mathcal{G}_{\text{DINO}}$, to overcome challenges from varied designs and styles. This model uses text prompts $\mathcal{P} = \{p_1, p_2, \ldots, p_n\}$ to identify character instances $\mathcal{C} = \{c_1, c_2, \ldots, c_m\}$ in comic panels $\mathcal{I}$ via mapping $\mathcal{G}_{\text{DINO}} : (\mathcal{I}, \mathcal{P}) \rightarrow \mathcal{C}$. Segmentation follows with the Segment Anything Model (SAM) [27], represented as $\mathcal{S}_{\text{SAM}}$, isolating each character $c_i \in \mathcal{C}$ by creating a binary mask $\mathcal{M}_i$, where $\mathcal{S}_{\text{SAM}} : c_i \rightarrow \mathcal{M}_i$. This process ensures precise character extraction while retaining intricate details for animation and processing.

2.5 Character Grouping and Mapping

We use a Vision Language Model (VLM), denoted $\mathcal{V}_{\text{LM}}$, to group and map unique comic characters. The characters detected $\mathcal{C} = \{c_1, c_2, \ldots, c_n\}$ are clustered based on appearance, costume, facial characteristics, body shape, color, and markers. Formally, $\mathcal{V}_{\text{LM}}$ maps the embeddings $\mathbf{e}_i \in \mathbb{R}^d$ of visual features $\mathbf{v}_i$ and textual prompts $\mathbf{t}_i$ into clusters $\mathcal{G} = \{\mathcal{G}_1, \mathcal{G}_2, \ldots, \mathcal{G}_k\}$:

$$\mathcal{V}_{\text{LM}} : (\mathbf{v}_i, \mathbf{t}_i) \rightarrow \mathbf{e}_i, \quad \text{and} \quad \text{Cluster}(\{\mathbf{e}_i\}) = \mathcal{G}, \tag{1}$$

, where each cluster $\mathcal{G}_j$ corresponds to a unique character. Spatial occurrences are then mapped across comic panels $\mathcal{P} = \{\mathcal{P}_1, \mathcal{P}_2, \ldots, \mathcal{P}_m\}$ using spatial coordinates (x_i, y_i). The mapping function $\mathbb{M}$ ensures continuity by linking clusters $\mathcal{G}_j$ to spatial occurrences across pages.

$$\mathbb{M} : (\mathcal{G}, \mathcal{P}) \rightarrow \Phi, \quad \text{where} \quad \Phi = \{\phi_1, \phi_2, \ldots, \phi_k\}, \tag{2}$$

with $\phi_j = \{\mathbf{s}_{j1}, \mathbf{s}_{j2}, \ldots\}$ tracking the spatial trajectory $\mathcal{G}_j$ of the clusters across panels. This pipeline maintains character continuity across panels and pages, thereby yields the desired results.

2.6 Character Speech Bubble Association

The character-speech bubble association algorithm (Algorithm 7) processes comic images $\{I_i\}$ with detected character boxes $\{C_i\}$ to generate mappings $\mathbb{M}$

Algorithm 7. Character-Speech Bubble Association

```
Input: Image paths {𝓘_i}, Character boxes {C_i}
Output: Mappings 𝕄, Comic frames 𝓕
for each image I_i do
    Detect panels: {R_p} ← 𝒫(I_i), sorted by (y, x)
    Detect speech bubbles: {B} ← 𝒮(I_i)
    Extract skeleton graph G and key points (junctions J, terminals T)
    for each panel R_p do
        Identify characters C_p and bubbles B_p inside R_p
        𝓜[i] ← {R_p, C_p, B_p}
    for each bubble b ∈ B do
        Find closest terminal t* ∈ T
        if Distance < θ_prox then
            Associate b with the nearest character in 𝕄
return 𝕄, 𝓕
```

and rendered frames $\mathcal{F}$. Firstly, panel regions $\{R_p\}$ are extracted by mask processing ($\mathcal{P}(I_i)$), while speech bubbles $\{H_b\}$ are detected by convex hull extraction ($\mathcal{S}(I_i)$). For each panel R_p, spatial analysis identifies the characters contained in $panel\_chars = \{c \in C_i | \mathcal{I}(c, R_p)\}$ and the bubbles $panel\_bubbles = \{b | \mathcal{I}(b, R_p)\}$. The proximity-based association constructs a skeleton graph G from bubble contours, then links each bubble b to the nearest character c when $\min_{t \in T} \|t_{pos} - o_b\|_2 < \theta_{prox}$, where T are terminal points in G and o_b is the anchor point of the bubble. Directional path tracing ($\mathcal{V}(G, t, J)$) resolves inter-panel overlaps, optimizing the global objective $\min \sum_{(b,c) \in \mathcal{M}} \|t_{pos} - c_{mouth}\|_2$ while preserving reading order constraints. Figure 5 shows the intermediate stages of processing and the final association for an example comic book page.

2.7 Forced Alignment for Text Based Animation

Forced alignment synchronizes spoken audio with text by identifying the exact start and end times of each word or phoneme in the audio. A systematic approach, as defined in Algorithm 8, is utilized to implement forced alignment for the synchronization of text and audio in the digitization of comics. Initially, the input dialogue text is converted to speech using a TTS model, while simultaneously masking the original text bubbles in the comic page image to create a

Algorithm 8. Forced Alignment for Comic Audio-Text Synchronization

```
Input: Comic page image I, dialogue text T, TTS model M
A ← TextToSpeech(T, M)                          ▷ Generate audio from text
I_masked ← MaskTextBubbles(I)                   ▷ Remove original text
W ← TokenizeToWords(T)                          ▷ Split text into words
{(w_i, t_start, t_end)} ← ForcedAlignment(W, A) ▷ Get word timings
Output: Masked image I_masked, audio A, word-timing pairs {(w_i, t_start, t_end)}
```

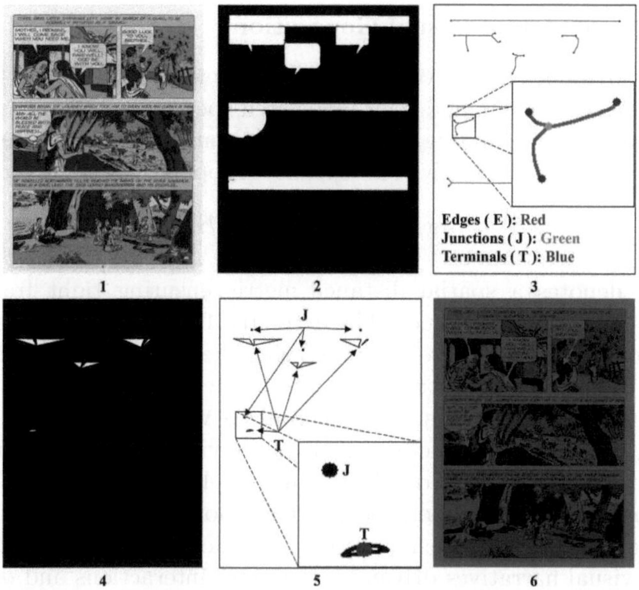

Fig. 5. An illustration of the character - Speech Bubble Association: 1) Input Page, 2) Speech Bubble Mask, 3) Skeletonized Speech Bubble as Graph, 4) Bubble Hull-Contour Difference, 5) Terminals and Junctions Selection, 6) Speech Bubble Pointing Directions.

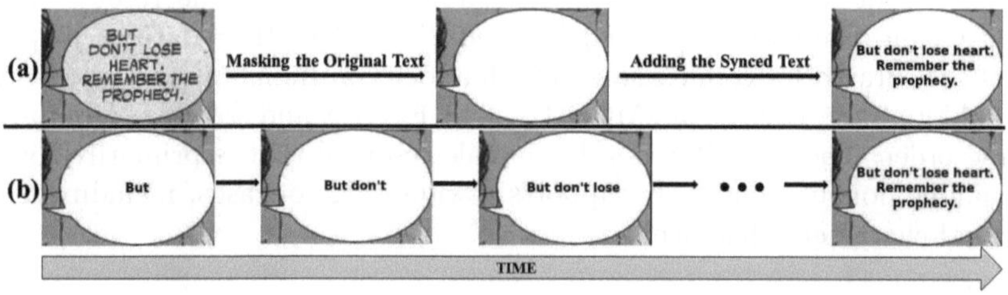

Fig. 6. An illustration of the forced alignment, animating text. (a) Approach, (b) Word by Word Text Rendering.

clean background. The dialogue text is then tokenized into individual words and forced alignment is applied to generate precise timing information (start and end timestamps) for each word in the generated audio. This process generates a triplet of the word, start time, and end time for each dialogue element, enabling synchronized text display with narration. The final output includes the masked comic image, generated audio, and word-timing pairs for harmonious text and audio presentation. The same is reflected in Fig. 6, where the approach and the illustration are shown in two different rows.

2.8 Non-linear Audio-Visual Generation

To construct dynamic narratives, we extract frames $\mathcal{F} = \{F_i\}_{i=1}^{N}$ from panels containing both characters and their associated speech bubbles. Given an image $\mathcal{I} \in \mathbb{R}^{m \times n}$, we define bounding boxes $\mathcal{B}_c$ and $\mathcal{B}_s$ around the detected character and speech bubble, respectively. The optimal cropping region $\mathcal{B}^*$ is obtained as:

$$\mathcal{B}^* = \arg\min_{\mathcal{B}} \left(d(\mathcal{B}_c, \mathcal{B}) + d(\mathcal{B}_s, \mathcal{B}) \right), \quad (3)$$

where $d(\cdot, \cdot)$ denotes a spatial distance metric ensuring tight framing while maintaining contextual coherence. The extracted frames are temporally ordered according to a function $\pi : \mathcal{F} \rightarrow \mathbb{N}$, preserving the narrative sequence based on the position of the panel and the continuity of the character.

The generated frames are then synchronized with the corresponding audio $\mathcal{A} = \{A_i\}_{i=1}^{N}$ to produce character-centric audio-visual. The duration τ_i of each frame F_i is calculated as $\tau_i = \alpha \cdot \ell(A_i)$, where $\ell(A_i)$ is the length of the speech audio segment, and α is a temporal scaling factor for smooth and natural rhythm. This pipeline enables the transformation of static comic book pages into non-linear audio-visual narratives driven by character interactions and dialogue.

3 Experimental Results and Analysis

3.1 Dataset

We experimented with two publicly available datasets: 1) the Indian Mythological Comic Database (IMCDB) [9], and 2) the COMICS dataset [13]. The IMCDB is curated for comic analysis with a focus on Indian mythology, history, and folklore. The comics in IMCDB are in English and follow a left-to-right reading order. The COMICS is a large-scale resource that is primarily sourced from public domain comics. It supports a wide range of tasks, including visual cloze and character coherence.

3.2 Panel Segmentation: Comparison with Baselines

Table 2. Comparison of proposed approach with others for panel segmentation.

Method	Dataset	IoU	Precision	Recall	Accuracy
Nguyen [20]	IMCDB	0.937	0.948	0.988	0.932
Gupta [9]	IMCDB	0.961	0.975	0.976	0.979
Proposed	IMCDB	0.967	0.981	0.985	0.972
	COMIC	0.954	0.983	0.970	0.960

Table 2 demonstrates the quantitative advantage of the proposed panel segmentation method over existing approaches. On the IMCDB dataset, our method achieves the highest Intersection over Union (IoU) of 0.967, outperforming previous works. In addition, it achieves superior precision (0.981) and recall (0.985), ensuring both higher accuracy in panel detection and reduced false positives. Furthermore, the proposed method exhibits strong generalizability by maintaining high performance in the COMIC dataset, achieving an IoU of 0.954 and precision of 0.983, despite variations in artistic style and panel structure. These results highlight the robustness of our approach in diverse comic datasets.

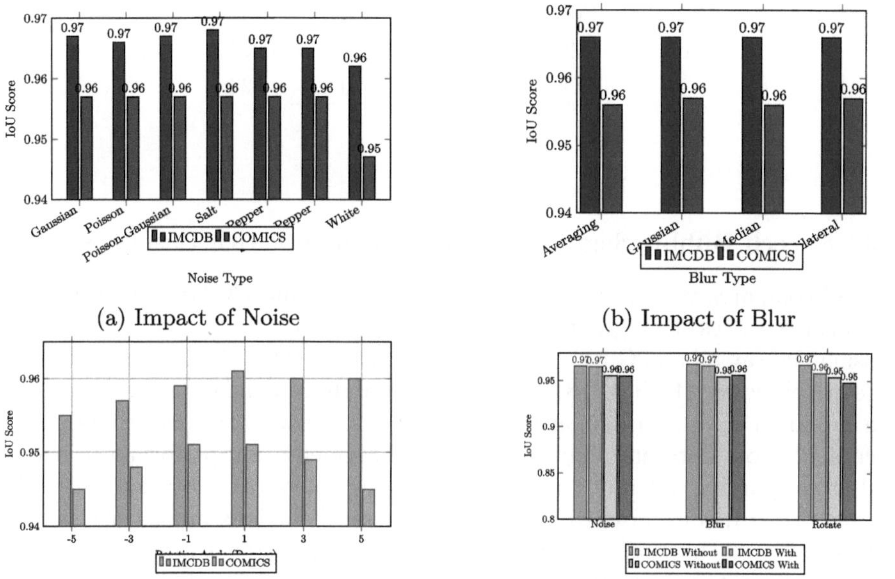

(a) Impact of Noise

(b) Impact of Blur

(c) Impact of Rotation

(d) IOU Drop Analysis Across Metrics

Fig. 7. Panel segmentation performance under different conditions.

Figure 7 presents the performance of the IoU under noise, blur, and rotation. The following parameters were used for each noise type: Gaussian ($\mu = 10, \sigma = 20$), Poisson (scale = 1), Poisson-Gaussian (scale = 2, $\mu = 20, \sigma = 15$), Salt (prob = 0.05), Pepper (prob = 0.01), Salt-Pepper (salt prob = 0.05, pepper prob = 0.01) and White ($\mu = 10, \sigma = 15$). For each blur type, the following parameters were used: Averaging ($k = 3$), Gaussian ($k = 3$), median ($k = 3$), and bilateral ($d = 5, \sigma = 75$), where k represents the size of the kernel and d denotes the diameter of each neighborhood of pixels used during filtering. The subfigures (a), (b), and (c) illustrate the IoU scores between data sets for different transformations. Figure 7(d) presents a grouped bar graph illustrating the IoU scores of the model on the IMCDB and COMICS datasets under various perturbation categories. The results demonstrate that noise-based perturbations have a negligible effect on IoU scores. The blur results in only a slight decrease in performance across both datasets, indicating that the model remains resilient even after the loss of sharp edges. However, a significant decline is observed under rotation-based perturbations, implying that spatial transformations notably impair the model's ability to recognize structures effectively.

Table 3. Speech Bubble segmentation comparison.

Method	Dataset	IoU	Precision	Recall	Accuracy
Dubray [6]	IMCDB	0.877	0.949	0.920	0.984
Gupta [9]	IMCDB	0.859	0.815	0.910	0.967
Proposed	IMCDB	0.898	0.955	0.845	0.977
	COMIC	0.718	0.867	0.821	0.924

3.3 Speech Bubble Segmentation: Comparison with Baselines

Table 3 compares various approaches in the IMCDB and COMIC datasets. Our approach achieves the highest IoU (0.898) and Precision (0.955) in IMCDB, surpassing Dubray [6] and Gupta [9]. However, in the COMIC dataset, performance is relatively lower (IoU: 0.718), indicating variability in the dataset. In general, the proposed approach demonstrates superior segmentation quality with high accuracy across datasets. Figure 8 shows the segmentation performance before and after applying different transformations: blur, noise, and rotation. The subfigures (a), (b), (c) and (d) illustrate how these distortions impact the performance. We used the same blur and noise types with the same parameters as in the panel segmentation task. However, each metric shows a noticeable performance decrease in this case. This can be explained by the fact that panel segmentation only requires detecting outer boundaries, while speech bubble segmentation depends on finer internal edges. Blur (Fig. 8b) introduces minor degradation, causing soft edges but preserving the overall structure. Noise (Fig. 8a) has a more disruptive effect, introducing irregularities that affect the sharpness of the boundary. Rotation (Fig. 8c) results in the most significant performance drop, as it misaligns the boundaries of the object, leading to segmentation errors.

3.4 Text Recognition: Comparison of Baselines

We evaluated three well-known OCR models: EasyOCR, PaddleOCR, and Pytesseract. For this comparison, we used common performance metrics, including Word Error Rate (WER), Character Error Rate (CER), and Match Error Rate (MER). The WER measures the percentage of words recognized incorrectly, CER checks errors at the character level, and MER indicates the portion of text regions not correctly recognized in the document. Lower scores in all these metrics mean the OCR model is more accurate. We also checked the average processing time per image in seconds to see how efficient each model is.

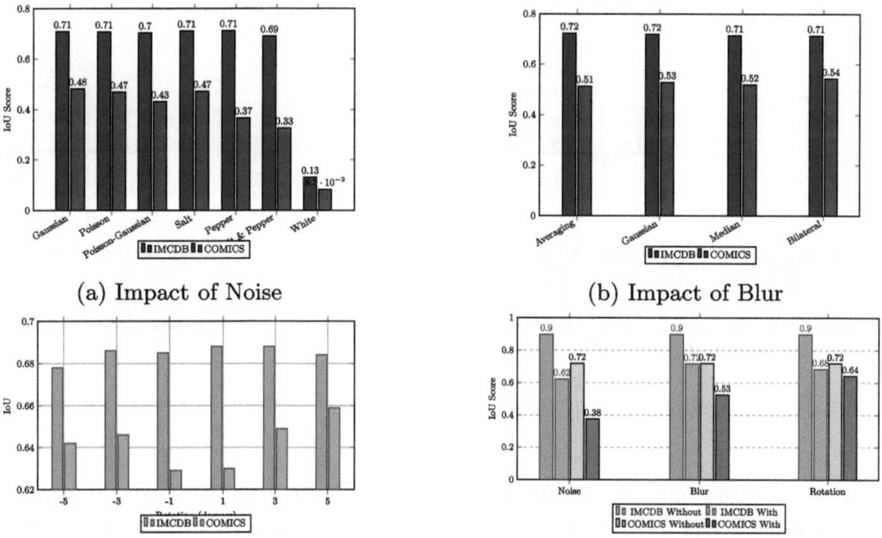

(a) Impact of Noise

(b) Impact of Blur

(c) Impact of Rotation

(d) IOU Analysis Across Metrics

Fig. 8. Speech segmentation performance under different conditions.

Table 4. OCR Model comparison

Model	Average Metrics (Per Image)			
	WER	CER	MER	Time (s)
EasyOCR	0.325	0.078	0.315	0.179
PaddleOCR	0.278	0.043	0.253	0.035
Pytesseract	0.406	0.282	0.371	0.149

According to Table 4, PaddleOCR outperforms the other models in all metrics. It achieves the lowest WER (0.278), lowest CER (0.043), and lowest MER (0.253), and it also has the fastest processing time at 0.035 s per image. EasyOCR performs moderately on all metrics, so it can be used as an alternative. Pytesseract, however, gives the highest error rates, with a WER of 0.406, CER of 0.282, and MER of 0.371, showing lower accuracy compared to others. Overall, these results show that PaddleOCR is the most reliable and efficient OCR system among the models we tested.

3.5 Non-linear Audio-Visual Generation

We employ a character-focused method for audio-visual synthesis, using non-sequential frame indexing based on character appearances in discontinuous segments (e.g., frames 1, 3, 20, 31). This technique targets characters and records their occurrences in dialogue scenes on comic pages. The resulting frames are

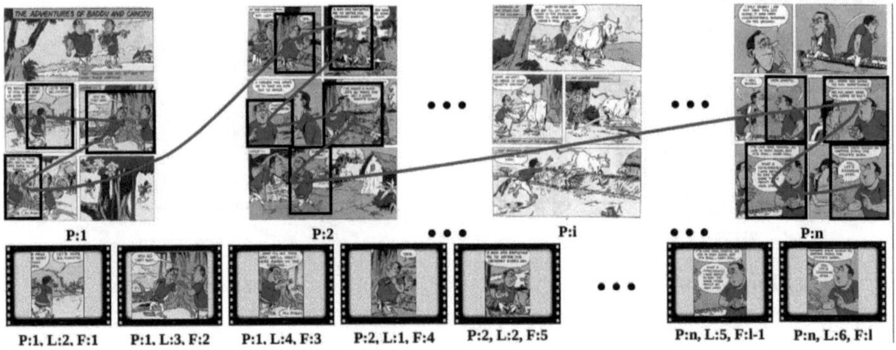

Fig. 9. Illustration of the non-linear audiovisual synthesis from scanned pages.

cropped to show only the character with the speech bubble. For example, in Fig. 9 Frame F:1 is from Page P:1, Panel L:2; Frame F:2 from Page P:1, Panel L:3; and Frame F:5 from Page P:2, Panel L:2. Panel's where the selected character is not saying anything (no speech bubble association), are not included in the final rendered audiovisual, e.g., the character is present in the second panel of page n, however, not rendered.

4 Conclusion and Future Work

We developed a fast and practical system for converting comics into interactive character-centered audiovisual scenes. Our approach performs well on two public datasets and gives better results than current methods for detecting panels and speech bubbles, even when images are blurry, noisy, or rotated. With a nonlinear storytelling style, users can experience more flexible and engaging comic stories. We also include a feature that shows the text together with audio, making the result more dynamic and enjoyable.

However, there are some limitations that we plan to improve in future work. Our method stops at the edge of panels and speech bubbles, but ground-truth labels often include these borders, which can lower recall when we compare results. Also, our system includes features like tails in speech bubbles, but the ground truth sometimes does not, which causes more false negatives. Our system works best in western comics, where panels and bubbles are well-structured, so it may not perform as well in other comic styles, such as manga. In the future, we will add emotion detection for speech bubbles, improve character focus using transformer models, add real-time style transfer for personalization, better handle complicated layouts for example objects spanning multiple panels, and improve how audio and text match. Introducing immersive features into static comics and use of modern methods to allow users to drive the story in their own way is also a potential research direction.

Acknowledgments. This work was partially supported by the iHub Drishti Foundation, IIT Jodhpur, under the project (TIH/iHubDrishti/Project/2023-24/47)) titled "Design and development of a software platform to create immersive content from digitally digitized and born digital documents".

References

1. Baek, Y., Lee, B., Han, D., Yun, S., Lee, H.: Character region awareness for text detection. In: Proceedings of the IEEE/CVF Conference on Computer Vision and Pattern Recognition, pp. 9365–9374 (2019)
2. Canny, J.: A computational approach to edge detection. IEEE Trans. Pattern Anal. Mach. Intell. **PAMI-8**(6), 679–698 (1986). https://doi.org/10.1109/TPAMI.1986.4767851
3. Cao, Y., Xufang, P., Chan, A., Lau, R.: DynamicManga: animating still manga via camera movement. IEEE Trans. Multimedia 1–1 (2016). https://doi.org/10.1109/TMM.2016.2609415
4. C. Correia, J.M., P. Gomes, A.J.: Balloon extraction from complex comic books using edge detection and histogram scoring. Multimedia Tools Appl. **75**(18), 11367–11390 (2015). https://doi.org/10.1007/s11042-015-2858-0
5. Du, Y., et al.: PP-OCR: A practical ultra lightweight OCR system. arXiv preprint arXiv:2009.09941 (2020)
6. Dubray, D., Laubrock, J.: Deep CNN-based speech balloon detection and segmentation for comic books. In: 2019 International Conference on Document Analysis and Recognition (ICDAR), pp. 1237–1243 (2019). https://api.semanticscholar.org/CorpusID:67787670
7. Durette, P.N., Contributors: gTTS: Google text-to-speech. https://gtts.readthedocs.io/. Accessed 2025
8. Dutta, A., Biswas, S., Das, A.K.: CNN-based segmentation of speech balloons and narrative text boxes from comic book page images. Int. J. Doc. Anal. Recogn. (IJDAR) (18), 49–62 (2021). https://doi.org/10.1007/s10032-021-00366-4
9. Gupta, V., Detani, V., Khokar, V., Chattopadhyay, C.: C2VNet: a deep learning framework towards comic strip to audio-visual scene synthesis. In: Lladós, J., Lopresti, D., Uchida, S. (eds.) ICDAR 2021. LNCS, vol. 12822, pp. 160–175. Springer, Cham (2021). https://doi.org/10.1007/978-3-030-86331-9_11
10. Halford, M.: Comic book panel segmentation (2022). https://maxhalford.github.io/blog/comic-book-panel-segmentation/
11. Ho, A.K.N., Burie, J.C., Ogier, J.M.: Panel and speech balloon extraction from comic books. In: 2012 10th IAPR International Workshop on Document Analysis Systems, pp. 424–428 (2012). https://doi.org/10.1109/DAS.2012.66
12. Iwata, M., Ito, A., Kise, K.: A study to achieve manga character retrieval method for manga images. In: 2014 11th IAPR International Workshop on Document Analysis Systems, pp. 309–313 (2014). https://doi.org/10.1109/DAS.2014.60
13. Iyyer, M., et al.: The amazing mysteries of the gutter: drawing inferences between panels in comic book narratives. In: IEEE Conference on Computer Vision and Pattern Recognition (2017)
14. Iyyer, M., et al.: The amazing mysteries of the gutter: Drawing inferences between panels in comic book narratives. CoRR abs/1611.05118 (2016). http://arxiv.org/abs/1611.05118

15. Li, Y., Aizawa, K., Matsui, Y.: Manga109Dialog: A large-scale dialogue dataset for comics speaker detection (2024). https://arxiv.org/abs/2306.17469
16. Li, Y., Hinami, R., Aizawa, K., Matsui, Y.: Zero-shot character identification and speaker prediction in comics via iterative multimodal fusion. In: Proceedings of the 32nd ACM International Conference on Multimedia, pp. 7366–7374. MM '24, Association for Computing Machinery, New York, NY, USA (2024). https://doi.org/10.1145/3664647.3681659
17. Liu, S., et al.: Grounding DINO: marrying DINO with grounded pre-training for open-set object detection. In: European Conference on Computer Vision, pp. 38–55. Springer (2024)
18. Nguyen, N.V., Rigaud, C., Burie, J.C.: Comic MTL: optimized multi-task learning for comic book image analysis. Int. J. Doc. Anal. Recognit. **22**(3), 265–284 (2019). https://doi.org/10.1007/s10032-019-00330-3
19. Ponsard, C.: Enhancing the accessibility for all of digital comic books. e-Minds **1** (2009)
20. Rigaud, C., Burie, J.C., et al.: What do we expect from comic panel extraction? In: 2019 International Conference on Document Analysis and Recognition Workshops (ICDARW). vol. 1, pp. 44–49. IEEE (2019)
21. Rigaud, C., Guérin, C., Karatzas, D., Burie, J.-C., Ogier, J.-M.: Knowledge-driven understanding of images in comic books. Int. J. Doc. Anal. Recogn. (IJDAR) **18**(3), 199–221 (2015). https://doi.org/10.1007/s10032-015-0243-1
22. Rigaud, C., Karatzas, D., Van de Weijer, J., Burie, J.C., Ogier, J.M.: An active contour model for speech balloon detection in comics. In: Proceedings of the 12th International Conference on Document Analysis and Recognition (ICDAR), pp. 1240–1244 (2013)
23. Rigaud, C., et al.: Speech balloon and speaker association for comics and manga understanding. In: 2015 13th International Conference on Document Analysis and Recognition (ICDAR), pp. 351–355 (2015). https://doi.org/10.1109/ICDAR.2015.7333782
24. Sachdeva, R., Shin, G., Zisserman, A.: Tails tell tales: chapter-wide manga transcriptions with character names. In: Proceedings of the Asian Conference on Computer Vision, pp. 2053–2069 (2024)
25. Soykan, G., Yuret, D., Sezgin, T.M.: Spatially augmented speech bubble to character association via comic multi-task learning. In: International Conference on Document Analysis and Recognition, pp. 231–256. Springer (2024)
26. Soykan, G., Yuret, D., Sezgin, T.M.: Identity-aware semi-supervised learning for comic character re-identification (2023). https://arxiv.org/abs/2308.09096
27. Zhao, X., et al.: Fast segment anything (2023)

Automatic Text Box Placement for Supporting Typographic Design

Jun Muraoka[1(✉)], Daichi Haraguchi[2], Naoto Inoue[2], Wataru Shimoda[2], Kota Yamaguchi[2], and Seiichi Uchida[1]

[1] Kyushu University, Fukuoka, Japan
jun.muraoka@human.ait.kyushu-u.ac.jp, uchida@ait.kyushu-u.ac.jp
[2] CyberAgent, Tokyo, Japan
{haraguchi_daichi_xa,inoue_naoto,wataru_shimoda, yamaguchi_kota}@cyberagent.co.jp

Abstract. In layout design for advertisements and web pages, balancing visual appeal and communication efficiency is crucial. This study examines automated text box placement in incomplete layouts, comparing a standard Transformer-based method, a small Vision and Language Model (Phi3.5-vision), a large pretrained VLM (Gemini), and an extended Transformer that processes multiple images. Evaluations on the Crello dataset show the standard Transformer-based models generally outperform VLM-based approaches, particularly when incorporating richer appearance information. However, all methods face challenges with very small text or densely populated layouts. These findings highlight the benefits of task-specific architectures and suggest avenues for further improvement in automated layout design.

Keywords: Document layout analysis · Structured document generation

1 Introduction

Layout design plays a crucial role in various forms of media, such as advertisements, posters, and web pages, by organizing information in a visually clear and comprehensible manner. In our digital society, where information overload is a constant challenge, effective layout design enhances both visual appeal and information delivery. However, achieving an optimal balance in the placement, size, and color harmony of design elements is a complex and time-intensive process, requiring both aesthetic sensibility and iterative refinements. Ineffective arrangements can hinder communication, highlighting the need for technologies that streamline the layout design process.

Modern design formats often employ a layered structure, allowing elements to be independently adjusted. This flexibility enables seamless modifications and refinements, making it particularly valuable for multimodal layouts that integrate text and images. While images provide visual impact, text conveys

L. Jin et al. (Eds.): ICDAR 2025 Workshops, LNCS 16225, pp. 175–191, 2026.
https://doi.org/10.1007/978-3-032-09368-4_11

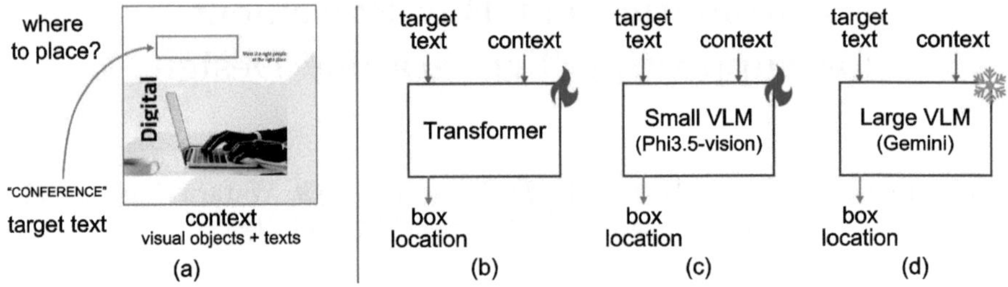

Fig. 1. (a) Overview of the text box placement task. (b)–(d) Three machine learning-based methods for solving the task.

detailed information, and their harmonious integration is crucial for clear communication. Automated layout generation has gained attention in research due to its potential to enhance efficiency and adaptability in design workflows.

The purpose of this study is to understand how recent methodologies can support typographic design through a simple but *very essential layout task*: that is, the optimal placement for incorporating a new text box into partially completed multimodal layouts. Figure 1(a) illustrates this estimation task. By leveraging the layered structure of modern layouts, elements can be dynamically adjusted to maintain visual coherence and structural integrity. Addressing this task is expected to improve both the efficiency and quality of the layout design process.

As possible methodologies, we implement and evaluate three methods for estimating optimal text box placements, as shown in Figs. 1(b), (c), and (d):

- A standard Transformer-based method (b), which directly trains the text placement task as a regression problem, considers relationships among a set of various elements, such as the whole image, text elements, and image elements.
- A small Vision and Language Model (VLM)-based method (c), which utilizes Phi3.5-vision [1], a lightweight VLM. This method relies on a single global image and estimates placements with a reduced set of input features compared to the standard Transformer-based method. Additionally, we examine whether placement estimation can be improved by controlling prompt inputs.
- A large VLM-based method (d), which leverages the large-scale pretrained model, Gemini [15], without additional training. For this method, we investigate its ability to perform estimation under the constraint of no additional training.

By comparing these three methods as fairly as possible, this study helps to understand which methodology is more promising for the optimal text box placement task. More importantly, this study helps to identify scenarios that are inherently easy or difficult for all three methods. A typical case that is easy for all methods suggests the presence of universal constraints that limit placement flexibility, leading to a relatively deterministic solution. Conversely, cases that are difficult for all methods indicate situations where placement freedom is extremely high, making the optimal solution ambiguous or even non-unique.

Understanding these inherent constraints and degrees of freedom in layout composition contributes to a deeper theoretical insight into automated text box placement.

This study can also be viewed as developing new application directions, such as appending additional text to AI-generated typographic layouts, supporting interactive text-manipulation processes, verifying the placement of individual text boxes, etc. Note that this study elucidates the effectiveness of a task-tailored training method, a fine-tuned VLM, and a fully-pretrained VLM in capturing spatial relationships and integrating new elements into existing designs.

2 Related Work

2.1 Layout Generation Using Basic Machine Learning Methods

Layout generation has been extensively studied in recent years. Content-agnostic approaches, which do not consider specific content, have been explored using Generative Adversarial Networks (GANs) [12], Variational Autoencoders (VAEs) [9], and diffusion models [6]. However, these methods face limitations in practical scenarios as they do not incorporate specific content like text or images into their design considerations.

More recently, content-aware layout generation, which accounts for background images or text, has gained attention. For instance, FlexDM [7] is a Transformer-based model that predicts masked elements and their attributes. Similarly, RADM [11] utilizes diffusion models to generate typed bounding boxes over background images, and RALF [4] enhances a Transformer framework with retrieval-augmented layout information.

2.2 Layout Generation Using Fine-Tuned VLMs for the Task

The use of fine-tuned VLMs for layout generation has garnered attention due to their scalability. COLE [8] introduced a pipeline to generate layouts from natural language instructions using a VLM to predict attributes like text placement. PosterLLaVA [17] also utilizes a multimodal LLM to position elements based on user prompts, while PosterLlama [14] employs a novel depth-based augmentation strategy to enhance robustness in data-scarce scenarios.

In this study, we leverage Phi3.5-vision [1], a lightweight VLM. Its smaller size and computational efficiency make it suitable for fine-tuning on our single text box placement task, while retaining sufficient expressive power to learn spatial relationships.

2.3 Layout Generation Using Large-Scale VLMs Without Training

Layout generation methods utilizing large-scale VLMs without additional training have also been proposed. While large-scale models, such as GPTs [2], pose challenges in terms of high computational costs and limited accessibility for fine-tuning, they have already been trained on vast and diverse knowledge sources.

As a result, they may have implicitly acquired substantial knowledge related to layout design. Particularly, large-scale VLMs like Gemini [15] can process both text and images, making them highly relevant for layout tasks that require an understanding of two-dimensional spatial structures. This potential for high performance without task-specific tuning makes them an important research direction.

LayoutPrompter [13] demonstrated the feasibility of performing various tasks solely through prompt engineering with pretrained LLMs. It achieved high levels of accuracy by incorporating mechanisms such as dynamic few-shot selection through retrieval and a Ranker module to evaluate multiple candidate outputs. Similarly, LayoutGPT [3] introduced a framework that generates not only 2D layouts but also 3D configurations flexibly, based on user instructions.

Let us conclude this review section with the reason why we need to conduct our research on the optimal single text box placement, although we can find several existing methods that estimate bounding boxes for all elements or perform multitask inferences for various attributes. The existing methods tend to evaluate overall accuracies, leaving the fundamental constraints and degrees of freedom in text placement underexplored. In contrast, we aim to clarify these constraints and better understand the inherent complexity of the task, by focusing on a single text box. Additionally, although various layout generation techniques exist, no comprehensive study has compared SOTA VLMs under a unified experimental setting. This work fills that gap by evaluating a Transformer-based model, a fine-tuned VLM, and a large-scale pretrained VLM, providing insights into their relative strengths, limitations, and applicability to automated layout design.

3 Three Machine Learning-Based Methods for Text Box Placement

This section describes the three methods selected for estimating the optimal placement of text boxes in layouts. As discussed in Sect. 2, layout generation has been approached through three major directions: (1) basic machine learning models tailored for layout tasks, (2) fine-tuned VLMs, and (3) large-scale pretrained VLMs used without additional training. To comprehensively evaluate their applicability to the single text box placement task, we select a representative method from each category:

- A standard Transformer-based method, which directly models spatial relationships between multimodal elements through task-specific training (Sect. 3.1).
- A fine-tuned VLM-based method, which adapts Phi3.5-vision [1] to process layout information as structured language input (Sect. 3.2).
- A large-scale pretrained VLM-based method, which leverages Gemini [15] without additional training to assess its zero-shot capabilities (Sect. 3.3).

By comparing these methods, we aim to assess not only their relative performance but also the fundamental challenges and constraints inherent in the single text box placement task.

All these methods estimate the optimal placement of a new text box within an existing layout. As illustrated in Fig. 2(a), the layout consists of N elements, where one element corresponds to the target text to be placed, and the remaining $N-1$ elements represent existing layout components as context. Note that one of $N-1$ contextual elements is the whole layout image (given by rendering all elements except for the target text).

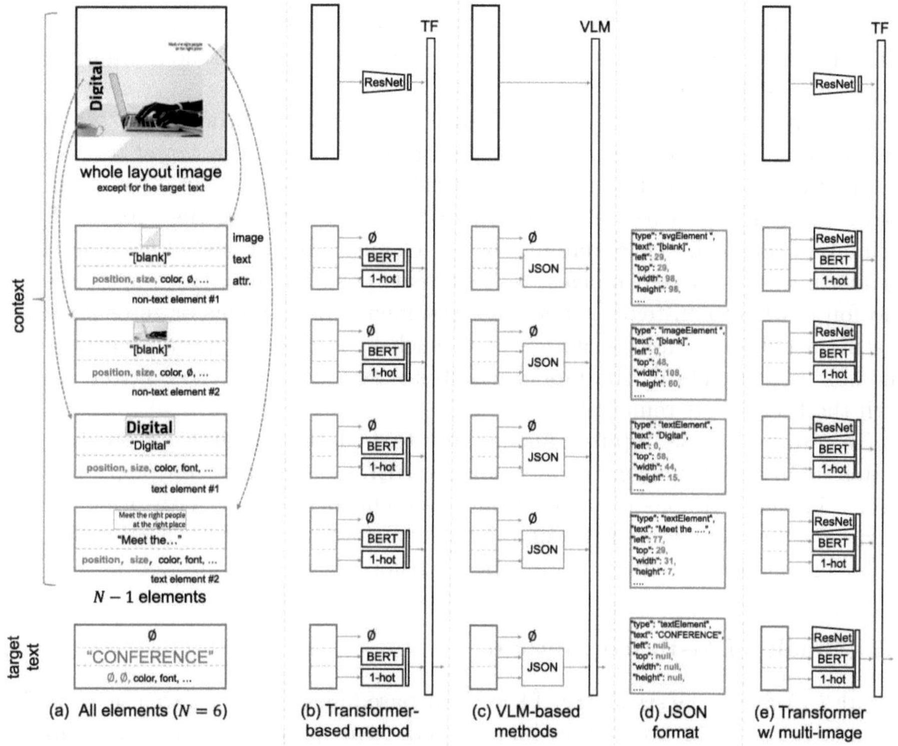

Fig. 2. (a) Details of N input elements. (b) and (c) Models for the optimal text box placement. (d) The details of the JSON format for VLM input. (e) Another version of (b) for utilizing a bitmap-based representation of individual input elements.

3.1 Standard Transformer-Based Method

The standard Transformer-based method estimates the placement of a target text box by processing the layout information of all N elements, as shown in Fig. 2(b). The model takes as input a set of feature representations for these elements and outputs the coordinates of the text box to be placed. The coordinates are given as a four-dimensional vector whose elements correspond to left, top, width, and height. Note that these values are normalized to the range $[0, 1]$, ensuring compatibility with various document sizes.

This method offers several advantages. First, the self-attention mechanism enables the model to learn spatial relationships between N elements, making it well-suited for layout tasks where interactions between text and surrounding objects are critical. Second, the Transformer architecture can handle an arbitrary number of input elements, allowing for greater flexibility in varying layout compositions. Third, by incorporating appropriate encoders, the model can seamlessly process diverse input types, including images, raw text, and categorical attributes represented as one-hot vectors, making it adaptable to multimodal layout representations.

As illustrated in Fig. 2(b), the Transformer encoder processes a set of N layout elements consisting of three types of inputs. First, the *target text* is represented by its rendered image, text content, character count, line count, angle, color, and font, but without location attributes, as these are the targets of our estimation. Second, the remaining $N - 1$ elements in the layout include similar attributes—rendered images, text content, position, size, type, angle, color, and font—where non-text elements are assigned empty strings or zeros for text-specific attributes, and categorical data such as font and type are encoded as one-hot vectors. Third, the model also takes as input the whole layout image with the target text removed.

To process these inputs, image and text data are first encoded using a ResNet50-based image encoder and a BERT-based text encoder, respectively. The embeddings obtained from these encoders, along with other layout attributes, are compressed through linear transformations and concatenated for each element before being fed into the Transformer encoder.

3.2 Small VLM-Based Method

Figure 2(c) shows the structure of the VLM-based method. Here, we use a "small" (light-weight) VLM to handle images and texts jointly. We employ Phi3.5-vision [1], which has already pretrained knowledge. Its lightweight nature facilitates fine-tuning for estimating the optimal text box placements even with limited training data. Its inputs (texts and one-hot attributes) are the same as in the standard Transformer-based method; however, instead of encoding them by BERT and linear transformations, the inputs are converted in JSON format, as illustrated in Fig. 2 (d), and fed to the Phi3.5-vision as a structured prompt. While the standard Transformer-based method predicts continuous values for text box placement, VLMs predict text box placement in a linguistic format, i.e., a sequence of tokens.

The JSON structure describes all the contextual elements and the target text using the fields: `type`, `text`, `left`, `top`, `width`, and `height`. The field `type` describes the type of element. In this paper, we use one type for texts and four types for images.[1] The field `text` describes the text sequence of each element; for non-text elements, this field is given as the special placeholder `{blank}`. The remaining `left`, `top`, `width`, and `height` specify the size and location of the bounding box for each element. Note that these four fields are set to `null` for the target text. The JSON descriptions of individual elements are sorted by their position and fed to the VLM. Finally, the VLM model generates output in JSON format, containing the estimated left, top, width, and height.

3.3 Large VLM-Based Method

Figure 2(c) also shows the method with a large VLM; the differences from the small VLM-based method are their model size. The large VLM-based method uses Gemini [15], which is much larger than Phi3.5-vision. Although Gemini is not trainable with our rather small dataset, it is already pretrained with vast knowledge. Therefore, even without any additional training, we can expect that Gemini already has sufficient knowledge about typography and layout patterns. We, consequently, evaluate Gemini with the same inputs and output as the small VLM-based method, illustrated in Figs. 2(c) and (d).

4 Experimental Results

4.1 Dataset

Professionally designed and layer-segmented typography datasets are scarce, and therefore, we used the Crello dataset [16], the best resource currently available. This dataset consists of professionally designed layouts, including social media posts, printed posters, and banner advertisements, all represented in a layered structure. Each design was created by typographic experts and is stored as a multi-layered file, similar to those used in Illustrator or other vector-based design software.

The dataset originally contains 19,095, 1,951, and 2,375 layouts for the training, validation, and test sets, respectively. Since our study focuses on estimating the position and size of a single text string, we excluded designs that do not contain textual elements. After this exclusion, the dataset includes 18,937, 1,933, and 2,314 layouts for training, validation, and testing, respectively. For the quantitative evaluation, we further divide the test set into 173 layouts with a single text element and 2,141 layouts with multiple text elements, following previous works [7,10].

[1] These five types are defined in the Crello dataset [16] as follows: `textElement` for texts, `imageElement` for bitmap images, `maskElement` for bitmap images with cropping by a mask, `svgElement` for shapes or textures in the SVG (i.e., vector) format, and `coloredBackground` for a constant-color background.

4.2 Training Details

Standard Transformer-Based Method. We used only the encoder of the Transformer model. The Transformer encoder consists of six layers. Each layer consists of eight attention heads, each with a dimension of 256, and a feed-forward network with a dimension of 512. The model was trained using the AdamW optimizer, with a learning rate of 0.0001 and a batch size of 64. After 150 epochs of training, we use the model with the lowest validation loss.

We use Complete IoU (CIoU) loss [18] to optimize the model. CIoU is an extended version of the standard IoU (Intersection over Union) that incorporates three key factors: the standard IoU between the predicted and ground-truth boxes, the distance between their centroids, and the consistency of aspect ratio. By accounting for centroid distance, CIoU mitigates the issue where standard IoU drops to zero when small boxes are slightly misaligned and do not overlap at all. Additionally, CIoU considers aspect ratio to ensure that the predicted bounding box maintains a shape similar to the ground-truth box. This is particularly important in text placement tasks, where bounding box proportions influence visual harmony and spatial alignment within the layout.

Small VLM-Based Method (Phi3.5-Vision). To fine-tune Phi3.5-vision [1], we employed LoRA (Low-Rank Adaptation) [5], a common technique for adapting LLMs and VLMs, with a rank of 128 and an alpha of 256. We trained the model using cross-entropy loss and optimized it with AdamW. The learning rate was set to 0.00005 and the batch size to 1. Training was conducted over five epochs, and we selected the model achieving the lowest validation loss for evaluation.

Large VLM-Based Method (Gemini). For the large-scale pretrained model, we employed Gemini 2.0 Flash (model identifier: gemini-2.0-flash-exp) [15] in a zero-shot manner, i.e., without any additional training. As already explained, we structured the model's inputs and outputs in JSON format, facilitating more reliable parsing and evaluation.

4.3 Evaluation Metrics

We employed the standard IoU and Boundary Displacement Error (BDE) [10] to evaluate how closely the predicted results align with the actual layout. BDE is similar to the centroid distance in CIoU. Specifically, BDE measures how far the edges of the predicted bounding box deviate from those of the ground-truth box. Specifically, it looks at the differences in position between the left, right, top, and bottom edges of both boxes and then averages those differences. A lower BDE value indicates closer alignment between the predicted and true boundaries, whereas a higher value suggests a larger mismatch.

Note that the VLM-based methods occasionally generate outputs that do not conform to the expected format, rendering them unsuitable for evaluation. Specifically, Phi3.5-vision produced 7 invalid outputs (out of 2,313), while Gemini produced 18. We excluded these cases from our analyses.

Table 1. Quantitative evaluation of text box placement accuracy. **Bold** and underline indicate the best and second-best results, respectively. Following prior works [7,10], we split the test data into layouts containing a single text element ("Single text") or multiple text elements ("Multiple text").

	Input image	Single text (173) IoU↑	Single text (173) BDE↓	Multiple text (2,141) IoU↑	Multiple text (2,141) BDE↓	All (2,314) IoU↑	All (2,314) BDE↓
Transformer	Single	**0.269**	**0.131**	**0.261**	**0.121**	**0.262**	**0.122**
Phi3.5-vision	Single	0.204	0.177	0.200	0.126	0.200	0.130
Gemini	Single	0.140	0.178	0.133	0.181	0.134	0.181
FlexDM [7]†	Multi⋆	0.357	0.098	0.110	0.141		
Transformer	Multi⋆	**0.365**	**0.096**	**0.295**	**0.104**	**0.300**	**0.104**

†: Values taken from results in [7]. ⋆: Discussed later in Sect. 5

4.4 Quantitative Evaluation

Quantitative Evaluation of Overall Performance. The results obtained using the aforementioned metrics are summarized in Table 1. As noted in Sect. 4.1, following prior work, we split the test set into "Single text" and "Multiple text" layouts. In the Single text subset, no other text elements are present, so the model is free from overlap concerns but also lacks the guidance that existing text might provide. Conversely, multiple text subsets provide more hints about where to place new text because other text elements are already laid out, but they add complexity to avoiding unnecessary overlaps between texts and maintaining a consistent layout.

Table 1 shows that the standard Transformer-based model consistently achieves the highest IoU and BDE scores across both Single-text and Multiple-text tasks. We attribute this superiority to two main factors. First, the Transformer is trained end-to-end specifically for text box placement, directly regressing continuous bounding-box coordinates and learning the spatial constraints needed for precise layout. Second, while Phi3.5-vision undergoes fine-tuning for this task and therefore outperforms Gemini, which is used in a purely zero-shot manner, both are general-purpose VLMs originally trained on broader objectives. Their token-based coordinate generation may also introduce discretization issues, causing slight format or numerical mismatches. Consequently, even the fine-tuned VLM struggles to capture the finer-grained relationships required for high-accuracy text positioning, highlighting the advantages of a specialized approach.

Figure 3 shows the histograms of IoU and BDE for each method, as well as an IoU-BDE scatter plot and its heatmap representation. The IoU histograms reveal that near-zero IoU values occur relatively frequently across all methods. However, this does not mean the placement task is intractable. In practice, IoU should be considered alongside BDE: a large number of cases with both near-zero IoU and small BDE often indicate small text boxes that barely fail to overlap due to minor misalignments. Indeed, cases with large BDE (and thus an IoU of

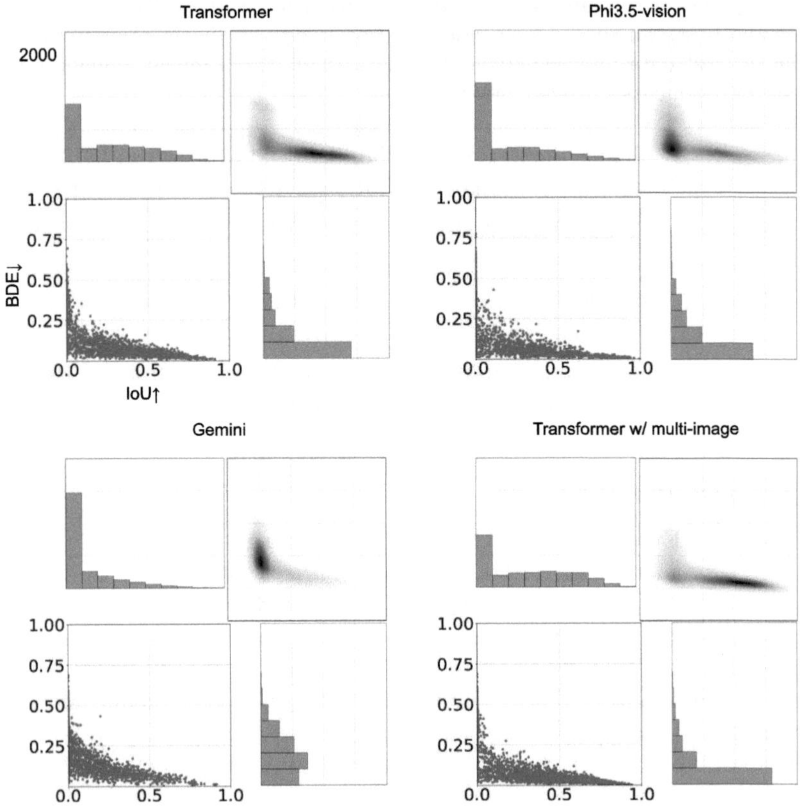

Fig. 3. Distributions of IoU and BDE in all the test set (single text + multiple text). We present IoU and BDE histograms, an IoU-BDE scatter plot, and a corresponding heatmap for each method.

zero) remain relatively rare for all methods, suggesting that text placement is still reasonably accurate in most cases.

When comparing methods, the Transformer and Phi3.5-vision methods exhibit pronounced peaks near zero in their BDE histograms, indicating that most predictions have only minor displacements. Meanwhile, Gemini's BDE distribution does not peak around zero, reflecting the model's inability to outperform those specialized or fine-tuned for text placement. Additionally, the frequency of high-precision placements (i.e., high IoU values) decreases monotonically for the two VLM-based methods. (In short, better IoU is less frequent.) By contrast, the Transformer's IoU histogram shows a secondary peak around IoU$\sim$ 0.3, deviating from the strictly monotonic trend.

Effect of the Target Text Area on Placement Accuracy. Figure 4 shows how the target text area affects placement accuracy for each method, visualized via Kernel Density Estimation (KDE). The purple curves represent the smallest

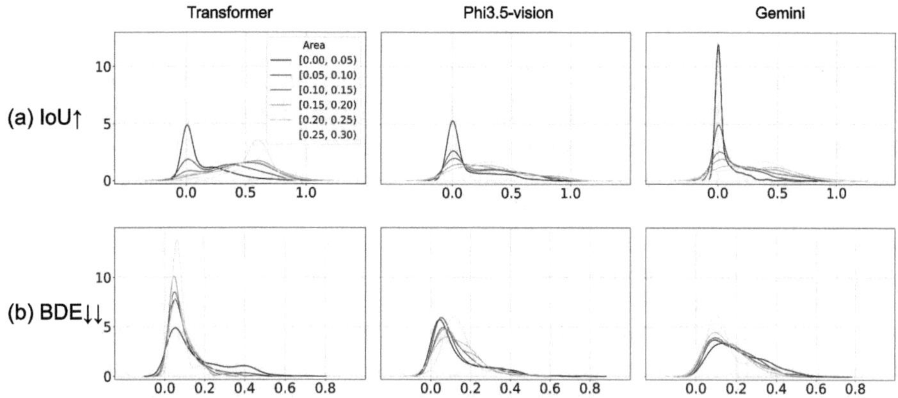

Fig. 4. Effect of the target text area on placement accuracy.

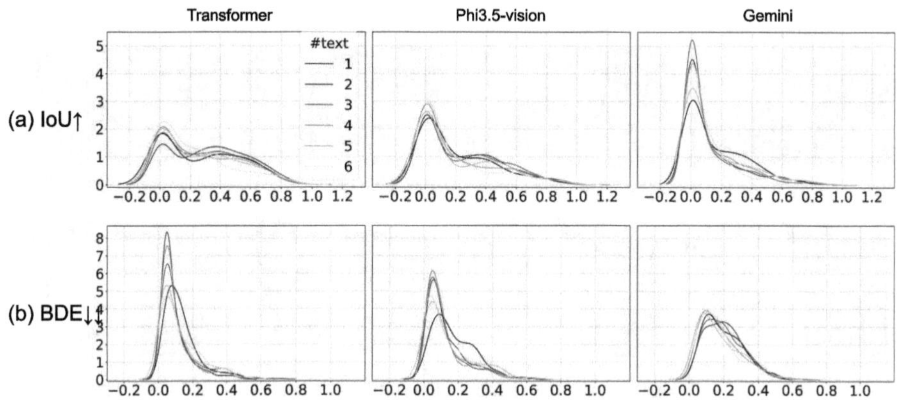

Fig. 5. Effect of the number of text elements on placement accuracy.

text areas, transitioning to yellow for progressively larger areas. Although IoU and BDE are each bounded within $[0, 1]$, KDE smoothing can produce density estimates that appear outside this interval.

Across all methods, smaller text areas lead to poorer placement accuracy in terms of both IoU and BDE. This is partly due to the nature of IoU, which inherently penalizes small bounding boxes, resulting in near-zero IoU values especially for Gemini. By contrast, larger text areas tend to improve both IoU and BDE; however, an exception emerges with Phi3.5-vision, which shows a tendency toward higher (i.e., worse) BDE as the target text area increases.

Effect of the Number of Text Elements on Placement Accuracy. Figure 5 shows KDE plots illustrating how the number of text elements in a layout influences placement accuracy for each method. The purple curve (#text = 1), represents a single-text layout with only the target text present. Cases where #text ≥2 indicate one or more additional text elements besides the target text.

Generally, as the number of text elements increases, IoU tends to decline because each text box becomes smaller in general. However, this decrease is not always monotonic. IoU starts off low at #text = 1 for the Transformer model, improves noticeably at #text = 2, and then exhibits more near-zero IoU as additional texts are added. A similar pattern is observed in the BDE metric, suggesting that having at least one other text element may help guide the placement of the target text.

In contrast, Phi3.5-vision's IoU remains relatively low and shows little change across different text counts. However, its BDE behavior mirrors that of the Transformer: the highest accuracy occurs when there are a few additional text elements (e.g., $2 \sim 4$), indicating that a moderate number of contextual texts can benefit the placement process.

Fig. 6. Representative examples where all three methods placed the text box successfully.

4.5 Qualitative Evaluation

Successful Cases. Figure 6 shows examples in which all three methods achieved consistently high-quality placements. In the top row ($N = 3$), the text "Make" seamlessly aligns with the existing text "SOAP at Home." In the second row

Fig. 7. Representative failure cases in all three methods. (Top) layouts where the text must be placed on a visually prominent area of the image, (middle) small text boxes requiring precise spacing, and (bottom) dense layouts that lead to unintended overlaps.

($N = 2$), the target text "classroom\n affects" (where a newline token is inserted between the two words) is accurately placed over a sufficiently large background image with clearly defined design elements. A satisfactory placement was frequently observed in such large background scenarios. This result also demonstrates the methods' ability to handle multi-line inputs effectively. Finally, the bottom row ($N = 5$) demonstrates that all three methods performed precise and contextually appropriate placements even when the target text must fit between existing text elements. Notably, even without task-specific fine-tuning, Gemini placed the bounding boxes well, showing a notable level of performance.

Failure Cases. Figure 7 shows representative failure cases observed in all three methods, focusing on three typical scenarios. The top row illustrates layouts where the text must be placed on a non-uniform background, yet models are typically trained to place text on flatter regions for better readability; limited data on such complex cases makes it difficult to learn appropriate placement. The middle row depicts small text boxes, which demand near-pixel-perfect alignment in tight whitespace. The bottom row presents dense layouts with limited free space, often forcing unintended overlaps. Even with dedicated or fine-tuned models, these edge conditions pose substantial challenges.

Fig. 8. Successful cases of Transformer with multiple image inputs. (Left) Context elements as images. (Middle) Results of Transformer with single whole image. (Right) Results of Transformer w/ multi-image.

5 Utilizing Multiple Image Inputs

Thus far, we have explored three methods, all of which take the whole layout as a single image input. However, as shown in Fig. 2(a), each element in a layout can also be rendered as an individual image (i.e., a bitmap). By providing each element's image as an additional input, it becomes possible to explicitly learn not only the overall layout but also the visual appearance of each component and its positional relationships. This approach increases computational complexity, but the additional cost remains manageable rather than intractable through suitable encoders and linear layers.

We extend the standard Transformer-based method by incorporating multiple images, as shown in Fig. 2(e). We refer to this approach as "Transformer w/ multi-image." Each layout element, such as texts, images, and other visual components, is provided as an additional image input in this setup. By encoding these individual images, the model can more clearly capture each element's appearance and its spatial relationships within the layout. Concretely, all text and non-text elements are rendered as images and passed through a ResNet50-based encoder. The resulting embeddings, along with text attributes and layout metadata, are concatenated for each element, and then fed into the transformer to predict bounding-box coordinates, following the same procedure as the single-image transformer method. In principle, providing multiple images as input to VLMs is also possible. However, VLMs have a maximum token limit, which practically restricts the extent to which multiple images can be utilized.

Figure 8 shows examples demonstrating the positive effect of providing multiple images as input. As shown, incorporating the appearance of each image element clarifies which elements function as "text containers." For example, such containers might be uniformly colored rectangles or speech-bubble-like graphics, where text tends to be frequently placed. The Transformer model effectively learns this tendency and positions text on these containers.

The bottom row of Table 1 shows the performance of "Transformer w/ multi-image," which achieves a substantial improvement over the version that uses only a single whole layout image. The gain is larger under the "single text" setting, where no other text elements besides the target text are present. This suggests that explicitly incorporating the appearance information of each image element, rather than just its coordinates or size, provides a stronger contextual cue for accurate placement. Figure 3 also visualizes the performance of Transformer w/ multi-image, which achieves more near-zero BDE cases than the other methods.

Although FlexDM [7] is capable of positioning multiple boxes, the experiment reported in its original paper evaluates only a single-box configuration under the multiple image input condition. Therefore, we present the corresponding numerical results in Table 1, which shows the standard Transformer outperforms FlexDM.

6 Conclusion, Limitation, and Future Work

We compared three approaches for automated text box placement: a specialized Transformer-based model, a small fine-tuned VLM (Phi3.5-vision), and a large pretrained VLM (Gemini). Overall, the Transformer-based method achieved the highest accuracy, as evidenced by both IoU and BDE. Notably, BDE values often remained low across the dataset, indicating that despite occasional IoU drops – especially with small text or complex layouts – most placements deviated only slightly from the ground truth. This result suggests that text box placement, while challenging, is tractable with carefully tailored models.

When comparing the VLM-based methods, Phi3.5-vision generally outperformed Gemini, demonstrating the advantage of task-specific fine-tuning over a purely zero-shot approach. Still, it did not match the specialized Transformer's ability to capture spatial relationships through continuous-coordinate regression consistently. Extending the Transformer to accept multiple image inputs further lowered BDE, highlighting the benefit of incorporating richer appearance information from each layout element.

For future work, several promising avenues exist. First, although we focused on a single text box, extending the proposed methods to jointly optimize multiple boxes could further enhance real-world layout design. Second, because many layouts allow more than one valid placement, representing text locations as a probability map, rather than a single deterministic coordinate, may offer a richer view of possible solutions. Third, some layouts simply do not have suitable whitespace or spacing; detecting such infeasible scenarios and suggesting alternative design adjustments would make these methods more practical. Finally, assembling multiple placement models – e.g., specialized Transformers, VLMs, or heuristics – into an ensemble could improve coverage and reliability, especially in challenging or unusual layout conditions.

Acknowledgements. This work was supported by JSPS KAKENHI-JP25H01149 and JST CRONOS-JPMJCS24K4.

References

1. Abdin, M., et al.: Phi-3 technical report: a highly capable language model locally on your phone. arXiv preprint arXiv:2404.14219 (2024)
2. Achiam, J., et al.: GPT-4 technical report. arXiv preprint arXiv:2303.08774 (2023)
3. Feng, W., et al.: LayoutGPT: compositional visual planning and generation with large language models. In: NeurIPS, vol. 36 (2024)
4. Horita, D., Inoue, N., Kikuchi, K., Yamaguchi, K., Aizawa, K.: Retrieval-augmented layout transformer for content-aware layout generation. In: CVPR, pp. 67–76 (2024)
5. Hu, E.J., et al.: Lora: low-rank adaptation of large language models. In: ICLR (2022)
6. Inoue, N., Kikuchi, K., Simo-Serra, E., Otani, M., Yamaguchi, K.: LayoutDM: discrete diffusion model for controllable layout generation. In: CVPR, pp. 10167–10176 (2023)
7. Inoue, N., Kikuchi, K., Simo-Serra, E., Otani, M., Yamaguchi, K.: Towards flexible multi-modal document models. In: CVPR, pp. 14287–14296 (2023)
8. Jia, P., et al.: Cole: a hierarchical generation framework for graphic design. arXiv preprint arXiv:2311.16974 (2023)
9. Jyothi, A.A., Durand, T., He, J., Sigal, L., Mori, G.: LayoutVAE: stochastic scene layout generation from a label set. In: ICCV, pp. 9895–9904 (2019)
10. Li, C., Zhang, P., Wang, C.: Harmonious textual layout generation over natural images via deep aesthetics learning. IEEE TMM **24**, 3416–3428 (2021)
11. Li, F., et al.: Relation-aware diffusion model for controllable poster layout generation. In: CIKM, pp. 1249–1258 (2023)
12. Li, J., Yang, J., Hertzmann, A., Zhang, J., Xu, T.: LayoutGAN: generating graphic layouts with wireframe discriminators. In: ICLR (2019)
13. Lin, J., Guo, J., Sun, S., Yang, Z., Lou, J.G., Zhang, D.: Layoutprompter: awaken the design ability of large language models. In: NeurIPS, vol. 36 (2023)
14. Seol, J., Kim, S., Yoo, J.: Posterllama: Bridging design ability of language model to contents-aware layout generation. In: ECCV (2024)
15. Team, G., et al.: Gemini: a family of highly capable multimodal models. arXiv preprint arXiv:2312.11805 (2023)
16. Yamaguchi, K.: CanvasVAE: learning to generate vector graphic documents. In: ICCV (2021)

17. Yang, T., Luo, Y., Qi, Z., Wu, Y., Shan, Y., Chen, C.W.: PosterLLaVA: constructing a unified multi-modal layout generator with LLM. arXiv preprint arXiv:2406.02884 (2024)
18. Zheng, Z., Wang, P., Liu, W., Li, J., Ye, R., Ren, D.: Distance-IoU loss: faster and better learning for bounding box regression. In: AAAI. vol. 34, pp. 12993–13000 (2020)

Visual Document Matching for Zero-Shot Document Classification

Lucas De Almeida Bandeira Macedo(✉), Joao Paulo Vieira Costa(✉), Joao Pedro Felix De Almeida, Pedro Garcia Freitas, and Li Weigang

University of Brasília, UnB - Brasília, Federal District, Brasilia 70910-900, Brazil
lucasabmacedo@hotmail.com, jpcosta1990@gmail.com

Abstract. Accurate document identification is crucial for ensuring compliance and maintaining consistency across various applications. Consequently, document classification has been extensively studied. However, the dynamic nature of documents often renders simple classification insufficient, necessitating frequent model retraining. To mitigate this constant maintenance, comparing incoming documents against known references is a more suitable approach. To address the zero-shot classification problem, we introduce Visual Document Matching (VDM). VDM focuses on verifying whether two documents share an identical visual layout structure, a particularly effective approach in scenarios where training classes do not align with those encountered during inference. Despite its significant potential, the Zero-Shot Learning (ZSL) approach remains largely underexplored in document layout understanding. To support our study, we introduce Layout-Aware Complex Document Information Processing (LA-CDIP), a dataset comprising 4,993 documents across 144 classes. We reorganized this dataset from the RVL-CDIP database to emphasize visual structure over semantic information. Our approach is benchmarked using a siamese network within a contrastive learning framework across multiple backbone architectures, including ResNet, EfficientNet, and Vision Transformer (ViT). In zero-shot scenarios, our method achieves an Equal Error Rate (EER) below 5% in 1-vs-1 verification with cross-validation. Furthermore, our VDM approach outperforms lighter Large Language Models (LLMs) and rivals GPT-4o. These findings highlight the superiority of specialized VDM techniques over general-purpose multimodal models, demonstrating high accuracy with significantly fewer parameters, making our approach more practical for real-world applications.

Keywords: Zero-shot learning · Metric Learning · Document Understanding · Document Image Classification · Visual Document Matching

1 Introduction

Documents are central to many organizations, acting as official records for contracts, identities, and evidence, and playing a crucial role in legal, financial, and

L. Jin et al. (Eds.): ICDAR 2025 Workshops, LNCS 16225, pp. 192–208, 2026.
https://doi.org/10.1007/978-3-032-09368-4_12

operational processes. Consequently, document classification, which involves categorizing documents into predefined classes [1], is an essential task for many businesses. Document Image Classification (DIC) is a more specialized task where textual information is not directly available, and input often consists of photos or scans of physical documents. Artificial Intelligence (AI) and Deep Learning (DL) techniques can automate DIC by leveraging traditional classification networks.

The challenge arises when previously accepted document forms change or when entirely new, unconsidered document classes must be handled and categorized. In such scenarios, models typically require retraining, as traditional classification networks often struggle to generalize to novel document layouts or entirely new classes. Therefore, a more robust alternative involves employing ZSL techniques. ZSL techniques are a machine learning paradigm that require models to generalize to tasks that were not seen during training [2,3]. This paradigm enables a single model to address multiple document classification problems. To build a reliable ZSL system, it is essential to identify a set of robust features that serve as anchors, ensuring documents share features if, and only if, they belong to the same class. This capability allows ZSL models to group classes, even those completely unseen during training. Therefore, a dataset for this task must be meticulously categorized with this requirement in mind.

1.1 Key Challenges in Document Classification

To achieve a ZSL document image classification system, there are several obstacles that researchers in this field must overcome, for instance:

1. the lack of a standardized dataset: although the most commonly used benchmark dataset categories are semantically broad and include visually heterogeneous sub-types. This intra-class diversity limits its suitability for tasks such as zero-shot learning and unsupervised clustering [4].
2. the absence of an established state-of-the-art methodology: unlike text classification and extraction, zero-shot classification lacks a widely accepted framework or baseline approaches.

Bridging these gaps is essential for developing scalable and efficient solutions for ZSL document classification in real-world applications.

1.2 Contributions of This Work

To address the aforementioned challenges, this work proposes VDM, a novel framework for tackling ZSL visual-based DIC. By leveraging metric learning techniques, documents with similar visual patterns can be identified and grouped, even if their class was entirely unseen during the training stage. The VDM problem is reduced to a binary classification problem [5] by calculating the similarity between two documents and defining a similarity metric (often based on a threshold). This classifies whether two documents belong to the same class or not. Although a reference anchor must be chosen for each class in a real-world application, this approach is still considered ZSL, as document classes can be excluded from the training set. This matching-based formulation draws inspiration from biometric verification systems.

Taking this into account, this paper makes several key contributions: it introduces a novel dataset designed for ZSL, focusing on visual information rather than document semantics, targeting a greater generalization across unseen classes. This enables the development and evaluation of models in this domain; proposes a VDM approach based on image similarity, leveraging ZSL techniques to generalize across unseen document layouts; presents an evaluation protocol to unify and standardize further research on the same dataset; and delivers a systematic evaluation of well-established neural network backbones in the context of zero-shot document layout matching, offering valuable benchmarks for the research area.

The remainder of this article is structured as follows: Sect. 2 reviews the state of the art in zero-shot learning and visually-rich document understanding, highlighting existing methods and their limitations. Section 3 presents the proposed VDM and it's supporting dataset, detailing its architecture, components, and underlying learning paradigm. Section 4 describes the experimental setup, including datasets, evaluation metrics, and baseline comparisons, followed by a comprehensive analysis of the results on Sect. 5. Finally, Sect. 6 concludes the article by summarizing the contributions, discussing potential applications, and outlining future research directions.

2 Related Work

Two primary themes prevalent in the existing literature are analyzed: (1) dataset availability and their suitability for ZSL opportunities and (2) current document processing frameworks, including an evaluation of their capabilities and constraints concerning ZSL.

2.1 Document Understanding Databases

The IIT-CDIP dataset [6], introduced in 2006, was a large-scale repository used for document classification and information retrieval. It originated from the Tobacco Documents Library, from UCSF Industry Documents Library collection, and contains millions of scanned documents. RVL-CDIP dataset [7] was introduced in 2015 as a dataset for image-based document classification, organizing 400,000 document images into 16 predefined categories such as letters, forms, and emails.

Several other benchmarks have been proposed for document understanding, though they target different tasks than document image classification. DocVQA [8] focuses on visual question answering, pairing over 12,000 document images with 50,000 questionâĂŞanswer pairs, and has become a standard for evaluating multimodal reasoning in LLMs. Other datasets, such as FUNSD [9], SROIE [10], CORD [11], and XFUND [12], address structured information extraction, especially key-value retrieval in forms, receipts, and administrative documents.

Recent studies have highlighted limitations in existing benchmarks datasets for DIC. RVL-CDIP includes semantically defined classes with high intra-class variability [4]. Other datasets such as MULTIEURLEX-DOC and WIKI-DOC address semantic granularity, but contain mostly static layouts, limiting their use for layout-aware models [13]. This reinforces the need for benchmarks that combine semantic consistency with layout diversity—precisely the goal of LACDIP.

2.2 Document Processing Works

Youssef et al. [14] propose a method to use Variational Autoencoders to detect deviations from a standard document template. Their approach is particularly suited for industrial compliance verification, where identifying visual discrepancies such as stamps, handwritten annotations, and misplaced signatures is crucial. By learning the distribution of compliant documents, the model detects anomalies as out-of-distribution samples, making it effective for scenarios with highly imbalanced datasets.

Sinha et al. [15] introduce CICA, a framework that enhances CLIP's performance in zero-shot classification by improving textual-visual feature alignment through content-injected contrastive learning. Both studies emphasize data-efficient approaches to document classification based on RVL-CDIP dataset.

Another study by Van Landeghem et al. [16] explores for visually rich document (VRD) applications, particularly Document Layout Analysiss (DLAs) and DICs. The study evaluates multiple architectures, including ResNet, Vision Transformer and Document Image Transformer. Their evaluation DICs was performed on the original categories of the RVL-CDIP dataset.

Collectively, these studies contribute to advancements in document understanding through anomaly detection, DLA and DIC. Liu et al. [1] provide a comprehensive survey of two decades of research in DIC. They highlight that although Siamese networks have shown strong performance in general image similarity tasks, they remain underexplored in the context of document image classification, particularly for capturing image appearance variations.

Given the limitations of RVL-CDIP [4], ZSL document classification [17] has shown lower performance compared to other computer vision tasks. In this context, our work contributes to the field of document understanding by introducing the LA-CDIP dataset and applying a protocol, which applies Siamese networks for VDM.

2.3 LLMs for Document Understanding

LLMs have demonstrated their ability to perform zero-shot classification in various domains, including document understanding. Scius-Bertrand et al. [17] investigate the application of LLMs, such as GPT-4 and RoBERTa, for zero-shot prompting and few-shot fine-tuning in DIC. Their study demonstrates that LLMs can achieve competitive performance with minimal labeled data, challenging the traditional reliance on large annotated datasets.

With the need to process multimodal information, there has been an advancement in models with visual capabilities. Llama 3.2 Vision [18], for example, was introduced as an open source alternative with multimodal reasoning capabilities.

InternVL 2.5 was introduced as an evolution of its predecessor, InternVL 2.0, incorporating enhancements that resulted in performance gains. InternVL 2.5 achieved 95.1% accuracy on the DocVQA benchmark, demonstrating document understanding capabilities [19]. Moreover, its performance is competitive with leading proprietary models, such as OpenAI's GPT-4o [20,21].

Qwen2.5-VL was introduced as an evolution of the Qwen2-VL series, surpassing its predecessor. According to its evaluation [22], Qwen2.5-VL demonstrated good performance in OCR-related tasks, chart interpretation, and document understanding. In the reported benchmarks, models such as Qwen2.5-VL 72B achieved top results in key tasks including DocVQA and ChartQA, while proprietary models like Claude-3.5, Gemini 1.5 Pro, and GPT-4o performed competitively across several categories.

Given their growing capabilities and accessibility, evaluating LLMs in the context of VDM serves both as a baseline and as a reference for general-purpose systems.

3 Methodology

Developing an effective solution for VDMs requires addressing key challenges related to dataset availability, document structure variability, and generalization to unseen layouts. This section provides a detailed description of these components and the experimental procedures used to evaluate their effectiveness.

3.1 Dataset Construction

The LA-CDIP proposed dataset is built as a reorganization of the RVL-CDIP [7] dataset. While RVL-CDIP's classes are divided by their meaning and textual content, in our dataset, classes are divided by the visual information. This is illustrated in Fig. 1, where both groups of documents are classified as the same class under RVL-CDIP, both being forms, but are re-arranged to be different classes under LA-CDIP, as they are visually distinct.

The construction of the database followed 2 stages: preliminary clustering, followed by a manual reorganization. The clustering is made with a private metric learning model, trained with a private document dataset, using a similar framework as the one presented here. The clustering is made possible by the metric learning nature of the solution, as the output of the trained model is a feature value in a feature space. The clustering algorithm used was Hierarchical Agglomerative Clustering, with Ward [23] strategy . The clustering served exclusively as suggestions, as the cluster themselves were not homogeneous, neither were unique. Then we manually reorganized and labeled the clusters, grouping by the visual layout information that often reflects the purpose of a document.

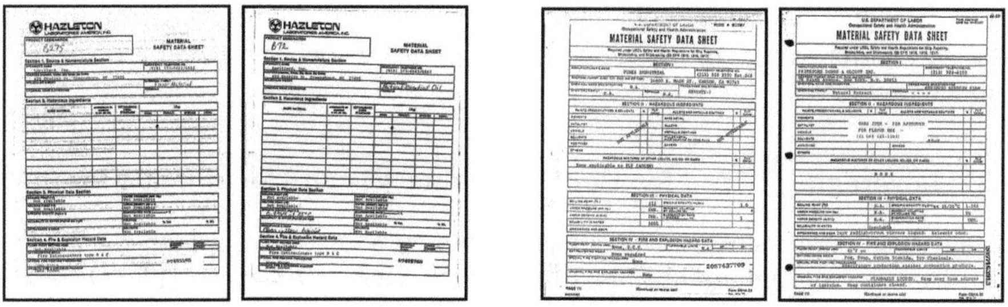

Fig. 1. Examples of two distinct classes under the proposed LA-CDIP dataset. Those classes originally belonged to the same class under RVL-CDIP organization.

Next, we conducted a second independent validation of the dataset, reviewing both intra-class and inter-class consistency.

To maintain train-test data consistency, data is split following the ZSL and Generalized Zero-Shot Learning (GZSL) [24]. These protocols define distinct methods for splitting training and test data:

1. ZSL: Represents a complete separation of training and test classes, ensuring no overlap between the splits.
2. GZSL: Simulates a more realistic scenario by allowing partial overlap, where half of the test or validation set consists of seen classes and the other half comprises unseen classes.

For both protocols, the data is initially divided into training and test sets. The training data is then further partitioned using 5-fold cross-validation [25].

The data for the test scenario and the data for each split are chosen randomly, and the same for every experiment. The test scenario is picked as a sixth split for the cross-validation, so it follows the same rules for the other 5 splits. For the ZSL scenario, a sixth of the classes are randomly chosen for each split, with no overlapping. Naturally, this scenario creates splits with variable sizes, as the classes themselves have variable sizes. For the GZSL scenario, half the classes in the whole dataset are chosen as classes that do not overlap between splits, and other half are diluted between the splits. While constructing a split, first, we choose the non overlapping classes for the split, then we fill the remaining slots with the overlapped classes, achieving splits with a constant size for this scenario.

This dataset can be used for a traditional visual classification problem, but since we tackle it with metric learning, additional considerations must be taken into account. While using a Siamese network, running inference on a single point of data returns no information, as the architecture requires some sort of comparison between different data elements. Randomly selecting pairs every test can fluctuate the results, therefore, to maintain the test consistency, we wrote a test protocol that assures the same pairs are tested every time. For every document in the set, we chose two other documents: one that shares the same class, and one that does not. This document is chosen randomly by its class, therefore, every class has the same probability of being picked. A protocol exists for the test set and for every cross-validation split. The dataset[1] prepared for VDMs is publicly available.

3.2 Leveraging LLMs for Benchmarking Visual Document Matching

Given their performance in document understanding benchmarks [18,22], we evaluate the multimodal LLMs LLaVA 3.2 Vision, InternVL 2.5, Qwen2.5-VL, GPT-4o (2024-11-20), and GPT-4o-mini (2024-07-18). These models were selected due to their demonstrated capabilities in tasks involving visual-text reasoning, OCR, and layout-based document analysis.

To benchmark VDMs, we designed a visual comparison prompt that instructs LLM to compare two document images, evaluating visual similarity. Models were asked to assign a similarity score from 0 to 100, categorized into five levels: Nearly Identical, Highly Similar, Moderately Similar, Weak Similarity, and Completely Different. This evaluation was carried out using Google Colab [26], under constraints in the size of the model and available computational resources. Table 1 illustrates the full inputâĂŞoutput interaction for a representative pair of documents.

To ensure a fair evaluation, both models are tested on 1,036 protocol documents from one dataset and 1,652 from another, assessing their VDM ability. This comparison aims to determine the feasibility of using LLMs for ZSL VDM and serves as a baseline for evaluating the performance and efficiency of our proposed method.

3.3 Modeling

Traditional classification methods, such as entropy learning, locks the generality of the model to the set of classes it has been trained on. Therefore, we use Zero Shot Learning techniques to be able to classify document layouts that have not been seen on the training set. To experiment on our dataset, we use a wide variety of Backbones for our trainings, but they all follow the same metric learning architecture with Siamese Networks. The overall architecture is shown on Fig. 2.

We benchmark our dataset choosing traditional, well established vision Neural Networks as our Backbones. They are ResNet [27], MobileNetV3 [28], EfficientNet [29], VGG [30] and ViT [31]. To adapt the architectures to a siamese network, we change only the last linear layer of each model—the classification layer—into a new linear layer with an arbitrary size n, suitable for our problem. Intuitively, the model learns to draw a representation of a input document in a feature space with n dimensions, such that documents that share the same class are represented clustered together, and documents that have different classes are far apart.

[1] Avaible at: https://github.com/jpcosta90/LA-CDIP.

Table 1. LLMâĂŞbased similarity assessment: condensed task instructions (top), the input document images (middle), and the model's JSON response (bottom).

Prompt Summary	`Image-1: <image> Image-2: <image>` Compare both images and return a JSON with `similarity_score`, `category`, and `justification`, based on layout, visual features, and content type. Scoring: (90–100) Nearly Identical, (70–89) Highly Similar, (50–69) Moderately Similar, (30–49) Weak Similarity, (0–29) Completely Different. Respond only with a JSON object.

Input Images

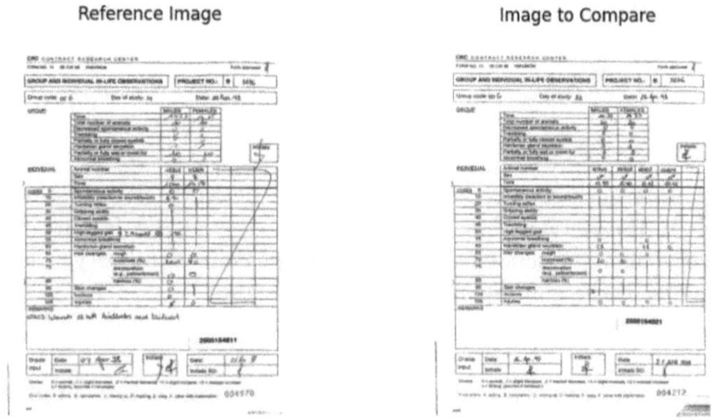

Output

```
{
  "similarity_score": 95,
  "category": "Nearly Identical",
  "justification": "Both documents share a nearly
identical layout with consistent headers, sections, and
tables, differing only in minor alignments."
}
```

We train computer vision models, therefore, their input data is the (R, G, B) document image matrix. First, we resize the input image to a shape compatible with each neural architecture. For most models, this is a $(224, 224)$ height and width. EfficientNet is the only architecture that does not follow this rule, as the different model versions increase their input size at the same time they increase the network depth and width. Then, every value of the image matrix is scaled from 0–255 to 0–1, and normalized with the mean and standard deviation of the current training split. During a training epoch, for each document in the dataset, a random document is chosen to close a pair. Every class has the same odds of being chosen to mitigate overfit in predominant classes.

We train these models within a supervised contrastive learning framework, utilizing the Contrastive Loss [32] as our loss function. Its objective is to minimize

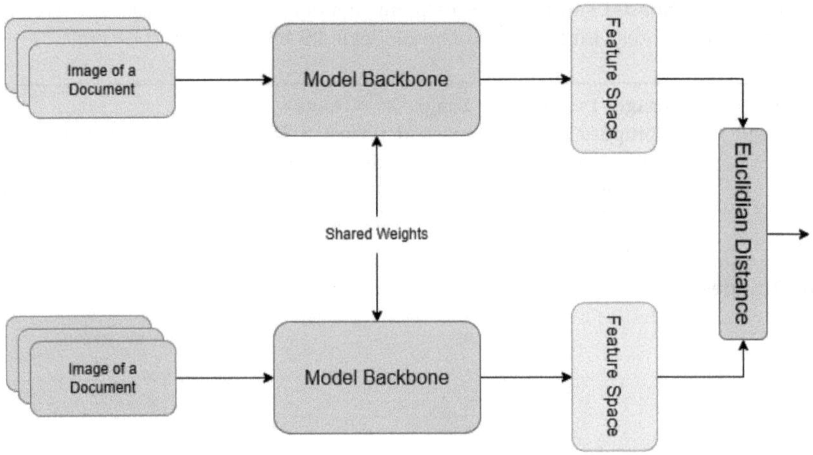

Fig. 2. Inference architecture of a siamese network. Two distinct documents are processed by the network, in parallel. Each side yields a feature vector, which then are compared using Euclidean Distance, to measure similarity. The model backbones in both sides of the siamese network share weights.

the distance between feature vectors belonging to the same class while maximizing the distance otherwise. Given a sample pair of feature vectors $\{(a, b)\}$, and a label y with a value of 1 if a and b share the same class and 0 otherwise, the loss is defined as:

$$Loss = y * d(a, b)^2 + (1 - y)) * abs(m - d(a, b))^2, \tag{1}$$

where d represents a metric function and m a hyperparameter defining the lower bound distance between samples of different classes. In this context of this paper, m=0.5 and d is the Euclidean distance between two points $P = (x_1, x_2, \cdots, x_n)$ and $Q = (y_1, y_2, \cdots, y_n)$ in an n-dimensional space defined mathematically as:

$$d(P, Q) = \sqrt{(x_1 - y_1)^2 + \cdots + (x_n - y_n)^2} = \sqrt{\sum_{i=1}^{n} (x_i - y_i)^2}. \tag{2}$$

4 Experimental Setup

4.1 LA-CDIP Dataset

The achieved dataset is composed of 4993 documents, divided in 144 different classes. Each class has at least two documents, the biggest class has 497 and the median size of the classes is 13, exposing an extra challenge on dealing with a unbalanced dataset.

4.2 Evaluation Criteria

Since a metric learning model yields a distance between two data points (in our case, two documents), we need to define a threshold to declare wether the two data points belong to the same class or not, to calculate how accurate the model is. For this, we use EER as metric [33]. EER is the point at which the False Acceptance Rate (FAR) and False Rejection Rate (FRR) are equal. The EER is calculated by setting the FAR equal to the FRR and finding the corresponding threshold at this point [34].

In other words, the EER is the value of error rate when the threshold value τ_{EER} gives

$$FAR(\tau_{EER}) = FRR(\tau_{EER}), \tag{3}$$

where

$$FAR(\tau) = \frac{\text{Number of false acceptances at threshold } \tau}{\text{Total number of negative samples}} \tag{4}$$

is the probability that a negative sample is incorrectly classified as positive [35]. On the other hand,

$$FRR(\tau) = \frac{\text{Number of false rejections at threshold } \tau}{\text{Total number of positive samples}} \tag{5}$$

is the probability that a positive sample (e.g., a genuine user) is incorrectly classified as negative.

For each chosen backbone, we train 10 models: a 5-fold cross-validation for each of the 2 scenarios, ZSL and GZSL. As shown in Sect. 3.1, each split follows a fixed validation protocol to ensure consistency. We also evaluate the trained models from each split on the independent test set, to compare them with the LLMs. We report the mean EER from all splits in a scenario.

4.3 Implementation Details

All the code for this work is written in Python, using PyTorch as the main library for neural network development. All the trained models were crafted with the TorchVision architecture collection. Our chosen default parameters are, Stochastic Gradient Descent (SGD) [36] as our optimization technique, a learning rate of 0.01, a momentum of 0.9, weight decay of 1^{-4}, 90 epochs of training and step learning rate decay of factor 0.1 each 30 epochs. Some models were trained with different hyperparameters, following the recommendations of their own work, with some adaptaions: ViT is trained with AdamW [37] optimization technique, 1^{-5} Weight Decay and Cosine Anneling [38] learning rate decay with a initial learning rate of 1^{-3}. MobileNetV3 is trained with weight decay of 1^{-5} and a 0.973 learning rate decay every 2 epochs. We use the provided ImageNet [39] weights as initial training weights, as they make an easier convergence on larger models.

5 Results

This section delves into the results of our research, primarily focusing on Table 2, which presents the performance of every model trained. The table details the mean EER from cross-validation in both scenarios, alongside the mean performance of each fold on the independent test set. While the differences between the ZSL and GZSL scenarios significantly influence the performance of visual models, this distinction is not relevant for the LLM test results, as these models were not fine-tuned for this specific task. Even if they were originally trained with some documents from the RVL-CDIP database, both scenarios effectively function as ZSL for these models. The results of the trained visual models and compare the effects of different backbones are discussed in Sect. 5.1. Subsequently, Sect. 5.2 will present the results from the chosen LLMs. Finally, Sect. 5.3 offers a comparative analysis of both approaches.

5.1 Visual Models

Among the visual models evaluated, smaller and more computationally efficient architectures demonstrably outperformed their larger counterparts. This trend is particularly evident across the ResNet variants: ResNet-18 and ResNet-34 exhibited robust performance, whereas larger versions (ResNet-50, ResNet-101, and ResNet-152) showed progressively diminishing results. Furthermore, despite its strong performance on datasets such as ImageNet [39], the ViT was the lowest-performing model for this specific task. The overall best-performing models are ResNet-34, EfficientNet-0, and EfficientNet-1. These architectures strike an effective balance in size, achieving robust generalization while effectively learning the intricate demands of VDM task. These results are visually supported by the t-SNE [40] projection shown in Fig. 3, where embeddings produced by ResNet and EfficientNet form clearly separated clusters according to their classes, while the Vision Transformer embeddings appear significantly more entangled.

Comparing the ZSL and the GZSL results, we can see that the GZSL consistently gives better results. This is expected, since half of the validation set on each fold is composed of classes seen in training. The biggest discrepancies between the two modalities can be seen in the largest models, where the overfitting damage is mitigated by the easier scenario. Even so, the best performance on the GZSL scenario is usually connected with a good performance on the ZSL scenario. For the test scenarios, reported values represent the mean performance across all folds, evaluated on the same data split. Given the zero-shot nature of the data split, there is inherently high variance across different splits, as each fold tests on entirely distinct patterns. We observe that, by chance, the ZSL test split proved more challenging than average, leading to higher error rates for most models relative to the mean cross-validation error rate. Conversely, the test GZSL scenario exhibited the opposite trend.

Table 2. Comparative performance between different visual backbones and Large Language Models. Following the columns: the architecture name; the architecture edition, if exists; the number of parameters of given architecture; cross-validation over the ZSL scenario; cross-validation over the GZSL scenario; test performance on the ZSL scenario; and test performance over the GZSL scenario. Every value is a mean EER (%) value over the CV folds. The best results are highlighted in bold for both the visual and language models separately.

Architecture	Edition	Params	ZSL	GZSL	Test ZSL	Test GZSL
VGG	11	129M	7.47	5.01	14.24	3.95
	13	129M	7.03	4.79	9.30	3.95
	16	134M	8.29	5.23	14.74	4.82
	19	139M	7.30	4.57	17.08	3.90
ResNet	18	11M	5.03	**1.54**	4.98	1.51
	34	21M	4.32	2.10	**4.13**	1.53
	50	23M	6.90	3.39	10.34	2.21
	101	42M	8.20	2.72	11.31	1.98
	152	58M	9.44	3.38	12.70	2.39
MobileNetV3	Small	1M	7.98	5.06	12.74	5.26
	Large	4M	8.16	4.27	8.45	4.43
EfficientNet	0	4M	4.41	2.27	6.02	**0.95**
	1	6M	**3.93**	3.54	8.88	2.70
	2	7M	5.73	2.61	7.29	2.14
	3	10M	5.65	3.64	7.37	2.34
ViT	Base	87M	12.43	7.97	19.72	5.19
	Large	305M	13.16	7.57	19.88	5.26
Llama	3.2	11B	–	–	13.95	21.90
InternVL	2.5	8B	–	–	8.58	10.40
Qwen-VL	2.5	7B	–	–	6.61	4.20
GPT 4o mini	2024-07-18	*	–	–	4.70	4.07
GPT 4o	2024-11-20	*	–	–	**2.75**	**1.33**

* The parameter count of GPT-4o has not been publicly disclosed.

5.2 Large Language Models

All models were evaluated under consistent conditions but further performance gains could be achieved through refined prompt engineering. Consequently, the presented results should be viewed not as definitive comparisons of model capabilities but rather as a baseline for assessing the viability of our proposed method. Despite these limitations, LLM results align with existing benchmarks for document understanding tasks.

Since the LLMs were not fine-tuned on the training data, the primary difference between the ZSL and GZSL scenarios lies in the number of classes for com-

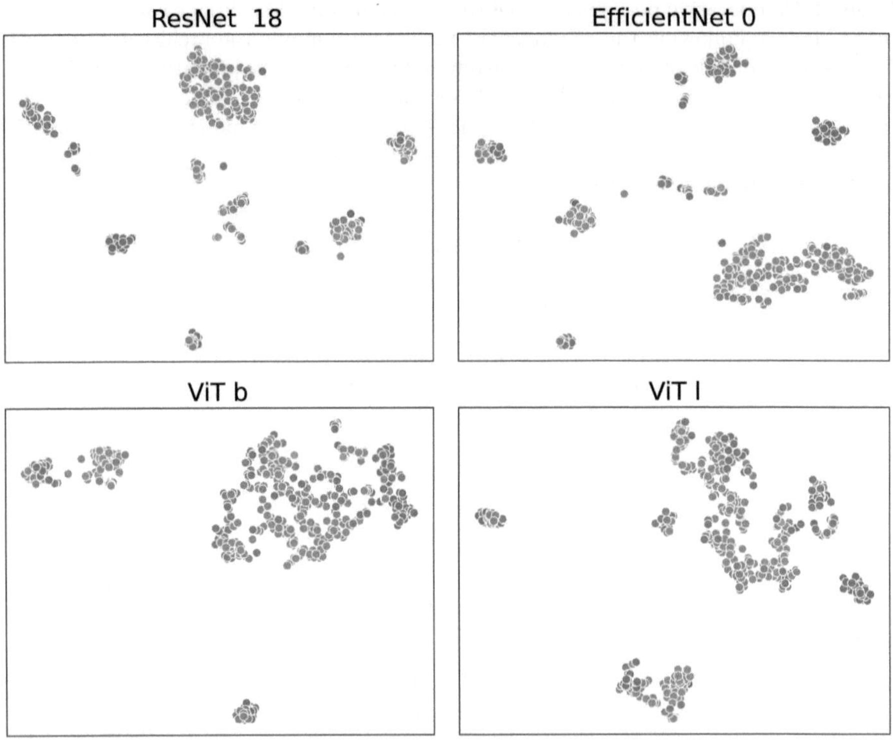

Fig. 3. t-SNE visualization of the Test ZSL scenario. These models have been trained on the same cross-validation fold.

parison. ZSL involves exactly one-sixth of the classes, whereas GZSL includes all classes, making GZSL a more comprehensive test. Interestingly, the two models exhibited divergent performance trends across these scenarios: InternVL excelled in the ZSL split, while GPT performed better in the GZSL split.

As shown in Table 2, GPT-4o demonstrated the best overall performance, achieving the lowest error rates in both the ZSL and GZSL scenarios. The GPT-4o-mini variant performed similarly but exhibited slightly lower accuracy, reflecting its inherent trade-off between efficiency and capability. In contrast, Llama 3.2 Vision encountered challenges handling multiple document inputs, necessitating image concatenation into a single input before processing. This limitation restricted its ability to directly compare layouts across distinct documents, differentiating its approach from other evaluated models. InternVL 2.5 and Qwen2.5-VL emerged as strong open-source alternatives for ZSL VDM.

5.3 Comparison

Is important to remind that GZSL is an advantageous scenario for vision models, since there is a overlap between train and test classes. However, LLMs have not been fine-tuned for this task, which means that both settings effectively serve as complete zero-shot evaluations. Therefore, the comparison was conducted in the ZSL scenario, where neither the visual nor the language models holds a particular advantage. When compared to most LLMs, VDM enabled visual models to handle ZSL while maintaining a significantly lower parameter count. This relationship between performance and model efficiency is illustrated in Fig. 4.

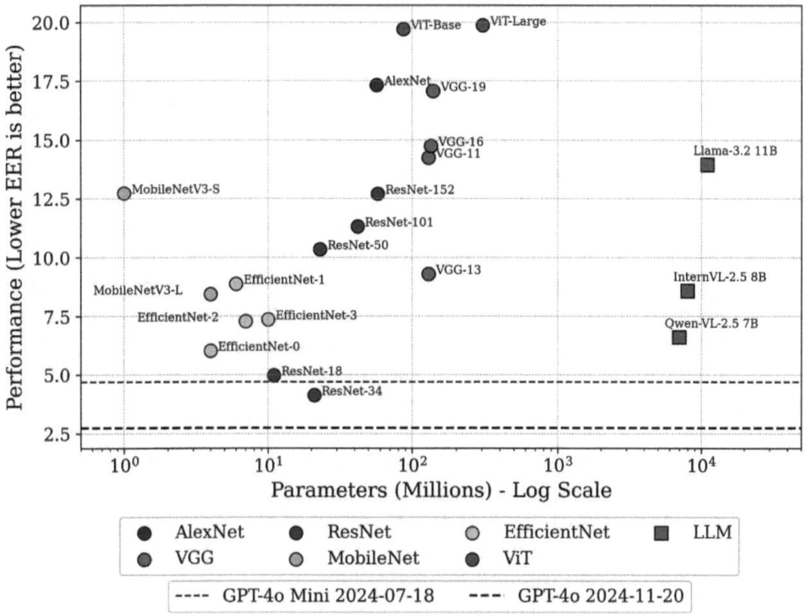

Fig. 4. Performance vs Parameters (ZSL)

The results highlight a trade-off between model complexity and effectiveness. Smaller visual backbones, such as ResNet-18 and EfficientNet-0, achieve competitive performance while using significantly fewer parameters than large multimodal models. Notably, ResNet-18 and ResNet-34 exhibit lower error rates than several larger architectures, reinforcing the efficiency of lightweight vision models in document matching tasks.

6 Conclusion and Future Work

In this work, we introduce LA-CDIP, a novel document image dataset uniquely categorized by its layout information. This dataset facilitates VDM research, offering a zero-shot alternative to the DIC problem. We benchmark LA-CDIP, by training siamese networks using the VDM approach on a diverse catalog of well-established visual backbones. We then compare their performance against popular, pre-trained LLMs. Our observations indicate that LLMs generally underperform visual models. While GPT-4o achieves a slight performance gain over visual models, its inference cost is hundreds of times higher, rendering its advantage impractical For future work, we plan to significantly expand the LA-CDIP dataset by reorganizing a larger portion of RVL-CDIP and incorporating additional document sources. Furthermore, we aim to explore advanced modeling techniques to further enhance zero-shot generalization.

References

1. Liu, L., Wang, Z., Qiu, T., Chen, Q., Yue, L., Suen, C.Y.: Document image classification: progress over two decades. Neurocomputing **453**, 223–240 (2021)
2. Xian, Y., Lampert, C.H., Schiele, B., Akata, Z.: Zero-shot learning - a comprehensive evaluation of the good, the bad and the ugly. IEEE Trans. Pattern Anal. Mach. Intell. **41**(9), 2251–2265 (2019)
3. Radford, A., et al.: Learning transferable visual models from natural language supervision. In: Meila, M., Zhang, T. (eds.) Proceedings of the 38th International Conference on Machine Learning, ICML 2021, 18-24 July 2021, Virtual Event. Proceedings of Machine Learning Research, vol. 139, pp. 8748–8763. PMLR (2021)
4. Larson, S., Lim, G., Leach, K.: On evaluation of document classification with RVL-CDIP. In: Vlachos, A., Augenstein, I. (eds.) Proceedings of the 17th Conference of the European Chapter of the Association for Computational Linguistics, Dubrovnik, Croatia, pp. 2665–2678. Association for Computational Linguistics (2023)
5. Jha, A., Samavedhi, A., Rakesh, V., Chandrashekar, J., Reddy, C.K.: Transformer-based models for long-form document matching: challenges and empirical analysis. In: Vlachos, A., Augenstein, I. (eds.) Findings of the Association for Computational Linguistics: EACL 2023, Dubrovnik, Croatia, May 2-6, 2023, pp. 2300–2310. Association for Computational Linguistics (2023)
6. Lewis, D.D., Agam, G., Argamon, S., Frieder, O., Grossman, D.A., Heard, J.: Building a test collection for complex document information processing. In: Efthimiadis, E.N., Dumais, S.T., Hawking, D., Järvelin, K. (eds.) SIGIR 2006: Proceedings of the 29th Annual International ACM SIGIR Conference on Research and Development in Information Retrieval, Seattle, Washington, USA, August 6-11, 2006, pp. 665–666. ACM (2006)
7. Harley, A.W., Ufkes, A., Derpanis, K.G.: Evaluation of deep convolutional nets for document image classification and retrieval. In: 13th International Conference on Document Analysis and Recognition, ICDAR 2015, Nancy, France, August 23–26, 2015, pp. 991–995. IEEE Computer Society (2015)
8. Mathew, M., Karatzas, D., Jawahar, C.V.: DocVQA: a dataset for VQA on document images. In: IEEE Winter Conference on Applications of Computer Vision, WACV 2021, Waikoloa, HI, USA, January 3–8, 2021, pp. 2199–2208. IEEE (2021)
9. Jaume, G., Ekenel, G.K., Thiran, J.-P.: FUNSD: a dataset for form understanding in noisy scanned documents. In: 2019 International Conference on Document Analysis and Recognition Workshops (ICDARW), pp. 1–6 (2019)
10. Huang, Z., et al.: ICDAR2019 competition on scanned receipt OCR and information extraction. In: 2019 International Conference on Document Analysis and Recognition (ICDAR), pp. 1516–1520. IEEE (2019)
11. Park, S., Shin, S., Kim, B., Cha, J., Lee, H.: CORD: a consolidated receipt dataset for post-OCR parsing. In: Document Intelligence Workshop at NeurIPS 2019 (2019)

12. Xu, Y., et al.: LayoutXLM: multimodal pre-training for multilingual visually-rich document understanding. CoRR, abs/2104.08836 (2021)
13. Fujinuma, Y., Varia, S., Sankaran, N., Appalaraju, S., Min, B., Vyas, Y.: A multimodal multilingual benchmark for document image classification. In: Bouamor, H., Pino, J., Bali, K., (eds.) Findings of the Association for Computational Linguistics: EMNLP 2023, Singapore, December 6–10, 2023, pp. 14361–14376. Association for Computational Linguistics (2023)
14. Youssef, A., Valvano, G., Veneri, G.: Document layout analysis with variational autoencoders: an industrial application. In: Ceci, M., Flesca, S., Masciari, E., Manco, G., Ras, Z.W. (eds.) ISMIS 2022. LNCS, vol. 13515, pp. 477–486. Springer, Cham (2022). https://doi.org/10.1007/978-3-031-16564-1_46
15. Sinha, S., Khan, M.S.U., Sheikh, T.U., Stricker, D., Afzal, M.Z.: CICA: content-injected contrastive alignment for zero-shot document image classification. In: Smith, E.H.B., Liwicki, M., Peng, L. (eds.) ICDAR 2024, Part IV. LNCS, vol. 14807, pp. 124–141. Springer, Cham (2024). https://doi.org/10.1007/978-3-031-70546-5_8
16. Landeghem, J.V., et al.: DistilDoc: knowledge distillation for visually-rich document applications. In: Smith, E.H.B., Liwicki, M., Peng, L. (eds.) ICDAR 2024, Part IV. LNCS, vol. 14807, pp. 195–217. Springer, Cham (2024). https://doi.org/10.1007/978-3-031-70546-5_12
17. Scius-Bertrand, A., Jungo, M., Vögtlin, L., Spat, J.-M., Fischer, A.: Zero-shot prompting and few-shot fine-tuning: revisiting document image classification using large language models. In: Antonacopoulos, A., Chaudhuri, S., Chellappa, R., Liu, C.-L., Bhattacharya, S., Pal, U. (eds.) ICPR 2024, Part XIX. LNCS, vol. 15319, pp. 152–166. Springer, Cham (2024). https://doi.org/10.1007/978-3-031-78495-8_10
18. Meta AI. Llama 3.2: From cloud to edge, now with vision (2024). https://ai.meta.com/blog/llama-3-2-connect-2024-vision-edge-mobile-devices/
19. Chen, Z., et al.: Expanding performance boundaries of open-source multimodal models with model, data, and test-time scaling. CoRR, abs/2412.05271 (2024)
20. OpenAI. Hello gpt-4o (2024). https://openai.com/index/hello-gpt-4o/
21. OpenAI. Gpt-4o mini: advancing cost-efficient intelligence (2024). https://openai.com/index/gpt-4o-mini-advancing-cost-efficient-intelligence/
22. Bai, S., et al. Qwen2.5-vl technical report. arXiv preprint arXiv:2502.13923 (2025)
23. Ward Jr., J.H.: Hierarchical grouping to optimize an objective function. J. Am. Stat. Assoc. **58**(301), 236–244 (1963)
24. Xian, Y., Lorenz, T., Schiele, B., Akata, Z.: Feature generating networks for zero-shot learning. In: 2018 IEEE Conference on Computer Vision and Pattern Recognition, CVPR 2018, Salt Lake City, UT, USA, June 18–22, 2018, pp. 5542–5551. Computer Vision Foundation/IEEE Computer Society (2018)
25. Wong, T.-T., Yeh, P.-Y.: Reliable accuracy estimates from k-fold cross validation. IEEE Trans. Knowl. Data Eng. **32**(8), 1586–1594 (2020)
26. Google Research. Google colaboratory (2024). https://colab.research.google.com/
27. He, K., Zhang, X., Ren, S., Sun, J.: Deep residual learning for image recognition. In: 2016 IEEE Conference on Computer Vision and Pattern Recognition, CVPR 2016, Las Vegas, NV, USA, June 27-30, 2016, pp. 770–778. IEEE Computer Society (2016)
28. Howard, A., et al.: Searching for mobilenetv3. In: 2019 IEEE/CVF International Conference on Computer Vision, ICCV 2019, Seoul, Korea (South), October 27 - November 2, 2019, pp. 1314–1324. IEEE (2019)

29. Tan, M., Le, Q.V.: Efficientnet: rethinking model scaling for convolutional neural networks. In: Chaudhuri, K., Salakhutdinov, R. (eds.) Proceedings of the 36th International Conference on Machine Learning, ICML 2019, 9-15 June 2019, Long Beach, California, USA, Proceedings of Machine Learning Research, vol. 97, pp. 6105–6114. PMLR (2019)
30. Simonyan, K., Zisserman, A.: Very deep convolutional networks for large-scale image recognition. In: Bengio, Y., LeCun, Y. (eds.) 3rd International Conference on Learning Representations, ICLR 2015, San Diego, CA, USA, May 7–9, 2015, Conference Track Proceedings (2015)
31. Dosovitskiy, A., et al.: An image is worth 16x16 words: transformers for image recognition at scale. In: 9th International Conference on Learning Representations, ICLR 2021, Virtual Event, Austria, May 3–7, 2021. OpenReview.net (2021)
32. Chopra, S., Hadsell, R., LeCun, Y.: Learning a similarity metric discriminatively, with application to face verification. In: 2005 IEEE Computer Society Conference on Computer Vision and Pattern Recognition (CVPR 2005), 20–26 June 2005, San Diego, CA, USA, pp. 539–546. IEEE Computer Society (2005)
33. Tolosana, R., et al.: ICDAR 2021 competition on on-line signature verification. In: Lladós, J., Lopresti, D., Uchida, S. (eds.) ICDAR 2021. LNCS, vol. 12824, pp. 723–737. Springer, Cham (2021). https://doi.org/10.1007/978-3-030-86337-1_48
34. Hofbauer, H., Uhl, A.: Calculating a boundary for the significance from the equal-error rate. In: 2016 International Conference on Biometrics (ICB), pp. 1–4 (2016)
35. Agrawal, P., Kapoor, R., Agrawal, S.: A hybrid partial fingerprint matching algorithm for estimation of equal error rate. In: 2014 IEEE International Conference on Advanced Communications, Control and Computing Technologies, pp. 1295–1299 (2014)
36. Robbins, H., Monro, S.: A stochastic approximation method. Ann. Math. Stat., 400–407 (1951)
37. Kingma, D.P., Ba, J.: Adam: a method for stochastic optimization (2015)
38. Liu, Z.: Super convergence cosine annealing with warm-up learning rate. In: CAIBDA 2022, 2nd International Conference on Artificial Intelligence, Big Data and Algorithms, Nanjing, China, 17-19 June 2022, pp. 1–7. VDE/IEEE (2022)
39. Deng, J., Dong, W., Socher, R., Li, L.-J., Li, K., Fei-Fei, L.: Imagenet: a large-scale hierarchical image database. In: 2009 IEEE Computer Society Conference on Computer Vision and Pattern Recognition (CVPR 2009), 20-25 June 2009, Miami, Florida, USA, pp. 248–255. IEEE Computer Society (2009)
40. van der Maaten, L., Hinton, G.: Visualizing Data using t-SNE. J. Mach. Learn. Res. **9**(86), 2579–2605 (2008)

Evaluating Popular Scene Text Detection and Recognition Methods on Tombstones

Mathias Seuret[1(✉)], Oliver Traub[1], Ning Guo[2], Florian Kordon[1], Thomas Gorges[1], and Vincent Christlein[1]

[1] Friedrich-Alexander-Universität Erlangen-Nürnberg, Martensstr. 3, 91058 Erlangen, Germany
{mathias.seuret,oliver.traub,florian.kordon,thomas.gorges, vincent.christlein}@fau.de
[2] International Audio Laboratories Erlangen, Erlangen, Germany
ning.guo@audiolabs-erlangen.de

Abstract. The preservation and analysis of tombstones is important for historical, cultural, and artistic research. However, manually extracting information from photographs of these tombstones is laborious, which makes the use of automatic analysis methods desirable. We consider three types of tasks: text detection, text recognition, and end-to-end detection-recognition. For each of these tasks, we select two popular methods and evaluate their performance on two tombstone datasets. We use a dataset of German tombstones, featuring a wide variety of styles, and another dataset consisting of Dutch tombstones with more uniform designs. Prior to evaluation, we fine-tune each model on the training subset of the corresponding dataset. The results show that current methods can handle simple datasets such as the Dutch one, achieving detection and recognition error rates below 5%. In contrast, more complex data still pose as challenging, requiring improvements in model architecture, pre-training, or additional fine-tuning data. Furthermore, the performance of end-to-end systems can be considered as insufficient, with character recognition error rates around 20% even in the best-case scenarios. As such, while they offer acceptable results on some of the easiest data, current methods are not yet mature enough for reliable automated text extraction from these types of images.

Keywords: tombstone · ocr · text recognition · historical document

1 Introduction

Recognition of written characters was one of the earliest applications of machine learning [1]. Initial research on Optical Character Recognition (OCR) focused primarily on automatic transcription of printed and handwritten text, such as postal item processing [2]. OCR systems are now widely available as commercial and open-source solutions. They typically combine two steps: localizing text within an image and transcribing it by recognizing the individual characters.

L. Jin et al. (Eds.): ICDAR 2025 Workshops, LNCS 16225, pp. 209–225, 2026.
https://doi.org/10.1007/978-3-032-09368-4_13

However, most such systems are designed for text that is intentionally easy to read, such as in documents, letters, signs, or advertisements. In contrast, many real-world texts, such as those in brand logos, restaurant signage, or decorative displays, prioritize aesthetics or visibility rather than readability. These texts are often affected by poor contrast, uncommon fonts, oblique angles, varying lighting conditions, or other distortions, that are common in outdoor environments.

Scene Text Detection (STD) and Scene Text Recognition (STR) methods address the challenges of detecting and transcribing such text "in the wild". The characteristics of this type of text are similar to those found on tombstones. Tombstone inscriptions often suffer from weathering, erosion, and the use of outdated fonts not present in modern training data. Automatically detecting and transcribing such inscriptions is essential for preserving the historical information these stones contain before it becomes illegible or is lost due to grave replacements.

Text present on tombstone may be valuable for future historical, genealogical, or demographic research. Given the large number of existing tombstones, manual transcription is not feasible. In this work, our objective is to evaluate state-of-the-art methods for STD and STR, as well as end-to-end systems that integrate both tasks. We also compare these methods with a widely used OCR system, EasyOCR[1], with the goal of identifying their strengths and weaknesses in this specific, yet important, application context.

1.1 Related Work

A project focusing on digitization of Taiwanese tombstones to preserve their information before stone degradation makes it illegible has been running for more than a decade [3]. Besides storing image data, they also preserve information (metadata and transcriptions) in the TEI[2] format. Another platform, Cemo-Memo [4] aims at collecting tombstone images and uses crowdsourcing to extract information from these images. More recently, Scius-Bertrand *et al.* worked on layout analysis [5]. They first applied pixel-wise labeling using a Tiramisu network [6], and then applied a seam carving-based method [7] for the segmentation of text columns. They later developed an annotation-free character detection method [8] using an iterative process to refine the detection candidates. Bos *et al.* [9] proposed a system to extract text and semantical information from a dataset of Dutch tombstones [10] (different from the one we used). They utilized a YOLOv5 detector, PaddleOCR, [11] followed by a set of rules for semantic interpretation.

2 Datasets

In this work, we use two datasets: the first consists of images of German tombstones (hereafter referred to as the German dataset), and the second consists

[1] https://www.jaided.ai/easyocr/.
[2] https://www.tei-c.org/.

Fig. 1. Samples from the German tombstone dataset.

of images of Dutch tombstones (referred to as the Dutch dataset). Each image in those two datasets was annotated manually using CVAT[3], with word-level annotations, consisting of sequences of bounding boxes, each containing a single word, and the respective contained text. These annotations are stored and used as the ground truth in the training and evaluation.

2.1 The German Dataset

The German dataset consists of 306 images. They were collected from three different graveyards in Bavaria. The images feature a great variety of tombstones, ranging from tombstones erected in the last ten years to tombstones that are already a century old. The data has been split into three subsets for training, validation, and testing, with ratios of 55%, 10%, and 35%. Each subset contains images of tombstones featuring a wide variety of styles. As illustrated with a few samples in Fig. 1, these tombstones have different materials, textures, fonts, colors, font backgrounds, various contrasts and luminosities. Additionally, the light and photography angles are not always optimal, which introduces perspective and image quality challenges. Moreover, the preservation of the text is not always optimal, especially for older stones, as they are frequently degraded due to weathering.

[3] https://www.cvat.ai/.

2.2 The Dutch Dataset

The second dataset, referred as the Dutch dataset, consists of 151 images. It contains pictures from various graveyards, as the data was collected from an online database of Jewish tombstones.[4] After downloading the images from this repository, we annotated them in the same manner as for the German dataset. In these images, only the Dutch text was annotated; all Hebrew script was ignored.

The appearances of the tombstones in this dataset are less diverse than in the German dataset, with most of them displaying black text on a grey or white background. Some representative samples are shown in Fig. 2.

There is little variation in the visual appearance of the text, which is typically rendered in either a standard bold font or a cursive font imitating handwriting. The structure of the text is also consistent: it often begins with "Hier rust" (meaning "here rests"), followed by the name of the deceased, the date of birth, and the date of death. The dates are sometimes given in the Gregorian calendar, but often in the Jewish calendar; in some cases, both formats are present.

A few images deviate from these patterns. These are usually older, black-and-white photographs or images with a yellowish tint. Some of these older images depict tombstones with different fonts and visible erosion, which has degraded the legibility of the text over time.

We split this dataset into three subsets for training, validation, and testing, with ratios of 45%, 10%, and 45%, respectively.

2.3 The Synthetic Dataset

We also generate a synthetic dataset consisting of German text samples. To achieve this, we used a fork[5] of the code from [12], a method to insert text into natural scenes using inpainting. The fork allows different text databases and languages to be used. In this way, we created a synthetic dataset containing 10000 images, with up to four text instances inserted into each image. The text is written in various orientations, and with multiple different fonts. We then split the dataset into two subsets for training and validation, with ratios of 95% and 5%, respectively. We used this dataset during the fine-tuning phase of the models, and did not include it in evaluations.

3 Scene Text Analysis Methods

European tombstones typically contain one or several names, as well as dates. Some contain images (e.g., portraits or visual elements linked to the person's life), however they are not very frequent. Thus, a majority of the relevant information can be obtained solely by extracting the text. In scene text, this is typically done either in two steps (text detection, then text recognition) or at once in an end-to-end manner. In this section, we present the evaluation of different methods, grouped by the task they solve: scene text detection, scene text recognition, and end-to-end methods encompassing both tasks.

[4] https://nld.stenenarchief.nl/het-project.

[5] https://github.com/LaJoKoch/SynthTextGerman.

Fig. 2. Samples from the Dutch tombstone dataset.

3.1 Scene Text Detection

MixNet. MixNet [13] proposes an alternative to the typical Vision Transformer (ViT) architecture [14], called Feature Shuffle Network (FSNet). FSNet is combined with a Transformer block [15] referred to as the Central Transformer Block (CTBlock), which is designed to detect the center line of the text in an image, i. e., the line running through the middle of the text from the first to the last character.

FSNet consists of multiple paths combining convolutional layers, downsampling operations, and shuffle blocks. The shuffle block divides an input features X_i into n segments (where n is the number of input feature maps). Each input is split into n parts X_i^k, upsampled or downsampled to match the dimensions of the i-th input, and then reassembled by stacking all k-th splits to form the k-th output.

The downsampling block consists of a convolutional layer with kernel size 3 and stride 2, followed by batch normalization and a ReLU activation. The convolutional block comprises two consecutive convolutional layers with kernel size 3, each followed by batch normalization and ReLU activation.

The outputs of all FSNet branches are upsampled to the same dimensions and concatenated. From these combined feature maps, text contours are extracted. A set of contour points is sampled from these contours and passed, along with the feature maps, to the CTBlock.

The CTBlock consists of two Transformer layers. The first predicts a set of center line points from the contour features. These predicted points, along with the contour points, are then refined in the second Transformer block to produce the final text line prediction.

SRFormer. The SRFormer [16] model adopts a different strategy for text detection by combining a segmentation-based approach with a regression step to refine the control points defining the polygons of detected text instances. The model employs a backbone network to extract initial feature representations from the input image. These features are then passed to a Transformer encoder, together with positional embeddings, in a manner similar to the ViT.

From the encoded features, the model samples the Top-K bounding boxes, which are used to construct the corresponding polygons. These polygons, together with the initial Top-K queries, serve as input to form the decoder queries. The decoder receives these queries along with the previously extracted features.

The decoder first performs a coarse segmentation to generate initial masks for the text instances. A subsequent regression module then refines the control points of these masks. Finally, the model outputs a set of predicted control points for each detected text instance, along with a confidence score.

3.2 Scene Text Recognition

PARSeq. PARSeq [17], short for Permutated Auto-Regressive Sequence, is a model architecture that combines a ViT for image encoding with a Transformer-based decoder named the Visio-lingual Decoder. The architecture uses attention masks to enforce learning of relationships between arbitrary subsets of sequence elements, rather than relying solely on sequential (left-to-right) dependencies.

The ViT encoder consists of 12 layers and uses non-standard patch dimensions of 8×4 pixels instead of the more commonly used squares (e.g., 16×16). Rather than selecting a single representation, the classification head is removed, and all output tokens from the final Multi-Head Attention (MHA) block are retained.

The Visio-lingual Decoder applies attention masks that vary between training and inference. These masks allow the decoder to model relationships between different positions in the output sequence.

During training, a combination of standard left-to-right masks and randomly permuted attention masks is used.

During inference, three masking modes are available:

(1) Left-to-right mask for autoregressive decoding,
(2) No mask for non-autoregressive decoding,
(3) Cloze mask for iterative refinement.

Empirically, AR decoding achieves the highest accuracy, though it is computationally more expensive since it requires up to N decoding steps, where N is the maximum supported sequence length.

The decoder operates in two stages:

(1) Self-attention is computed between the input context X and positional queries (which function as positional embeddings), modulated by the chosen attention mask.
(2) Cross-attention follows between the output of the self-attention layer and the ViT-encoded image features, using an additional MHA layer.

The final output is generated by a multilayer perceptron followed by a linear layer, yielding a prediction tensor $y \in \mathbb{R}^{(N+1)\times(S+1)}$, where S is the size of the character set and the additional symbol represents the end-of-sequence token. The length $N+1$ ensures that the model can always produce an explicit marker at the end of the sequence.

CLIP4STR. The CLIP4STR [18] architecture builds upon the PARSeq model, with the main difference being the replacement of the ViT by a pre-trained CLIP [19] vision encoder. The goal is to leverage the general visual-textual knowledge learned by CLIP from large-scale training on hundreds of millions of image–text pairs, which likely include more varied textual content than what is found in typical STR datasets comprising only a few tens of thousands of images.

Instead of using a single decoder, CLIP4STR introduces two parallel decoders with identical architecture but different inputs: a visual decoder and a cross-modal decoder.

- The visual decoder takes as input a set of positional queries p (serving a similar role as positional embeddings) and the visual features f_V from the CLIP vision encoder. It outputs a prediction y_V.
- The cross-modal decoder receives a concatenation of the visual features f_V and the output y_V of the visual decoder, along with the features f_T from the CLIP text encoder and the positional queries p, producing the final output y.

During training, both decoders contribute to the objective function, using a combined cross-entropy loss:

$$L = \mathrm{CE}(y_V, \hat{y}) + \mathrm{CE}(y, \hat{y}) \quad (1)$$

This dual-decoder strategy improves performance on several STR benchmarks compared to prior models. However, it comes at the cost of significantly increased inference time – up to 5× slower for the largest variant of the model.

3.3 End-to-End Methods

SWINTS. The SwinTextSpotter [20] architecture, or shorter SWINTS, is inspired by MaskTextSpotter [21] and employs a modified Swin Transformer [22] as its backbone for feature extraction. The model consists of four main components: the backbone (a Dilated Swin Transformer), the Text Detector, the Recognition Conversion module, and the Recognizer.

The Dilated Swin Transformer extends the original Swin Transformer by applying two dilated convolution layers with residual connections to each of its outputs. Based on an initial bounding box prediction $bbox_0$, RoI Align [23] is performed, producing feature maps f_0, which are passed to the Text Detector.

The Detector module includes a Transformer encoder, a flattening operation on its outputs, and a detection head that produces a refined bounding box $bbox_k$ along with a mask prediction $mask_k$. This detection process is repeated K times, with RoI features recomputed at each iteration using the latest bounding box $bbox_k$.

The Recognition Conversion module takes the final RoI features and the output of the Transformer encoder from the last detection step. It fuses this information into a new feature representation, which is then passed to the Recognizer. The Recognizer employs a Transformer architecture consisting of a Focal Transformer encoder [24] and a decoder identical to that used in MaskTextSpotter.

DeepSolo. DeepSolo [25] takes another approach by combining a backbone of either a ResNet [26], SwinTransformer or a Vision Transformer with Aggregation and Enhancement (ViTAE) with a typical transformer consisting of an encoder and decoder to predict the text instances in the image, with their associated text by predicting the Bézier points that form the center line of the text.

Another variant entitled DeepSolo++ [27] exist, which supports multiple types of scripts by adding an output layer for each script and a script classification to the output. The image first gets processed by the backbone, and the resulting features are passed to the encoder with a positional embedding added to them. The encoder's output features are used to sample some K sets of Bézier curve points. From these Bézier points, points on the curve, they create are sampled, which are then combined with another positional embedding and passed to an MLP to then be used as queries for the decoder. The key and values for the decoder are taken from the encoder output features. The decoder then predicts the outputs for the different subtasks using individual heads. The subtasks are the confidence score predictions, the character classifications, and the center and boundary points predictions.

4 Experimental Setup

OCR results on a dataset are typically better if transfer learning is applied. I.e., the model is already trained on some (similar if possible) data, and then fine-tuned for the considered dataset, rather than training from scratch [28]. For this reason, and also as our datasets comprise relatively small numbers of samples, we opted for this approach. Another advantage of pre-training is a shorter training time. By using pre-trained models, we were able to reach training time of a few hours only.

Table 1. Data augmentation methods used during training by each method.

	Applied augmentation								
Model	Crop.	Rot.	Shear.	Trans.	Resizing	Contrast	Brightness	Light	Satur.
MixNet	✓	✓	✓	✓	✓				
SRFormer	✓	✓			✓	✓	✓	✓	✓
SWINTS	✓	✓			✓	✓	✓		✓
DeepSolo	✓	✓			✓	✓	✓		✓
PARSeq		✓	✓	✓	✓				
CLIP4STR		✓	✓	✓	✓				

4.1 Training Methodology

For the methods used in this work, except for EasyOCR, pre-trained models and source codes for training and evaluation are available. Whenever possible, we used models pre-trained on the "TotalText" [29] dataset. These models were initially trained on large-scale synthetic datasets, containing hundreds of thousands to millions of images, and subsequently fine-tuned on standard benchmark datasets.

We applied different types of data augmentations: 1) cropping, 2) rotations, 3) shearing, 4) linear translation, 5) resizing, 6) contrast, 7) brightness, 8) lighting, and 9) saturation change. However, the publicly available training codes which we used did not use all of these augmentations. We assume that augmentation selections for these methods have already been optimized. Details are given in Table 1.

We adopted the original training parameters provided for each pretrained model and modified only the number of epochs to ensure convergence. The batch size was adjusted to fully utilize the available memory during training. In addition, several models were adapted to use a different recognition head, allowing for a character set tailored to our needs rather than the default one. This custom character set includes all uppercase and lowercase letters, digits, German umlauts, as well as punctuation such as brackets, periods, spaces, and dashes. The recognition head was replaced for all models performing text recognition, regardless of the dataset they were originally trained on, to ensure fair conditions for subsequent evaluations.

For evaluation purpose, we save the models when their smallest validation loss during training was reached. As early stopping was not easy to integrate in the various frameworks, we did not employ this method.

4.2 Metrics

For text detection, we useTedEval [30]. This tool computes an optimal matching between detections and the ground truth using not only the intersection over union, but also taking into account the difference of angles between bounding boxes. This allows to compute the precision and the recall.

In the case of the text recognition, we use the Normalized Edit Distance (NED). It is defined as:

$$\mathrm{NED}\,(p,t) = \frac{L\,(p,t)}{\mathrm{len}\,(t)} \tag{2}$$

where p is the predicted text, t is the ground truth, and L is the Levenshtein distance [31].

The evaluation of the end-to-end (E2E) methods was performed using End-toend Evaluation Metric (EEM) [32], which combines bounding box matching with text label scoring. First, ground truth and detected boxes are paired by selecting for each G_i the D_j with the largest intersection, and vice versa. Unmatched boxes are represented as (G_i, NULL) or (NULL, D_j). These pairs are then merged into sets by checking for shared elements to identify one-to-one, one-to-many, or many-to-many matches.

Each set is scored using the Normalized Score (NS), defined as:

$$\mathrm{NS} = 1 - \frac{\sum_i^N L(p_i, t_i)}{\sum_i^N \max(\mathrm{len}(t_i), \mathrm{len}(p_i))} \tag{3}$$

where N is the number of box sets. For sets with multiple boxes, labels are merged by ordering boxes left to right (by x-axis) and concatenating their texts.

This merging and scoring are done for all sets, and the final score is aggregated over the full dataset. Additionally, edit operations (insertions, deletions, substitutions) are analyzed per method to identify character-level errors. For further comparison, recognition-only methods and EasyOCR were evaluated using the bounding boxes predicted by each E2E method.

5 Experiments

For our experiments, the results are split by the type of tasks and by the dataset used. We first go over the results of the detection, then the recognition, and lastly, the E2E text recognition.

5.1 Text Detection

We measured the precision and the recall before and after the fine-tuning phase. The results are shown in Fig. 3. For this task, we see a clear improvement in the precision and/or recall for all models. While the recall sometimes drops slightly, the F1-score does always improve.

The recall did not change much, except for SRFormer and DeepSolo, however, the number of false detections dropped significantly, which improved the precision. We can see similar trends for both datasets, although MixNet has a slightly lower recall for the German dataset after fine-tuning. We believe this is due to some overfitting during the training, as MixNet struggles with overlapping bounding boxes. The results on the Dutch dataset, especially with SRFormer,

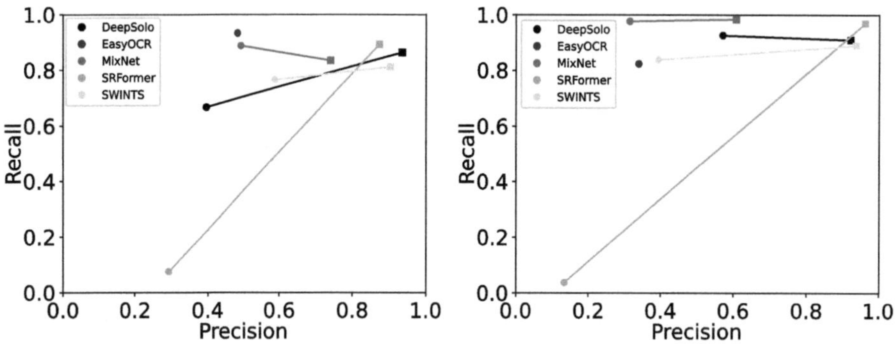

Fig. 3. Evaluation on the effect of fine-tuning the different models. The left plot is for the German dataset, and the right plot for the Dutch one. Circles correspond to the performance before fine-tuning, while squares are the final results, i.e., after fine-tuning.

are almost perfect. However, there is still room for improvement in the case of the German dataset, especially for the recall.

During our experiments, we saw slightly lower precision and recall for dates compared to the rest of the text. We think that although dates are more strictly formatted and should be easier to detect, their relative lower quantity has a negative impact on the detection.

5.2 Text Recognition

The NED is calculated on the test sets of the two datasets. Moreover, as we want more granular details on the evaluation, we grouped characters into several categories, and computed the contribution to the NED for each of these categories. The categories are upper case, lower case, German umlaut (such as *ä*) in the case of the German dataset, and the rest (numbers, punctuation, etc.).

The results are shown in Fig. 4, where the total length of a bar corresponds to the NED for the corresponding model, and the different colors encode the type of character contributing to the error.

We can see that PARSeq and CLIP4STR vastly outperform EasyOCR, even when they are not trained on the corresponding data. Overall, the NED is quite good for the Dutch dataset, and slightly better than for the German dataset. We can also notice that fine-tuning PARSeq degrades its performance. Using the synthetic dataset during the training had either no effect or a slightly negative impact. We think this is due to the nature of this synthetic dataset being too different from real world data.

During our investigation, we also saw that CLIP4STR did sometimes hallucinate words or letters, leading to a NED being higher than 5 for some samples[6].

[6] For example, an instance of *wie* was transcribed as *vvieeeeeeeeeeeeeeeeeeeeee*, while other boxes from the tombstone were properly transcribed.

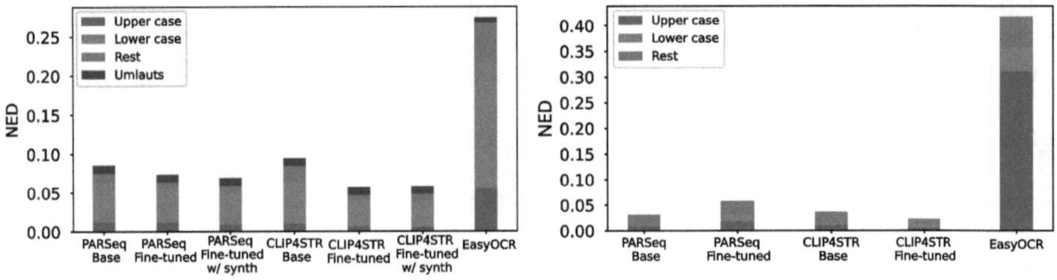

Fig. 4. Text recognition NED, with details of the contribution of different character types. Results for the German dataset are on the left, and for the Dutch one on the right.

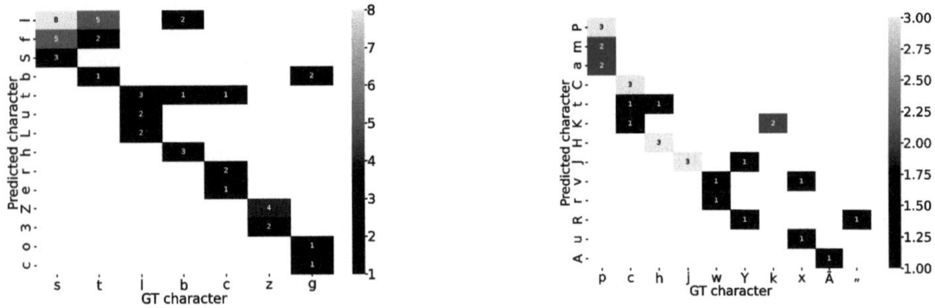

Fig. 5. Confusion matrix for some of the characters which get the most commonly mistaken by CLIP4STR. Left side for the German dataset, right side for the Dutch one.

We also computed confusion matrices for the most common character substitutions, i.e., when one character is recognized as another one. While this ignores needed deletions and insertions, it provides an interesting overview of how the methods perform and where they face difficulties. The confusion matrices for CLIP4STR are shown in Fig. 5.

We can see that many of the confusions come from mistakes between lower and upper cases. Some letters, such as *p*, have little differences between lower and upper case, and the variety of fonts makes their distinction challenging.

We can also note some mistakes which seem more difficult to understand at first glance, such as the *s* often mistaken for a *l*. We think this is due to the presence of *long s* in some of the training data.

We also saw that a large fraction of the German umlaut are not recognized properly, which can most likely be linked to the data quality. Some tombstones are covered by small degradation which look similar to dots, and this makes the recognition of the real dots more challenging.

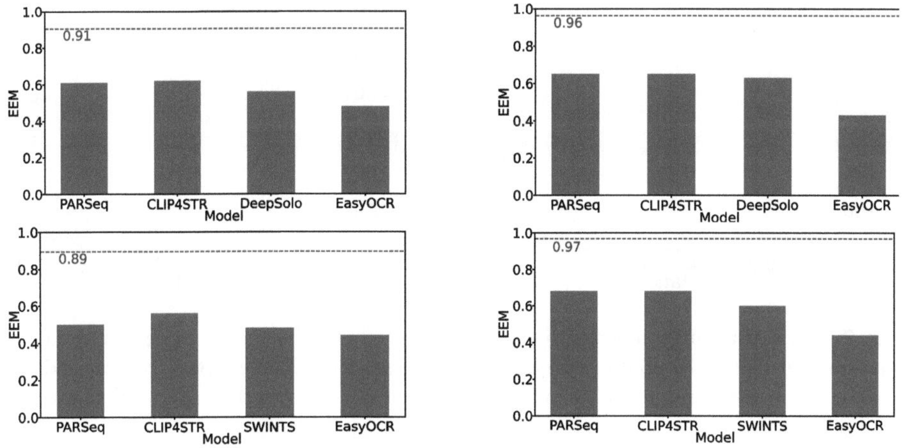

Fig. 6. EEM for DeepSolo (top) and SWINTS (bottom), for the German (left) and Dutch (right) datasets. The other methods are evaluated on the E2E's detections. The red line is the theoretical maximum EEM for the given detections.

5.3 End-to-End Methods

For the E2E methods, we measured three types of values for each dataset. First, the EEM obtained by the method itself. Second, the EEM obtained by giving the E2E method's detection result to the STR methods mentioned above. And, third, we computed the theoretical highest EEM which could be reached using the E2E method's detections, assuming that the detected bounding boxes match perfectly the ground truth. The results are given in Fig. 6.

We can see that the E2E methods have moderate results compared to the other ones, while there is a large room for improvement (see the dashed red line). The gap between these methods and what the STR methods obtained is rather small. EasyOCR, here again, obtains lower results.

Our investigations have shown, as illustrated in Fig. 7, that DeepSolo and SWINTS produce bounding boxes that cover typically between 80% and almost 100% of the matched ground truth box. Very few predicted boxes have a lower overlapping than 80%, with an exception of a fair amount which are marked with an overlapping of 0%. These are predicted boxes without match in the ground truth, i.e., false detections. The results are similar for both datasets, thus we show it only for the German dataset.

6 Discussion

In this paper, we evaluated the performance of various STD and STR models on tombstone images and compared them to EasyOCR, an off-the-shelf solution.

We evaluated two detection models, MixNet and SRFormer, and the detection part of two E2E models, DeepSolo and SWINTS, using precision and recall

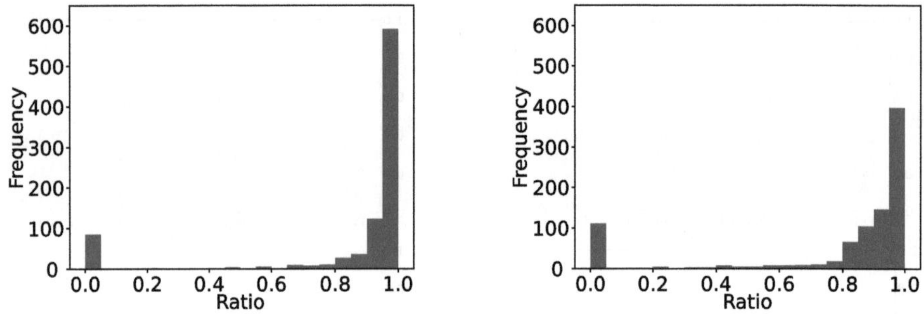

Fig. 7. Histogram of ground truth coverage by predicted bounding boxes for the German dataset using DeepSolo (left) and SWINTS (right). The left peak corresponds to undetected boxes, and most detected boxes have an intersection over union greater than 80%.

measurements. We also evaluated two recognition-only models, PARSeq and CLIP4STR, using NED, analyzing per-character error contributions and confusion patterns. We also ran experiments on the two E2E methods, DeepSolo and SWINTS, and additionally replaced parts of their pipeline by the recognition models.

On the Dutch dataset, most detection models achieved precision and recall near 99, except MixNet, which underperformed due to overlapping bounding boxes. All models outperformed EasyOCR, which showed high recall but low precision. On the German dataset, precision and recall dropped to 80–85%, with MixNet again performing worst. All models still exceeded EasyOCR's precision.

Regarding the task of text recognition, PARSeq and CLIP4STR achieved NED $\approx$ 10% on German data, improved by $\approx$ 1% with pre-training on synthetic text. Both outperformed EasyOCR (NED $\approx$ 26%). Umlauts and uppercase errors persisted; lowercase, digits, and punctuation improved. Umlaut confusions likely stem from rarity and visual similarity. Potential remedies include weighted loss functions or corrections using language models, although these approaches are limited when dealing with proper names, particularly given the historical variation in their orthography.

CLIP4STR slightly outperformed PARSeq but requires more resources. It occasionally hallucinates long outputs. Most NED values are below 0.5, with 75% of the samples recognized perfectly.

Dutch data showed better recognition: NED ranged 0.02–0.05, versus 0.4 for EasyOCR. Error categories were minimal; most samples had few or no errors. STR models handled this domain well despite smaller dataset size, suggesting potential improvements for German with additional data. Current recognition models are suitable for Dutch, not German.

E2E recognition using detected boxes yielded poor results: only 60% character accuracy, with false detections lowering maximum achievable rates. On Dutch data, best-case recognition was 96–97%; on German, 89–91%. Even matched detections showed 20–30% recognition error, making them Pimpractical. We expect that manually correcting data with such high error rates would take as long as a fully manual transcription of the inscriptions.

7 Conclusion

We evaluated the feasibility of using modern STD and STR methods to transcribe tombstone text. We selected six models: two each for detection-only, recognition-only, and E2E tasks, and fine-tuned them on annotated images of real tombstones.

Detection models performed well on the Dutch dataset, with error rates around 1%, but struggled with the more complex German dataset, achieving about 80% precision and recall. MixNet underperformed across both datasets, likely due to training issues caused by overlapping bounding boxes.

Recognition models (CLIP4STR and PARSeq) achieved 2–5% error on the Dutch dataset and 8–10% on the German dataset, with CLIP4STR performing slightly better but prone to hallucination in rare cases. Pre-training with synthetic German text reduced the error by 1%.

E2E recognition performed significantly worse, achieving EEM scores of only 0.6 to 0.7 on data where 0.9 was expected. The performance drop is likely due to overlapping and fragmented bounding boxes, leading to incorrect label associations.

In conclusion, current methods are insufficient for reliable full transcription of diverse tombstone images with small datasets. However, they may assist manual annotation by automatically detecting text regions, accelerating dataset creation for future model training.

References

1. Fukushima, K.: Cognitron: a self-organizing multilayered neural network. Biol. Cybern. **20**(3), 121–136 (1975)
2. LeCun, Y., et al.: Backpropagation applied to handwritten zip code recognition. Neural Comput. **1**(4), 541–551 (1989)
3. Streiter, O., Goudin, Y., Huang, C.J.: Thakbong, digitalizing Taiwan's tombstones for teaching, research and documentation. In: TELDAP 2010-The International Conference on Taiwan e-Learning and Digital Archives Program, pp. 146–157 (2010)
4. May, M.J., Kantor, E., Zror, N.: CemoMemo: making more out of gravestones (with help from the crowd). J. Comput. Cultural Heritage (JOCCH) **14**(4), 1–22 (2021)
5. Scius-Bertrand, A., Voegtlin, L., Alberti, M., Fischer, A., Bui, M.: Layout analysis and text column segmentation for historical Vietnamese steles. In: Proceedings of the 5th International Workshop on Historical Document Imaging and Processing, pp. 84–89 (2019)
6. Jégou, S., Drozdzal, M., Vazquez, D., Romero, A., Bengio, Y.: The one hundred layers tiramisu: fully convolutional densenets for semantic segmentation. In: Proceedings of the IEEE Conference on Computer Vision and Pattern Recognition Workshops, pp. 11–19 (2017)

7. Alberti, M., Vögtlin, L., Pondenkandath, V., Seuret, M., Ingold, R., Liwicki, M.: Labeling, cutting, grouping: an efficient text line segmentation method for medieval manuscripts. In: 2019 International Conference on Document Analysis and Recognition (ICDAR), pp. 1200–1206. IEEE (2019)
8. Scius-Bertrand, A., Jungo, M., Wolf, B., Fischer, A., Bui, M.: Annotation-Free Character Detection in Historical Vietnamese Stele Images. In: Lladós, J., Lopresti, D., Uchida, S. (eds.) ICDAR 2021. LNCS, vol. 12821, pp. 432–447. Springer, Cham (2021). https://doi.org/10.1007/978-3-030-86549-8_28
9. Bos, J., Marocico, C., Tatar, A.E., Mzayek, Y.: Automatically interpreting Dutch tombstone inscriptions. Comput. Linguist. Netherlands J. **12**, 253–267 (2022)
10. Bos, J.: A semantically annotated corpus of tombstone inscriptions. Int. J. Digit. Hum. **3**(1), 1–33 (2022)
11. Cui, C., et al.: Paddleocr 3.0 technical report (2025)
12. Gupta, A., Vedaldi, A., Zisserman, A.: Synthetic data for text localisation in natural images. In: Proceedings of the IEEE Conference on Computer Vision and Pattern Recognition, pp. 2315–2324 (2016)
13. Zeng, Y.-X., Hsieh, J.-W., Li, X., Chang, M.-C.: Mixnet: toward accurate detection of challenging scene text in the wild. arXiv preprint arXiv:2308.12817 (2023)
14. Dosovitskiy, A., et al.: An image is worth 16x16 words: Transformers for image recognition at scale. arXiv preprint arXiv:2010.11929 (2020)
15. Vaswani, A., et al.: Attention is all you need. In: Advances in Neural Information Processing Systems, vol. 30 (2017)
16. Qingwen, B., Park, S., Khang, M., Cheng, Y.: Srformer: text detection transformer with incorporated segmentation and regression. In: Proceedings of the AAAI Conference on Artificial Intelligence **38**, 855–863 (2024)
17. Bautista, D., Atienza, R.: Scene text recognition with permuted autoregressive sequence models. In: Avidan, S., Brostow, G., Cissé, M., Farinella, G.M., Hassner, T. (eds.) ECCV 2022. LNCS, pp. 178–196. Springer (2022). https://doi.org/10.1007/978-3-031-19815-1_11
18. Zhao, S., Wang, X., Zhu, L., Yang, Y.: Clip4str: a simple baseline for scene text recognition with pre-trained vision-language model. arXiv preprint arXiv:2305.14014 (2023)
19. Radford, A., et al.: Learning transferable visual models from natural language supervision. In: International Conference on Machine Learning, pp. 8748–8763. PmLR (2021)
20. Huang, M., et al.: Swintextspotter: scene text spotting via better synergy between text detection and text recognition. In: proceedings of the IEEE/CVF Conference on Computer Vision and Pattern Recognition, pp. 4593–4603 (2022)
21. Lyu, P., Liao, M., Yao, C., Wu, W., Bai, X.: Mask textspotter: an end-to-end trainable neural network for spotting text with arbitrary shapes. In: Ferrari, V., Hebert, M., Sminchisescu, C., Weiss, Y. (eds.) ECCV 2018. LNCS, vol 11218, pp. 67–83. Springer, Cham (2018). https://doi.org/10.1007/978-3-030-01264-9_5
22. Liu, Z., et al.: Swin transformer: hierarchical vision transformer using shifted windows. In: Proceedings of the IEEE/CVF International Conference on Computer Vision, pp. 10012–10022 (2021)
23. He, K., Gkioxari, G., Dollár, P., Girshick, R.: Mask R-CNN. In: Proceedings of the IEEE International Conference on Computer Vision, pp. 2961–2969 (2017)
24. Yang, J., et al.: Focal self-attention for local-global interactions in vision transformers. arXiv preprint arXiv:2107.00641 (2021)

25. Ye, M., et al.: Deepsolo: let transformer decoder with explicit points solo for text spotting. In: Proceedings of the IEEE/CVF Conference on Computer Vision and Pattern Recognition, pp. 19348–19357 (2023)
26. He, K., Zhang, X., Ren, S., Sun, J.: Deep residual learning for image recognition. In: Proceedings of the IEEE Conference on Computer Vision and Pattern Recognition, pp. 770–778 (2016)
27. Ye, M., et al.: Deepsolo++: let transformer decoder with explicit points solo for multilingual text spotting. arXiv preprint arXiv:2305.19957 (2023)
28. Reul, C., Springmann, U., Wick, C., Puppe, F.: Improving OCR accuracy on early printed books by combining pretraining, voting, and active learning. J. Lang. Technol. Comput. Linguist. **33**(1), 3–24 (2018)
29. Ch'ng, C.K., Chan, C.S., Liu, C.: Total-text: towards orientation robustness in scene text detection. Int. J. Doc. Anal. Recogn. (IJDAR) **23**, 31–52 (2020)
30. Lee, C.Y., Baek, Y., Lee, H.: Tedeval: a fair evaluation metric for scene text detectors. In: 2019 International Conference on Document Analysis and Recognition Workshops (ICDARW), vol. 7, pp. 14–17. IEEE (2019)
31. Levenshtein, V.I., et al.: Binary codes capable of correcting deletions, insertions, and reversals. In: Soviet physics doklady, vol. 10, pp. 707–710. Soviet Union (1966)
32. Hao, J., et al.: EEM: an end-to-end evaluation metric for scene text detection and recognition. In: Lladós, J., Lopresti, D., Uchida, S. (eds.) ICDAR 2021. LNCS, vol. 12824, pp. 95–108. Springer, Cham (2021). https://doi.org/10.1007/978-3-030-86337-1_7

Deep Learning for Defect Detection in Answer Document Image

Jun Xie[1], Yiming Xia[2], Sailong Wu[3], Ruiqing Wu[1,4](✉), and Yirong Chen[1]

[1] School of Information and Communication Engineering, University of Electronic Science and Technology of China, Chengdu 611731, China

[2] School of Integrated Circuit Science and Engineering, University of Electronic Science and Technology of China,Chengdu 611731, China

[3] Research and Development Institute of Oilfield Technology, China Oilfield Services Limited, Langfang, China

wusl5@cnooc.com.cn

[4] National Key Laboratory on Blind Signal Processing, University of Electronic Science and Technology of China,Chengdu 611731, China

rqwu@uestc.edu.cn

Abstract. Defects, such as stains and scratches, in the answer document image can degrade the image quality and seriously reduce the accuracy of text and mark recognition. The existing methods to detect these defects are difficult due to weak ability to suppress interference from the complex background, and result in lower detection accuracy. This paper proposes a new method that deeply explores visual features to detect defects using deep learning. Firstly, the Oriented Bounding Box (OBB) is used to minimize the bounding rectangle for abnormal objects with flexible shapes. Secondly, a new convolution module combining the Space-to-Depth Convolution (SPDConv) is designed to improve the feature extraction ability for objects with complex backgrounds. In addition, the attention mechanism with Multi-Scale Context Aggregation (MSCA) is employed to effectively capture multi-scale information and establish dependent relationships between remote pixels. In order to detect long-line defect objects, the P6 detector head is introduced to expand the receptive field range. Finally, the proposed method is integrated into YOLOv8 to verify defect detection. The experimental results show that the proposed method can effectively detect the four types of defects in answer document images, and theÂămAP_{50}Âăreaches 92.1% on our datasetÂă(The dataset is available a https://github.com/micangdao/defectanswersheetimage).

Keywords: answer document image · defect detection · space-to-depth convolution · multiscale context aggregation · oriented bounding box · visual features

1 Introduction

In recent years, automated marking systems are gradually replacing the time-consuming traditional marking methods due to the development of artificial

L. Jin et al. (Eds.): ICDAR 2025 Workshops, LNCS 16225, pp. 226–244, 2026.
https://doi.org/10.1007/978-3-032-09368-4_14

intelligence, and have become an important means of improving marking efficiency. However, defects, including stains and scratches, are frequently observed in answer document images. These defects pose a great challenge to the quality of automated marking and scoring. Therefore, defect detection in answer document images is crucial for enhancing automatic marking accuracy, preventing errors, and enabling manual correction of misjudgments.

Based on visual morphology and structural characteristics, we classify the defects into four major categories: rubbing smudge, cloud blemish, pen scratch and scanned line. Rubbing smudges are residual marks caused by candidates repeatedly erasing and modifying filled-in answers during the examination process. Such defects generally occupy small regions (less than 30×30 pixels) and demonstrate notably low grayscale levels. Cloud blemishes arise due to various factors, such as the storage, scanning, compression or dragging of answer documents. These actions can cause the original ink or pen markings to smudge and contaminate other areas of the document. As a result, cloud blemishes exhibit variability in size and shape, ranging from small to medium objects on the answer document image. Pen scratches primarily result from meaningless and irregular lines left on the answer document during the writing process, often caused by the careless handling or use of the pen. Scanned lines are thin artifacts caused by misalignment of the answer document during scanning. These defects typically are characterized by exceptionally large aspect ratios and varying levels of darkness. Due to the stochastic shaking of the scanning process, scanned lines may appear at arbitrary locations throughout the document image. Examples of the four types of defects are shown in Fig. 1.

The shape variations of defects in answer document images, coupled with complex background interference, pose significant challenges to detection methods [1,2]. Furthermore, the industry typically deploys high-speed devices to achieve optimal production efficiency, performing quality assessment and defect detection concurrently with document image scanning and storage. A prime example is the Canon imageFORMULA DR-G1130, a widely adopted high-speed scanner capable of processing 130 pages per minute. Consequently, real-time defect detection and image preprocessing systems still face significant challenges in meeting ultra-low latency requirements. In order to improve the capability and speed of defect detection in complex backgrounds, in this paper, a new method for detecting defects based on YOLOv8 is proposed to enhance the ability to extract features of weak and multi-shaped targets. The method incorporates a novel Conv-SPD-Conv module that significantly improves low-resolution target feature extraction through a new spatial coding technique. To address diverse defect scales, we integrate an improved MSCA (Multi-Scale Context Aggregation) [3] attention mechanism into the detector, enhancing multi-scale feature learning and detection performance. To optimize scanned line detection, we integrate the P6 detector head into our detector. This modification expands the network's receptive field while significantly improving detection performance for ultra-long linear defects.

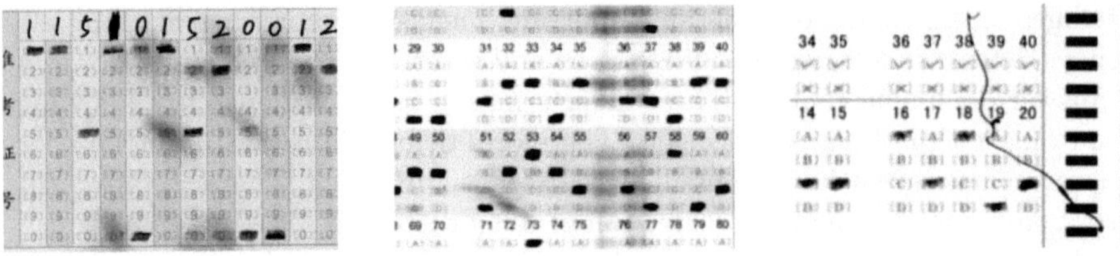

(a) rubbing smudge (b) cloud blemish (c) pen scratch

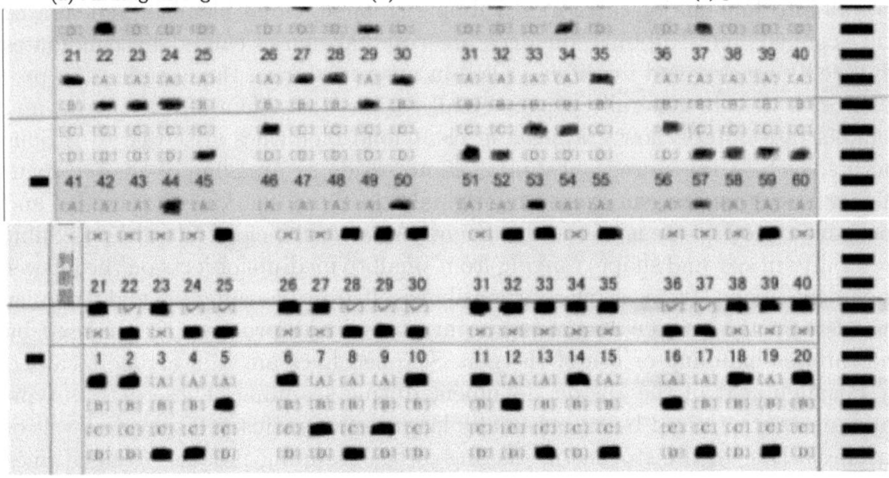

(d) scanned line

Fig. 1. Example images of the four types of defects.

2 Related Works

Target detection methods based on deep learning have become one of the effective ways to detect weak targets in complex environments due to their high-performance feature extraction. In recent years, target detection tasks have encountered challenges with objects of indeterminate shapes, particularly those that are narrow and elongated. The horizontal bounding boxes used for such targets often involve excessive background information. In order to achieve more accurate detection, some researchers designed the R2CNN model to support rotated target detection by improving the Faster R-CNN [4]. The rotated target detection is able to locate the target more accurately in the image by introducing an additional angle parameter. Since then, based on a variety of target detection methods, researchers have proposed some rotated target detection methods, such as Oriented R-CNN [5], Rotated Faster R-CNN [6], and YOLO-OBB [7].

In addition, to enhance the performance of target detection, researchers have proposed numerous feature enhancement modules or innovative network structures. For example, Pu et al. [8] propose an adaptive rotational convolution to extract target features with different orientations in an image. Compared with the traditional convolution, it has more accurate feature extraction capability.

Zhang et al. [9] proposed an efficient detector that combines feature enhancement, fusion and spatial context-awareness to address the problems of inadequate target feature representation and background confusion. On the other hand, reference [10] addresses the diversity and rich detail characteristics of document images by proposing a document tampering detection method based on spatial frequency domain and multi-scale feature fusion networks. Moreover, the application of Transformers [11] in computer vision has emerged as a significant research direction, with researchers developing several representative models: DETR [12], as the first end-to-end vision Transformer model, revolutionized object detection paradigms. Building upon this foundation, subsequent derivatives such as Swin Transformer [13], RT-DETR [14], and LW-DETR [15] have emerged as outstanding detectors. However, these models typically cannot match the real-time performance of YOLO-series models at smaller scales.

Many researchers have also applied deep learning-based target detection methods to the field of document image detection. C. Chen et al. [16] introduced an improved technique combining Frequency Domain Halftone Amplification (FHAG) with Band of Interest (BOI) localization, which enhances the generalization of document attack detection by optimizing frequency domain feature amplification. Addressing the challenges of irregular shapes, dense distribution, and significant aspect ratio variations in text lines, X. Hu et al. [17] proposed a universal framework, DB-SegHist, which significantly improves text detection accuracy through a segmentation reinforcement strategy. To tackle the structural complexity of ancient documents, A. Kundu et al. [18] innovatively combined the YOLO detection model with a rule-driven graph traversal algorithm, expanding the applicability of line segmentation methods in historical literature. M. Li et al. [19] proposed a Transformer-based document deblurring method for addressing stain interference generated during the digitization process of document images. This method precisely captures stain features at local, instance, and semantic levels while effectively preserving the integrity of document content. Furthermore, MSE. Kasem et al. [20] introduced a cutting-edge model based on the Swin-L Transformer backbone network combined with advanced Transformer mechanisms, further enhancing the performance of table detection in document images.

3 Principle of Proposed Method

3.1 Oriented Bounding Box Detection(OBB)

Defects in answer document images typically exhibit two key characteristics: arbitrary orientations due to scanning variations and cluttered backgrounds from handwritten content. These characteristics make oriented bounding boxes more effective than horizontal bounding boxes, as they provide tighter object encapsulation and reduce redundancy. Generally, an oriented bounding box is represented by five elements: center coordinates (x, y), width and height (w, h), and rotation angle θ.

In order to make the model better adapted to the regression task of oriented bounding box detection, we adopt probabilistic intersection over union (ProbIoU) [21] loss as the optimization objective to train on rotating rectangular frames. ProbIoU is a probability distribution-based IoU computation method, which, compared with the traditional IoU, is able to better handle the overlapping region between the rotated boxes, especially when the target has a large rotation angle or irregular shape.

ProbIoU transforms the ground-truth and predicted frames into a Gaussian distribution $\mathcal{G}(\mu, \Sigma)$, specifically:

$$\Sigma = \begin{bmatrix} \frac{1}{12}w^2cos^2\theta + \frac{1}{12}h^2sin^2\theta & \frac{1}{24}(w^2 - h^2)sin2\theta \\ \frac{1}{24}(w^2 - h^2)sin2\theta & \frac{1}{12}w^2cos^2\theta + \frac{1}{12}h^2sin^2\theta \end{bmatrix} \tag{1}$$

$$\mu = (x, y)^T \tag{2}$$

As shown in the above equations, by designing a network that regresses five parameters (x, y, w, h, θ), ProbIoU ensures that the shape and rotation angle of the predicted oriented bounding box more closely align with those of the ground truth box.

The similarity between two Gaussian distributions is obtained using the Bhattacharyya coefficient, and the Hellinger distance is introduced to bring the two distributions closer, thereby increasing the overlap between the ground-truth box and the predicted box. Therefore, ProbIoU and the final loss function ($L_{ProbIoU}$) are formulated as follows:

$$ProbIoU = 1 - \sqrt{1 - \int_{R_2} \sqrt{p(x)q(x)}dx} \tag{3}$$

$$L_{ProbIoU} = 1 - ProbIoU \tag{4}$$

In the expression, $p \backsim \mathcal{G}(\mu_1, \Sigma_1)$ and $q \backsim \mathcal{G}(\mu_2, \Sigma_2)$ represent the probability density functions of the Gaussian distributions for the ground truth box and the predicted box, respectively.

3.2 Enhanced Convolutional SPD-Conv

Most existing feature extraction networks usually use step-length convolution and pooling layers. However, step-length convolution and pooling layers are likely to cause the loss of fine-grained information when processing low-resolution small targets with complex backgrounds [22]. Answer document images typically exhibit homogeneous backgrounds, where defects are often deeply embedded among printed text, handwritten markings, filled-in boxes, form lines, and multi-colored background elements. This demands detection methods with robust feature extraction capabilities to accurately identify such defects.

SPD-Conv [23] combines a space-to-depth layer with a non-strided convolution layer. The space-to-depth layer transforms the feature map's spatial dimension into a depth dimension, preserving more information by increasing channel amount. This approach minimizes information loss and captures finer details, enhancing performance on complex tasks. The architecture of SPD-Conv is shown in Fig. 2, where Space-to-Depth Split rearranges the spatial blocks of the feature maps into four new feature maps, maintaining the same number of channels per map while reducing their spatial dimensions by a factor of two. Finally, the four feature maps are concatenated along the channel dimension, followed by a pointwise convolution to restore the original channel depth while preserving spatial resolution.

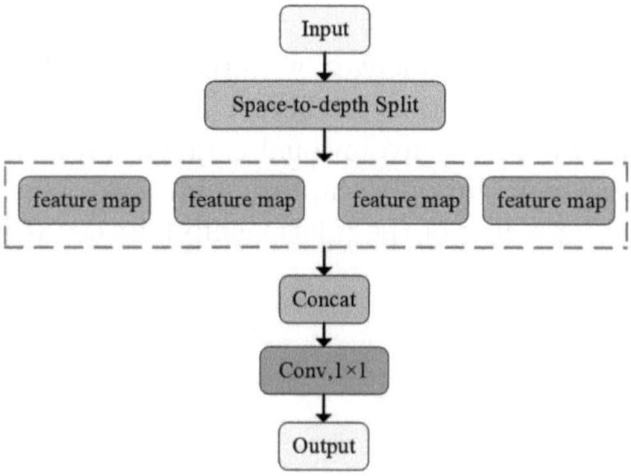

Fig. 2. The basic SPD-Conv.

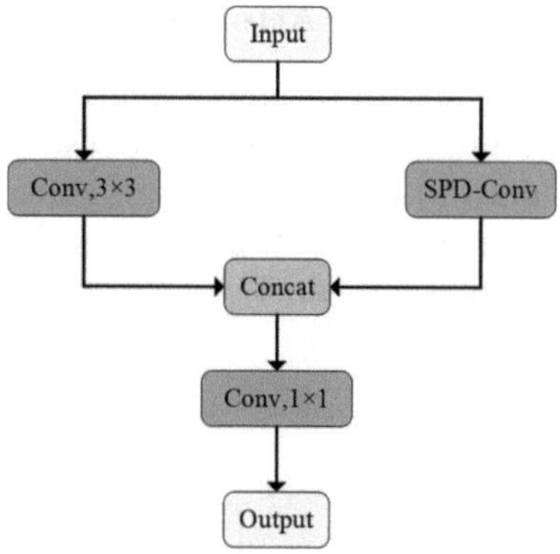

Fig. 3. Conv-SPD-Conv.

To improve feature extraction, we design a novel convolution module (Conv-SPD-Conv), which integrates a convolution block and SPD-Conv, and achieves effective fusion of local features and global context information through feature concatenation and channel adjustment. This module structure is depicted in Fig. 3.

First, the input feature map is processed through two branches. One branch is a 3×3 traditional convolution layer, which extracts local features, while the other branch employs the SPD-Conv module to capture more refined features. Next, the outputs of the two branches are concatenated to fuse local details with global context information. Finally, a 1×1 convolution is applied to adjust the channels and further integrate the features, producing the output feature map.

As shown in Fig. 2, for a feature map F_{in} with C channels and both height and width of S, after undergoing SPD-Conv, it is split to obtain four feature maps, each with a channel of C and both height and width of $\frac{S}{2}$, as described below:

$$\begin{aligned} f_{0,0} &= F_{in}[0:S:2, 0:S:2], \quad f_{1,0} = F_{in}[1:S:2, 0:S:2] \\ f_{0,1} &= F_{in}[0:S:2, 1:S:2], \quad f_{1,1} = F_{in}[1:S:2, 1:S:2] \end{aligned} \tag{5}$$

Subsequently, the four feature maps are concatenated and subjected to a non-strided convolution, resulting in a feature map $F_{SPDConv\_out}$ with a channel count of C and both height and width of $\frac{S}{2}$:

$$F_{SPDConv\_out} = Conv_{1\times1}([f_{0,0}, f_{1,0}, f_{0,1}, f_{1,1}]) \tag{6}$$

Thus, after passing through the enhanced convolution module Conv-SPD-Conv, the output of F_{in} is as follows, with its dimensions being $C \times \frac{S}{2} \times \frac{S}{2}$:

$$F_{out} = Conv_{1\times1}([Conv_{3\times3}(F_{in}), F_{SPDConv\_out}]) \tag{7}$$

3.3 MSCA Attention Mechanism

To cope with the above multi-scale detection challenges, we introduce the MSCA attention mechanism into the feature extraction backbone network. MSCA, shown in Fig. 4, consists of three parts: deep convolutional aggregation of local information, multi-branch depth-wise convolution to capture multi-scale context information, and 1×1 convolutional modeling of the relationship between different channels.

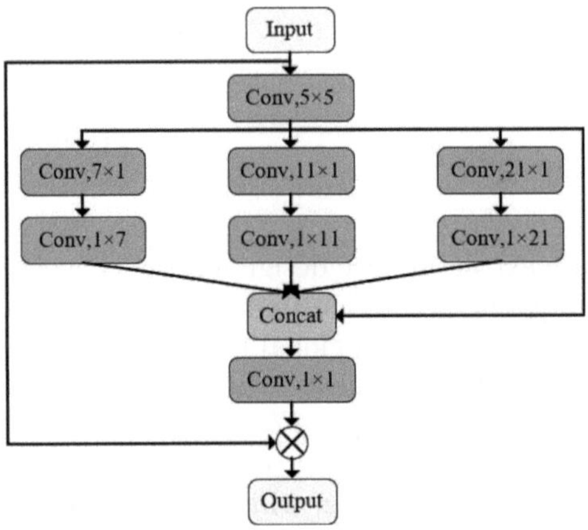

Fig. 4. Blocks of MSCA attention mechanism.

MSCA uses two deep strip convolutions to approximate the depth-wise convolution of a large convolution kernel, and the kernel sizes of each branch are set to 7, 11, and 21, respectively. Compared with the traditional convolution, the strip convolution has lighter weight and is more conducive to the detection of scratches and scanned lines, which are skewed towards elongated morphology. Furthermore, we concatenate the outputs of deep strip convolutions at multiple scales, preserving the complete information of multi-scale features. This approach ensures the retention of both fine details of smaller targets and the global context of larger targets, enhancing the overall detection capability.

3.4 Extended Detector Head

Among the defects in the answer document image, scanned lines are typical anomalous targets, which are characterized by elongated stripes and usually extend more than 800 pixels horizontally. The detection of such super-large-scale elongated targets faces two challenges. First, the conventional detection head is limited by the feature map resolution and receptive field range, which makes it difficult to completely capture the global structural features of the scanned lines. Second, the extreme aspect ratio of elongated targets introduces localization bias in conventional bounding box regression mechanisms [24].

In the YOLOv8 baseline architecture, the feature maps at each level of the feature pyramid are illustrated in Fig. 5, where P1 to P5 represent feature maps at different levels. The size of each feature map is half that of the previous layer. Therefore, for an input image of size $C \times S \times S$, the feature map size of P5 is $\frac{S}{32}$, with corresponding receptive fields of 32 pixels. However, for detecting ultra-long objects, such a receptive field range is often insufficient, making it difficult to fully capture the global feature information of large-scale targets.

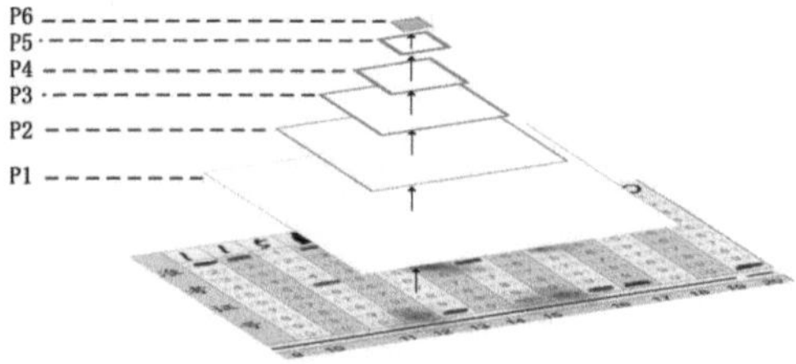

Fig. 5. The features pyramid in the YOLOv8 architecture.

Aiming at the above issues, we introduce a P6 feature layer with a size of $\frac{S}{64}$ into the feature pyramid, which enables the model's receptive field to expand up to 64 pixels. By expanding the model's receptive field and enhancing semantic abstraction in high-level features, the detector achieves more accurate detection of large-scale elongated targets with extreme aspect ratios.

As illustrated in Fig. 6, the P6 detection head comprises two branches, each consisting of two 3×3 convolutions and one 1×1 convolution. The extracted feature layers from P3 to P6 are passed through their respective detection heads to distill information, thereby obtaining the final object detection bounding boxes along with their corresponding categories.

3.5 Full Framework of Proposed Method

As shown in Fig. 7, the overall architecture of the answer document defect detection model proposed in this study is built on YOLOv8 as a benchmark framework, which is developed by Ultralytics and supports oriented bounding boxes with strong scalability and excellent performance.

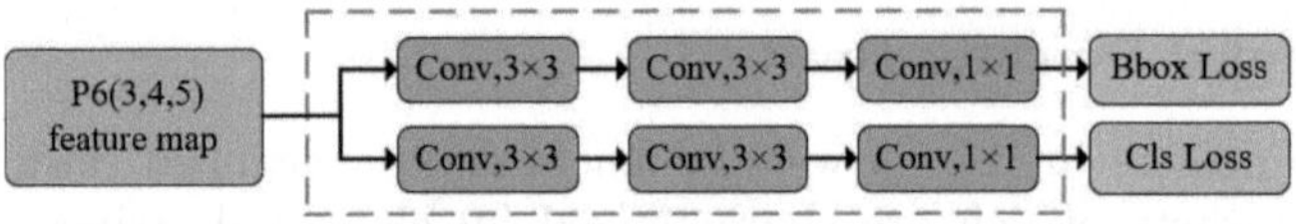

Fig. 6. The features pyramid in the YOLOv8 architecture.

Our proposed architecture is divided into three parts in total: feature extraction network, feature fusion network and rotating frame detection head. The feature extraction layer consists of four parts: the novel convolutional module Conv-SPD-Conv, the CSP Bottleneck with 2 Convolutions (C2f), the MSCA attention mechanism, and Spatial Pyramid Pooling Fast (SPPF). Among them, C2f can better aggregate multi-scale information by concatenating the outputs of different Bottleneck modules with the original feature maps. SPPF can effectively fuse feature information of different scales and improve the model's ability to recognize targets.The feature fusion uses a path aggregation network structure to fuse feature information at different scales through top-down and bottom-up approaches. The detection head uses a 1×1 convolutional layer to generate predictions on feature maps that have been downsampled by factors of 8, 16, 32, and 64, thereby producing prediction results across multiple scales.

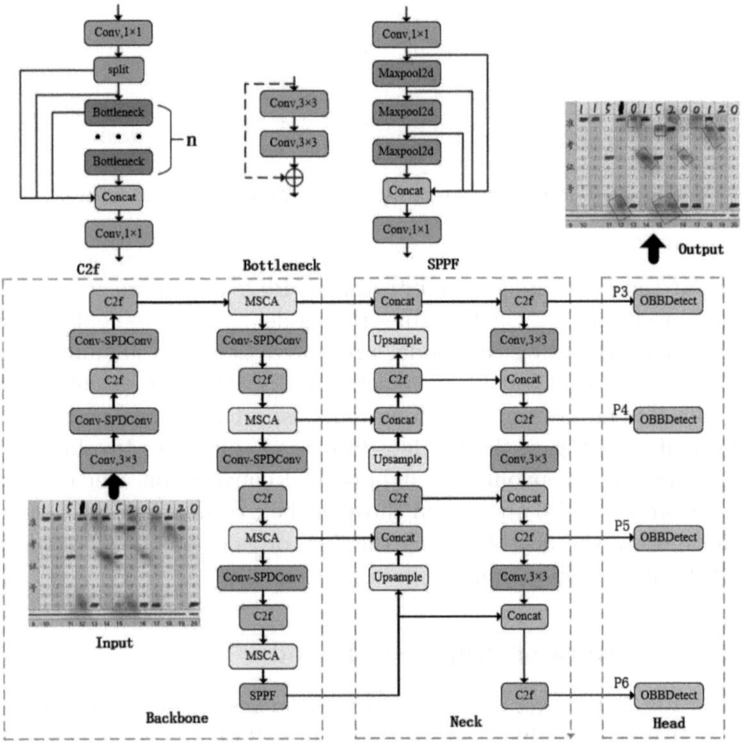

Fig. 7. Defect detection framework of answer document image.

4 Experiments and Analysis

4.1 Dataset

We conduct extensive experiments on three document image defect datasets, namely Document Defects Object Dataset (DocDefectOD), Synthetic Document Defects Object Dataset (SDocDefectOD), and SignverOD [25], to evaluate the effectiveness of our proposed defect detector.

DocDefectOD: We built this dataset comprising 11,162 authentic answer document images, with a total of 84,112 rotated bounding box annotations across four defects. Figure 8 plots the object size distribution of four defects in DocDefectOD. As can be seen, rubbing smudges are small in size, mostly within 30 × 30 pixels. Cloud blemishes exhibit a wide range of sizes, with the majority of concentrated in the lower-left quadrant of the image, and the pixel values covering a broad spectrum. The pen scratch and scanned line are more concentrated on the axes, with not only a high aspect ratio but also an extremely narrow and long shape. The specific number of images and labels for each type of defects are presented in Table 1 ("Images" represents the number of images containing the corresponding defects in dataset, and "Labels" represents the number of labels of the corresponding defects) (Fig. 8).

Table 1. Data Statistics of defects in DocDefectOD and SDocDefectOD

Defects	DocDefectOD		SDocDefectOD	
	Images	Labels	Images	Labels
rubbing smudge	1480	4388		
cloud blemish	8891	68208	1198	2317
pen scratch	992	1693	1198	2424
scanned line	7552	9823	781	1182

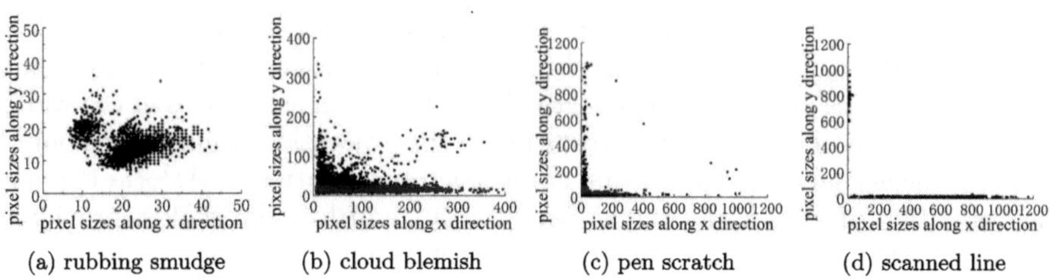

Fig. 8. Sizes distribution of the four types of defects in DocDefectOD.

SDocDefectOD: To enhance the diversity of defects and evaluate the robustness of the defect detector, we employed image processing algorithms to generate extensive simulated defects across over 100 styles of authentic document images. These synthetic defects demonstrate near-identical visual morphology and distribution characteristics compared to authentic defects. All synthesized document images were compiled into the SDocDefectOD. The specific number of images and labels for each type of defect are presented in Table 1.

SignverOD: This is an open-source dataset hosted on Kaggle, comprising a total of 2,576 scanned document images across four categories: signatures, initials, redactions, and handwritten dates. SignverOD covers a diverse set of document types including memos, emails, bank cheques, lease agreements, letters, and invoices. From SignverOD, we selected signatures, redactions, and handwritten dates as three types of defects in printed document images to evaluate the effectiveness of the detect detector.

DocDefectOD and SDocDefectOD will be randomly divided into 10% as the test set and the remaining data will be divided into the train set and validation set according to the ratio of 8:2 by using the cross-validation method.

4.2 Evaluation Metrics

Precision, Recall, mAP_{50}, and $mAP_{50:95}$ are employed as evaluation metrics. mAP_{50} means mean Average Precision at an intersection over union of 0.5, and $mAP_{50:95}$ denotes mean Average Precision over intersection over union values from 0.5 to 0.95 in steps of 0.05.

Precision measures the proportion of true positives among all predicted positives, while recall measures the proportion of true positives among all actual positives. These conflicting metrics make it hard to evaluate a model comprehensively. To address the imbalance between precision and recall, mAP_{50} provides a balanced evaluation metric by averaging precision at various recall levels at an IoU threshold of 0.5.

In particular,ÂămAP_{50}Âăevaluates the detection results through bounding boxes, allowing a certain degree of IoU tolerance. It is more tolerant of slight deviations in the width direction of linear targets, while focusing on the overall localization and recognition ability of the targets. This characteristic makes it especially suitable for the detection tasks of linear targets (pen scratches and scanned lines). In contrast, pixel-level metrics demand excessively high precision for the boundaries and are easily affected by factors such as changes in the width of the artifacts and edge blurring, resulting in the evaluation results being out of line with the actual detection effects. For a stricter evaluation, $mAP_{50:95}$ extends the evaluation process by averaging across a range of IoU thresholds, providing a more detailed assessment of detector performance.

4.3 Experiment Setting

Our proposed approach is implemented using PyTorch v2.0 and all experiments are performed on a workstation equipped with 2 Nvidia 1080 Ti GPUs (11

GBmemory). The models are optimized using the Adaptive Moment Estimation(Adam) algorithm with a batch size of 16, a momentum of 0.937, a weight decay of 0.0005, and 300 epochs. The initial learning rate is set to 0.001 and undergoes linear decay over training epochs, eventually reaching a minimum value of 1e-5. To enhance the detection performance for small and faint defects, all experimental inputs are standardized to 960×960 pixel resolution.

4.4 Ablation Analysis

To analyze the importance of each component in the defect detector, we progressively integrated the Conv-SPD-Conv, improved MSCA, and P6 head into the baseline to verify their effectiveness. The ablation experiment was conducted in DocDefectOD. Table 2 shows the impact of adding each module on evaluation metrics, where $\surd$ represents using the module and $\times$ represents not using the module.

Table 2. Ablation experiments results(%)

Yolov8-OBB	Conv-SPD-Conv	MSCA	P6	Precision	Recall	mAP_{50}	$mAP_{50:95}$
√	×	×	×	86.7	81.9	89.8	60.5
√	√	×	×	86.3	82.3	90.4	61.9
√	×	√	×	87.9	83.5	91.4	62.6
√	×	×	√	**89.1**	82.3	91.3	62.6
√	√	√	√	88.8	**84.0**	**92.1**	**64.8**

1) Conv-SPD-Conv: As shown in Table 2, the feature extraction network, using Conv-SPD-Conv, can improve mAP_{50} (from 89.8% to 90.4%) and $mAP_{50:95}$ (from 60.5% to 61.9%), which indicates the impact of the fine-grained features in the feature map and the fusion of local and global contextual information.

2) MSCA: Table 2 shows that adding MSCA can improve all evaluation metrics, especially in terms of $mAP_{50:95}$ (from 60.5% to 62.6%), which indicates that the MSCA can effectively trigger the model to dynamically adjust the weights of the features at each scale and focus more on the main features of defects.

3) P6: In Table 2, it shows that P6 can also improve all evaluation metrics, especially precision has the best increasement. In practice, incorporating the P6 head significantly enhances the detection performance for scanned lines, while providing more modest improvements for the remaining three defect types. This confirms that P6 head expands the detector's feature pyramid hierarchy, thereby improving its coverage of a broader range of target scales.

For the YOLOv8-OBB baseline model incorporating Conv-SPD Conv, MSCA, and P6 heads, mAP_{50} increased by 2.3%, and $mAP_{50:95}$ improved by 4.3%. The specific evaluation metrics of the four defects are shown in Table 3.

It can be noticed that the detection performance of scanned line is the best, with precision and recall above 92%, and the highest 97% of mAP_{50}. Secondly, the mAP_{50} of rubbing smudge and cloud blemish is above 95%, which indicates that the defect detector can basically detect these defects effectively. Pen scratches have the least amount of data, and the detector fails to fully learn its key features, resulting in its mAP_{50} being only 78.1%.

Table 3. Evaluation metrics of four defects in DocDefectOD

Class	Images	Instance	Precision(%)	Recall(%)	Map_{50}(%)	$Map_{50:95}$(%)
all	1862	16337	88.8	84.0	92.1	64.8
rubbing smudge	317	838	94.4	87.2	96.8	75.0
cloud blemish	1583	13297	91.5	87.7	95.8	74.6
pen scratch	188	357	77.3	68.1	78.1	47.5
scanned line	1350	1845	92.2	92.8	97.7	62.2

4.5 Comparison Among Different Methods

We conducted comprehensive comparative experiments across all three datasets to validate the superior defects detection capability of our proposed detector for document images. On both DocDefectOD and SDocDefectOD benchmarks, we compared our method against seven popular oriented object detection methods: Rotated Faster R-CNN, Oriented R-CNN, R3Det [26], S2ANet [27], YOLOv5-OBB, YOLOv8-OBB, and YOLOv11-OBB. Additionally, we conducted inference speed tests on all eight models (including ours) across both datasets. The speed metric was measured in FPS (Frames Per Second), which represents the number of document images processed per second. The results of different methods on DocDefectOD and SDocDefectOD are shown in Table 4.

1) DocDefectOD: As shown in Table 4, Rotated Faster R-CNN, Oriented R-CNN, and R3Det have lower mAP_{50}, none of which is more than 60%. S2ANet has a greater improvement than the previous three methods, with a mAP_{50} of 75.7%. In comparison, the YOLO variants achieve better detection of defects in the answer document image. Compared with YOLOv5-OBB and YOLOv11-OBB, the mAP_{50} of YOLOv8-OBB has exceeding improvement by 0.8% and 6.0%, respectively. Compared with other detection methods, the proposed defect detector reaches the highest mAP_{50} at 92.1%. Because of the introduction of Conv-SPD-Conv, MSCA and P6 detection head, the parameters and GFLOPs of our method are slightly higher than those of the baseline YOLOv8-OBB.

2) SDocDefectOD: Table 4 shows that the proposed defect detector achieves the best performance on the SDocDefectOD benchmark, achieving a remarkable mAP_{50} of 95.9%. Compared with Rotated Faster R-CNN, Oriented R-CNN, R3Det and S2ANet, the defect detector exhibits significant advantages in

mAP_{50}, model parameters, and FLOPs. Notably, it should be noted that the performance gap between the defect detector and YOLOv11-OBB on SDocDefectOD is marginal, with only a 0.4% lead in mAP_{50}.

Table 4. Results of different methods on DocDefectOD and SDocDefectOD

Method	DocDefectOD				SDocDefectOD			
	mAP_{50}(%)	Params(M)	GFLOPs	FPS	mAP_{50}(%)	Params(M)	GFLOPs	FPS
Rotated Faster R-CNN	55.9	41.1	198.4	4.1	56.9	41.1	198.3	4.1
Oriented R-CNN	59.7	41.1	198.5	4.1	78.3	41.1	198.4	4.1
R3Det	57.3	36.9	227.1	4.2	63.5	41.1	227.1	4.2
S2ANet	75.7	38.6	196.5	4.1	80.4	38.5	196.4	4.1
YOLOv5-OBB	89.0	**2.6**	7.4	51.2	89.5	**2.6**	7.4	**47.1**
YOLOv8-OBB	89.8	3.1	8.4	**51.3**	93.9	3.1	8.3	46.0
YOLOv11-OBB	83.8	2.7	**6.7**	47.9	95.5	2.7	**6.7**	43.5
Our Method	**92.1**	6.0	10.0	31.8	**95.9**	6.0	10.0	28.9

We replaced the oriented bounding box detection head with a horizontal bounding box head in the defect detector, and conducted comprehensive comparisons with seven popular horizontal object detection methods on the SignverOD benchmark. The evaluated algorithms include Dynamic R-CNN [28], DINO [29], RTMDet-tiny [30], DDQ-DETR [31], RT-DETR, YOLOv8, and YOLOv11. Detailed experimental results are presented in Table 5.

Table 5. Results of different methods in SignverOD

Method	mAP_{50}(%)	Params(M)	GFLOPs	FPS
Dynamic R-CNN	80.0	41.4	172.0	5.7
Dino	77.1	47.5	229.1	3.6
RTMDet-tiny	80.4	4.9	8.0	29.4
DDQ-DETR	77.5	43.2	183.4	3.2
RT-DETR	75.5	42.8	130.5	6.8
YOLOv8	91.4	3.0	8.2	**42.6**
YOLOv11	89.6	**2.6**	**6.4**	41.3
Our Method (not use OBB)	**93.2**	6.0	9.9	29.4

As shown in Table 5, our detector achieves the highest mAP_{50} of 93.2% on the SignverOD benchmark, outperforming the baseline YOLOv8 by 1.8%. This demonstrates that the proposed Conv-SPD-Conv module, enhanced MSCA mechanism, and P6 head can enhance the defect detector's detection capability in document images. Furthermore, the defect detector maintains significant advantages over other leading methods including Dynamic R-CNN, DINO, RTMDet-tiny, DDQ-DETR, and RT-DETR.

In Tables 4 and 5, the YOLO series models consistently achieve higher FPS across all three datasets, demonstrating their inherent advantage in meeting real-time processing requirements for document images. While the proposed detector exhibits a reduction in FPS compared to standard YOLO models due to the integration of the Conv-SPD-Conv module, enhanced MSCA mechanism, and P6 head, it still maintains practical real-time performance. Notably, this trade-off yields significant improvements in detection accuracy, achieving an optimal balance between precision and inference speed.

In order to evaluate the visual effect of our proposed defect detector, original document image samples are selected, and partial visualizations of the detection results under different methods are shown in Figs. 9, 10 and 11.

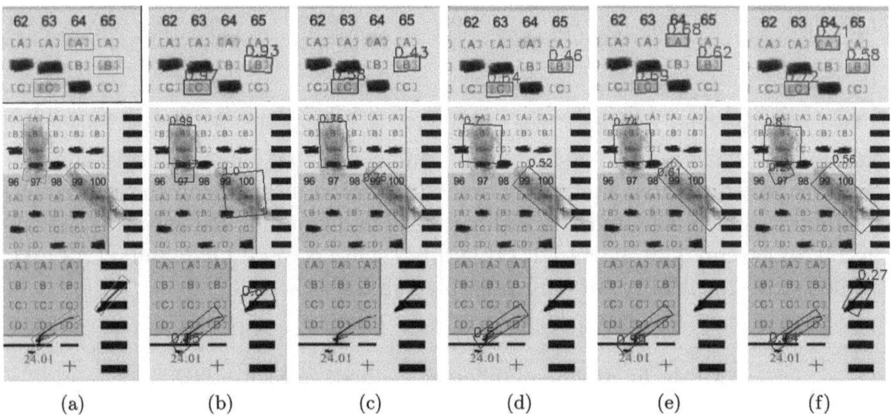

Fig. 9. Results of different methods for rubbing smudge, cloud blemish, and pen scratch in DocDefectOD. (a) Origin Image. (b) Rotated Faster R-CNN. (c) S2Anet. (d) YOLOv5-OBB. (e) YOLOv8-OBB. (f) Our Method.

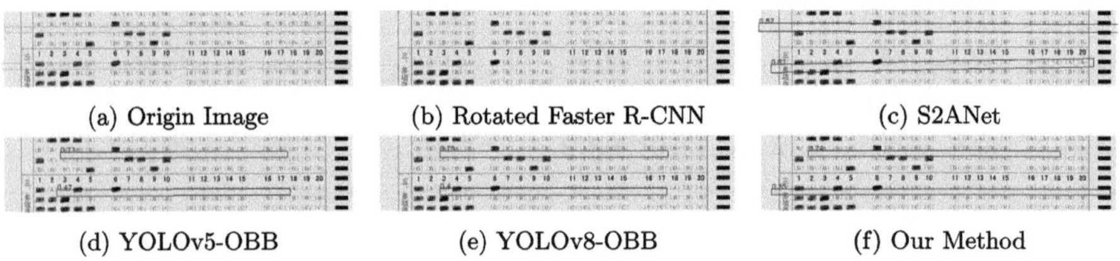

Fig. 10. Results of different methods for scanned line in DocDefectOD.

Figure 9 and Fig. 10 presents representative detection results from the DocDefectOD. In Fig. 9, the three detection result rows from top to bottom correspond to rubbing smudge (top row), cloud blemish (middle row), and pen scratch (bottom row) respectively, while Fig. 10 demonstrates the detection results for scanned line defects. All subfigures (a) in Fig. 9 and Fig. 10 indicate the location of defects in the original image; (b) to (f) show the detection results of Rotated Faster R-CNN, S2ANet, YOLOv5-OBB, YOLOv8-OBB, and our method for the original image, respectively. In addition, all detection results are visualized and marked with rotated rectangular boxes. Directly above each rotated rectangular box, a corresponding detection score is labeled, which intuitively reflects the assessment of the reliability of the current detection result by the model. The score ranges from 0 to 1. When the score approaches 1, it indicates that the model is more confident in the detection result, that is, the detection confidence is higher.

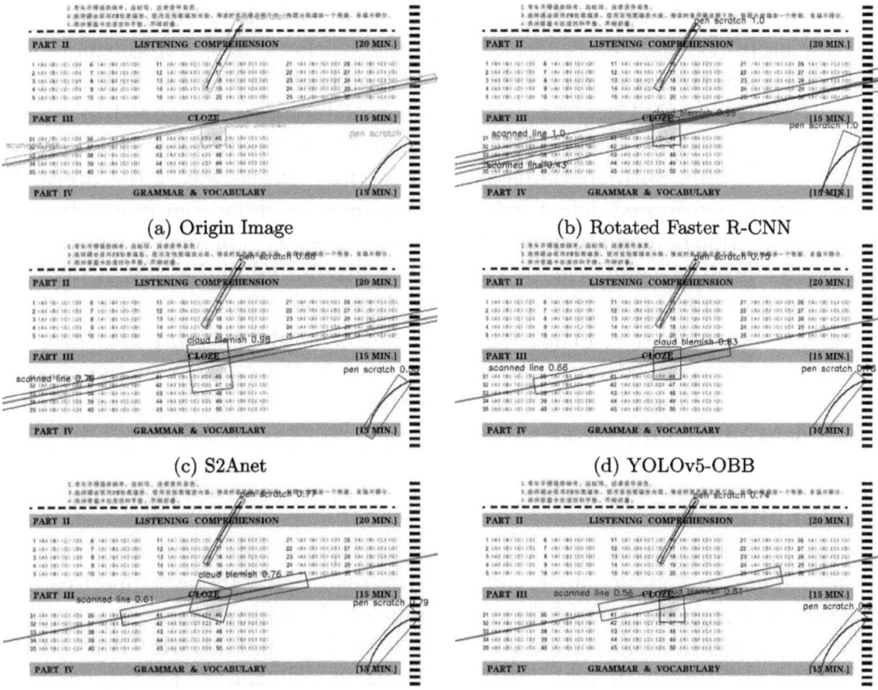

(a) Origin Image (b) Rotated Faster R-CNN

(c) S2Anet (d) YOLOv5-OBB

(e) YOLOv8-OBB (f) Our Method

Fig. 11. Results of different methods in SDocDefectOD.

As illustrated in Fig. 9, both YOLOv8-OBB and our method can effectively detect rubbing smudges, but the other three methods fail to detect the faintest rubbing smudge at the top. We attribute this to the insufficient feature extraction capabilities of these three models, which leads to relatively weaker detection performance on small and faint defects. All methods can effectively detect large cloud blemishes, but only Rotated Faster R-CNN and our method detect a small portion of cloud blemishes that are isolated by the filled-in boxes. Evidently, S2ANet missed two obvious pen scratches, while YOLOv5-OBB and YOLOv8-OBB failed to detect the rightmost pen scratch. This is attributed to the insufficient sensitivity of S2ANet's rotated bounding box parameterization to linear targets, combined with the limited receptive fields of YOLOv5 and YOLOv8, resulting in suboptimal detection performance for pen scratches.

Obviously, in Fig. 10, Rotated Faster R-CNN exhibits insufficient capability in learning the relevant features of scanned lines with extremely large aspect ratios that appear in complex background regions, making it difficult to detect such defects. S2Anet can detect two scanned lines, but the angle of the bottom scanned line is significantly inaccurate. Both YOLOv5-OBB and YOLOv8-OBB can only detect part of the scanned lines. In contrast, our method can detect the scanned lines relatively completely and accurately.

Figure 11 presents the results of various methods for some defects on SDocDefectOD. It is clearly observable that the synthesized cloud blemish and pen scratch closely resemble real defects, while the synthetic lines exhibit more distinctive features compared to actual defects. As shown in Fig. 11, both Rotated Faster R-CNN and S2ANet can relatively completely detect scanned lines, but Rotated Faster R-CNN yields a false detection. S2ANet and YOLOv8 can detect cloud blemish, but both methods tend to identify excessively large areas. In comparison, the proposed defect detector demonstrates more comprehensive and accurate detection of cloud blemish and pen scratches, with precise localization of scanned line positions and orientations. This is attributed to the MSCA module, which captures multi-scale features and enables more precise target detection. However, it should be noted that YOLO algorithm exhibits incomplete detection performance for scanned line. We believe this is because YOLO series models are constrained by their bounding box generation mechanism, making it difficult to fully detect extremely slender defects.

5 Conclusion

In this paper, we conduct an analysis of the distinctive shape properties and detection challenges in answer document images. To address these unique difficulties, we sequentially introduce the Conv-SPD-Conv module, MSCA attention mechanism and the P6 detection head. Compared with the traditional convolution, the proposed Conv-SPD-Conv can extract the fine-grained features in the feature map more efficiently, thus enhancing the model's capability to detect weak targets. MSCA can gain the more detailed information in complex scenes through multi-scale feature extraction and channel attention enhancement, thus improving the model's ability to focus on important features of targets of different sizes. The P6 detection head is based on a lower-resolution feature map with a larger receptive field, enabling it to capture a broader range of contextual information, which significantly improves the model's capability to detect very long objects. By integrating these three modules into the YOLOv8 baseline model, we design a novel defect detector. In addition, we have built two document image defect detection datasets (DocDefectOD and SDocDefectOD): one containing over 80,000 authentic defect instances, and another comprising synthetic defects across diverse document styles. Extensive experiments on SignverOD and our proposed datasets demonstrate that our defect detector achieves superior performance for defect detection in document image.

References

1. Wu, K., Fu, H., Li, W.: Handwriting text-line detection and recognition in answer sheet composition with few labeled data. In: 2020 IEEE 11th International Conference on Software Engineering and Service Science. IEEE ICSESS, pp. 129–132 (2020)
2. Shaikh, E., Mohiuddin, I., Manzoor, A., Automated grading for handwritten answer sheets using convolutional neural networks. In: 2nd International Conference On New Trends In Computing Sciences. IEEE, pp. 1–6 (2019)
3. Guo, M.H., Lu, C.Z., Hou, Q., et al.: Segnext: rethinking convolutional attention design for semantic segmentation. Adv. Neural. Inf. Process. Syst. **35**, 1140–1156 (2022)
4. Jiang, Y., Zhu, X., Wang, X., et al.: R2 CNN: rotational region CNN for arbitrarily-oriented scene text detection. In: 24th International Conference on Pattern Recognition, Beijing, pp. 3610–3615 (2018)
5. Xie, X., Cheng, G., Wang, J., et al.: Oriented R-CNN for object detection. In: Proceedings of the IEEE/CVF International Conference On Computer Vision, pp. 3520–3529 (2021)
6. Yang, S., Pei, Z., Zhou, F., et al.: Rotated faster R-CNN for oriented object detection in aerial images. In: Proceedings of the 2020 3rd International Conference on Robot Systems and Applications, pp. 35–39 (2020)
7. Feng, S., Huang, Y., Zhang, N.: An improved YOLOv8 OBB model for ship detection through stable diffusion data augmentation. Sensors **24**(17), 5850 (2024)
8. Pu, Y., Wang, Y., Xia, Z., et al.: Adaptive rotated convolution for rotated object detection. In: Proceedings of the IEEE/CVF International Conference on Computer Vision, pp. 6589–6600 (2023)
9. Zhang, Y., Ye, M., Zhu, G., et al.: FFCA-YOLO for small object detection in remote sensing images. IEEE Trans. Geosci. Remote Sens. **62**, 1–15 (2024)
10. Li, L., Bai, Y., Zhang, S., et al.: Document forgery detection based on spatial-frequency and multi-scale feature network. J. Visual Commun. Image Representation, 104393 (2025)
11. Vaswani, A., Shazeer, N., Parmar, N., et al.: Attention is all you need. Adv. Neural Inf. Process. Syst., 30 (2017)
12. Carion, N., Massa, F., Synnaeve, G., et al.: End-to-end object detection with transformers. In: European Conference on Computer Vision. Cham: Springer International Publishing, pp. 213–229 (2020)
13. Liu, Z., Lin, Y., Cao, Y., et al.: Swin transformer: hierarchical vision transformer using shifted windows. In: Proceedings of the IEEE/CVF International Conference On Computer Vision, pp. 10012–10022 (2021)
14. Zhao, Y., Lv, W., Xu, S., et al.: Detrs beat yolos on real-time object detection. In: Proceedings of the IEEE/CVF Conference on Computer Vision and Pattern Recognition, pp. 16965–16974 (2024)

15. Chen, Q., Su, X., Zhang, X., et al.: Lw-detr: a transformer replacement to yolo for real-time detection. arXiv preprint arXiv:2406.03459 (2024)
16. Chen, C., Li, B., Cai, R., et al.: Distortion model-based spectral augmentation for generalized recaptured document detection. IEEE Trans. Inf. Forensics Secur. **19**, 1283–1298 (2023)
17. Hu, X., Wei, B., Gao, L., et al.: SegHist: a general segmentation-based framework for chinese historical document text line detection. In: International Conference on Document Analysis and Recognition. Cham: Springer Nature Switzerland, pp. 391–410 (2024)
18. Kundu, A., Bhattacharya, U.: YOLO Assisted A* algorithm for robust line segmentation of degraded document images. In: International Conference on Document Analysis and Recognition. Cham: Springer Nature Switzerland, pp. 407–424 (2024)
19. Li, M., Sun, H., Lei, Y., et al.: High-fidelity document stain removal via a large-scale real-world dataset and a memory-augmented transformer. arxiv preprint arxiv:2410.22922 (2024)
20. Kasem, M.S.E., Mahmoud, M., Yagoub, B., et al.: HTTD: a hierarchical transformer for accurate table detection in document images. Mathematics **13**(2), 266 (2025)
21. Murrugarra Llerena, L.N., Kirsten, L.F., Zeni, et al.: Probabilistic intersection-over-union for training and evaluation of oriented object detectors. IEEE Trans. Image Process. **33**, 671–681 (2024)
22. Huang, D., Zhang, X., Zhang, R., et al.: DWM: a decomposable Winograd method for convolution acceleration. In: Proceedings of the AAAI Conference on Artificial Intelligence, vol. 34, no. 04, pp. 4174–4181 (2020)
23. Sunkara, R., Luo, T.: No more strided convolutions or pooling: a new CNN building block for low-resolution images and small objects. In: Joint European Conference on Machine Learning and Knowledge Discovery in Databases. Cham: Springer Nature Switzerland, pp. 443–459 (2022)
24. Zhang, R., Deng, B., Cheng, X., et al.: GCS-YOLOv8: a lightweight face extractor to assist deepfake detection. Sensors **24**(21), 6781 (2024)
25. Victor, D.: A dataset for handwritten signature objectdetection in scanned documents. victordibia.com (2022). https://victordibia.com/signverod.pdf
26. Yang, X., Yan, J., Feng, Z., et al.: R3det: refined single-stage detector with feature refinement for rotating object. In: Proceedings of the AAAI Conference on Artificial Intelligence, vol. 35, no. 4, pp. 3163–3171 (2021)
27. Han, J., Ding, J., Li, J., et al.: Align deep features for oriented object detection. IEEE Trans. Geosci. Remote Sens. **60**, 1–11 (2021)
28. Zhang, H., Chang, H., Ma, B., Wang, N., Chen, X.: Dynamic R-CNN: Towards High Quality Object Detection via Dynamic Training. In: Vedaldi, A., Bischof, H., Brox, T., Frahm, J.-M. (eds.) ECCV 2020. LNCS, vol. 12360, pp. 260–275. Springer, Cham (2020). https://doi.org/10.1007/978-3-030-58555-6_16
29. Zhang, H., Li, F., Liu, S., et al.: Dino: detr with improved denoising anchor boxes for end-to-end object detection. arxiv preprint arxiv:2203.03605 (2022)
30. Lyu, C., Zhang, W., Huang, H., et al.: Rtmdet: an empirical study of designing real-time object detectors. arxiv preprint arxiv:2212.07784 (2022)
31. Zhang, S., Wang, X., Wang, J., et al.: Dense distinct query for end-to-end object detection. In: Proceedings of the IEEE/CVF Conference on Computer Vision and Pattern Recognition, pp. 7329–7338 (2023)

ResNet-TPP: A Parallel PHOC-PHOS Framework for Zero-Shot Handwritten Word Recognition in Low-Resource Scripts

Aniket Gurav[1(✉)], Sukalpa Chanda[1], and Marius Pedersen[2]

[1] Department of Computer Science and Communication, Østfold University College, Halden 1757, Norway
aniketag@hiof.no , sukalpa@ieee.org
[2] Department of Computer Science, NTNU, Gjøvik, Norway
marius.pedersen@ntnu.no

Abstract. Handwritten word recognition in low-resource languages, such as historical Norwegian, faces significant challenges due to limited annotated data and complex script variations. This paper introduces a novel zero-shot learning (ZSL) architecture for recognizing seen/unseen words in historical documents, leveraging a pre-trained ResNet-50 backbone, enhanced Temporal Pyramid Pooling (TPP), and parallel Pyramidal Histogram of Characters (PHOC) and Shapes (PHOS) branches. Unlike prior frameworks like Pho(SC)Net and ResPho(SC)Net, our approach employs independent TPP paths for each PHOC level, improving sub-word alignment. This design enhances generalization to unseen classes, achieving improvements in seen and unseen class accuracies on Norwegian and George Washington (GW) datasets. By addressing data scarcity, our method supports the preservation of endangered scripts and facilitates access to cultural heritage. The architecture's efficiency and scalability make it a robust solution for low-resource document analysis, with potential applications in linguistic research and multilingual word recognition for historical texts.

Keywords: Zero-shot Learning · Handwritten Word Recognition · Low-resource Languages · Deep Learning · Document Analysis

1 Introduction

Historical manuscripts are essential for preserving cultural heritage, offering researchers, historians, and the public access to linguistic and societal records [22]. In low-resource languages like historical Norwegian, these documents provide critical insights into cultural evolution, yet digitization is challenged by scarce annotated data and complex script variations [4]. The Norwegian dataset, with 7,052 word images across 2,166 classes, features multi-writer variability and diacritical marks (e.g., æ, ø, å), while the GW dataset, with 4,894 images across 1,471 classes, exhibits a consistent writing style. Figure 1 shows a

L. Jin et al. (Eds.): ICDAR 2025 Workshops, LNCS 16225, pp. 245–259, 2026.
https://doi.org/10.1007/978-3-032-09368-4_15

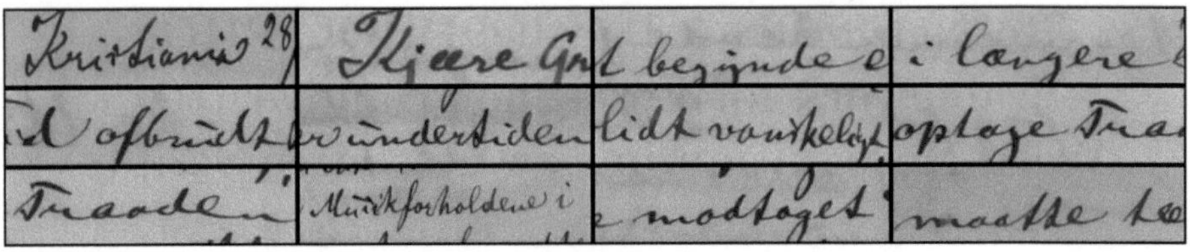

Fig. 1. Sample word image crops from the Norwegian dataset, showcasing multi-writer variability, diacritics, and historical degradation.

few examples from the Norwegian dataset. Multi-writer datasets introduce significant intra-class variability, hampering generalization [4]. Traditional Optical Character Recognition (OCR) struggles with cursive text and poor segmentation, often yielding unseen class accuracies below 60% [4] [14]. For instance, a historian searching for "Kristiania" in 19th-century Norwegian manuscripts may struggle to locate relevant documents without efficient automated tools [17]. Manual handling risks damaging fragile manuscripts, limiting access to historical knowledge and impeding robust OCR for non-English scripts [1].

It is not always feasible to synthetically generate data like [8,10,18]. ZSL addresses data scarcity by enabling recognition of unseen word classes [3]. Unlike object recognition, where semantic attributes like color are abundant, word recognition relies on sparser binary and structural features, making ZSL more challenging [3]. In ZSL, a model maps word images to an attribute signature space, such as the Pyramidal Histogram of Shapes and Characters (PHO(SC)), bridging seen and unseen classes [4]. Given a training set $S = \{(x_i, y_i, c(y_i))\}$, where $x_i \in X$ is an image, $y_i \in Y^S$ is a seen label, and $c(y_i) \in C$ is the attribute signature, the goal is to classify unseen labels $Y^U = \{u_1, \ldots, u_L\}$ using signatures $U = \{u_l, c(u_l)\}$ [4].
ZSL faces:

(i) limited training data;
(ii) high intra-class variability; and
(iii) the need for compositional embeddings [2,21].

Generalized ZSL (GZSL) [4] requires balanced performance across seen and unseen classes [4].

The PHO(SC) representation enhances ZSL by combining PHOC and PHOS per h segments, yielding $(1+2+3+4+5) \times 12 = 180$ for Norwegian and 165 for English [4]. The PHO(SC) vector (PHOC+PHOS) has lengths of $646+180 = 826$ for Norwegian and $604+165 = 769$ for English. High-dimensional PHOC vectors risk overfitting in low-data regimes, necessitating independent feature encoding [4]. No prior method has addressed sub-word-level feature disentanglement within PHOC using a multi-path ZSL strategy, a gap our work fills.

Prior ZSL frameworks like Pho(SC)Net [22] and ResPho(SC)Net [9] face limitations. Pho(SC)Net's shallow CNN and Spatial Pyramid Pooling (SPP) dilute sequential text structures, with GZSL unseen accuracies of 1.5%âĂŞ6.5% on Norwegian data [4,22]. ResPho(SC)Net's ResNet-18 and shared TPP improve performance but suffer from feature entanglement, limiting granularity for Norwegian diacritics (GZSL unseen accuracies: 31.75%âĂŞ39%) [9]. CTC-based models, reliant on sequence transduction, fail in multi-author ZSL as they cannot generalize to unseen words, unlike attribute-based models using compositional embeddings [4]. These shortcomings highlight the need for tailored low-resource architectures.

Focusing on low-resource datasets like Norwegian and GW is driven by the need to preserve endangered scripts [1]. Unlike high-resource datasets, these face severe data constraints, rendering traditional deep learning ineffective [15,16]. Norwegian's diacritics and multi-writer variability, and GW's degraded script, necessitate innovative ZSL solutions [9]. Word recognition, which identifies a word in a lexicon for indexing, differs from word spotting, which locates similar image regions for retrieval [4,22]. Our work focuses on zero-shot word recognition to support precise transcription.

Our Contributions. Our work, **ResNet-TPP** [1], advances zero-shot handwritten word recognition with:

- Parallel TPP paths (refer fig.2) for each PHOC level (2, 3, 4, 5, bigram), enabling sub-word-level feature disentanglement, unlike ResPho(SC)Net's shared TPP [4,24].
- A pre-trained ResNet-50 backbone, ensuring robust feature transfer and faster convergence (within 30 epochs), reducing data needs compared to Pho(SC)Net's shallow CNN and ResPho(SC)Net's non-pre-trained backbone [6,11].
- A parameter-efficient design with independent PHOC encoding, scalable to other low-resource languages, supporting cultural preservation and multilingual OCR [1,14].

Evaluated on Norwegian and GW datasets, ResNet-TPP improves ZSL and GZSL performance.

2 Related Work

Handwritten word recognition has progressed from manual indexing with handcrafted features to modern deep learning-based models capable of end-to-end visual understanding. Early approaches utilized features such as stroke patterns, contours, and projection profiles [17,20], but these methods failed to generalize across diverse handwriting styles and degraded historical documents [1]. Classical machine learning models, including Support Vector Machines (SVMs), offered marginal improvements but required extensive manual feature engineering, limiting their adaptability to low-resource scripts like historical Norwegian [2].

[1] Code available at: https://github.com/aniketntnu/ResNet-TPP/tree/main.

The introduction of deep learning, especially Convolutional Neural Networks (CNNs), marked a significant shift. CNNs enabled robust representation learning directly from raw word images, eliminating the need for hand-crafted features. Architectures like HWNet and PHOCNet exploited CNN backbones to learn semantic embeddings such as the PHOC, which encodes character-level presence across pyramidal splits [15,16,22]. These models showed high performance on well-annotated datasets but continued to struggle in low-resource settings, where labeled examples per class are scarce. Sequence modeling with Recurrent Neural Networks (RNNs) and CTC-based frameworks further improved recognition, but they are known to fail in multi-writer, zero-shot scenarios due to their reliance on training-time lexical patterns [4,12,19].

To address the lack of training data, ZSL has been adapted from object recognition to handwritten text [3]. ZSL enables models to classify previously unseen words by mapping them into a shared attribute space—most commonly using PHOC. However, PHOC alone struggles with visually similar but lexically different word images due to its lack of shape-level context. This limitation led to the development of PHOS, which encodes primitive visual features such as ascenders, loops, and diagonals [5]. The joint PHOC+PHOS representation, known as PHO(SC), became the basis of Pho(SC)Net [22], which used a shallow CNN with SPP. While effective on English benchmarks, Pho(SC)Net underperformed on low-resource languages, where generalization to unseen words was critical.

ResPho(SC)Net addressed these shortcomings by introducing a deeper ResNet-18 backbone and replacing SPP with a shared TPP layer, which better captures the left-to-right layout of handwritten words [4]. The shared representation improved convergence and recognition accuracy, particularly for multi-writer datasets like historical Norwegian. However, the model's use of shared TPP and fully connected (FC) layers for PHOC levels introduced feature entanglement, limiting its ability to model sub-word structures such as diacritics or affixes. Furthermore, the high dimensionality of PHOC embeddings (646-D for Norwegian) poses overfitting risks in low-data regimes.

Other approaches have explored attention-based architectures and sequence-to-sequence learning for unconstrained handwriting recognition [19]. While promising, these methods generally require large labeled corpora, making them impractical for historical scripts. Similarly, recent progress in large language models (LLMs) has enabled zero-shot classification in natural language tasks, but their use in image-based handwritten word recognition—especially for rare scripts—remains largely unexplored.

Our work builds on these insights by introducing a parallel PHOC encoding architecture with independent TPP paths for each pyramid level. This design avoids inter-level interference, improves sub-word alignment, and better models the compositional structure of complex scripts. Combined with a pre-trained ResNet-50 backbone for transferable visual features, our proposed ResNet-TPP model effectively addresses the limitations of earlier ZSL approaches, delivering improved performance in both seen and unseen class recognition for low-resource historical datasets like Norwegian and GW.

Handwritten word recognition has evolved from manual indexing with handcrafted features (e.g., stroke patterns, contours) to deep learning-based approaches [1,17]. Early methods using Support Vector Machines struggled with low-resource scripts like historical Norwegian due to extensive feature engineering [2]. Deep Convolutional Neural Networks (CNNs) shifted the paradigm, enabling robust feature extraction. HWNet and PHOCNet leveraged CNNs to learn PHOC embeddings, encoding character presence across pyramidal splits [15,16]. However, these models require large annotated datasets, limiting their efficacy in low-resource settings. Sequence models like Recurrent Neural Networks with CTC loss improved recognition but failed in zero-shot scenarios due to reliance on seen lexical patterns [4,12]. ZSL addresses data scarcity by mapping images to attribute spaces like PHOC [3]. Pho(SC)Net introduced PHOS alongside PHOC, forming PHO(SC) embeddings to capture visual and character attributes [5]. Despite its success on English datasets, Pho(SC)Net's shallow CNN and SPP diluted sequential text structures, yielding GZSL unseen accuracies of 1.5%–6.5% on Norwegian [4,22]. ResPho(SC)Net improved performance using a ResNet-18 backbone and shared TPP, but its shared TPP entangled PHOC level features, limiting sub-word granularity for Norwegian diacritics (GZSL unseen: 31.75%–39%) [4]. CTC-based models, effective for sequence transduction, underperformed in multi-writer ZSL due to poor generalization to unseen words [4,19]. Recent advances explored attention-based and transformer models for handwriting recognition [14,25]. However, these require large corpora, impractical for historical scripts. Generative approaches like GANs and diffusion models synthesize handwritten text but focus on style transfer, not ZSL recognition [7,13]. Our ResNet-TPP model builds on Pho(SC)Net and ResPho(SC)Net, introducing parallel TPP paths for each PHOC level to disentangle sub-word features, enhancing diacritic recognition in Norwegian and degraded text in GW [4]. The pre-trained ResNet-50 backbone ensures robust feature transfer, addressing data scarcity and outperforming non-pre-trained models [6,11]. This scalable design supports low-resource scripts, advancing cultural preservation [1].

3 Methodology

Our ResNet-TPP architecture advances zero-shot handwritten word recognition for low-resource scripts like historical Norwegian, overcoming limitations in Pho(SC)Net and ResPho(SC)Net [4,22]. We pursue three goals:

(i) enhance generalization to unseen classes using transferable features,
(ii) capture the sequential nature of handwritten words, and
(iii) eliminate architectural bottlenecks in PHOC/PHOS encoding that impede sub-word alignment. Figure 2 illustrates the architecture.

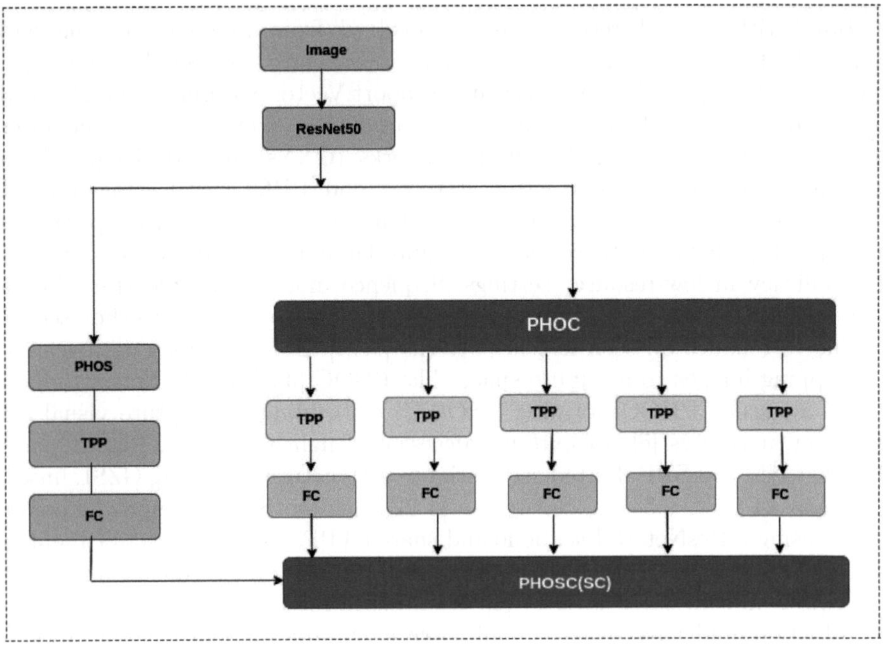

Fig. 2. ResNet-TPP architecture, featuring a pre-trained ResNet-50 backbone, a PHOS branch with shared TPP, and a PHOC branch with parallel TPP paths.

3.1 Problem Formulation

ZSL enables recognition of words not seen during training, crucial for low-resource scripts like historical Norwegian, where annotated data is scarce (e.g., 7,052 images for 2,166 classes) [4]. In ZSL, we define a training set $S = \{(x_i, y_i, c(y_i))\}_{i=1}^{N}$, where $x_i \in X$ is a word image (e.g., a handwritten "norske"), $y_i \in Y^S$ is a seen word label (e.g., "norske"), and $c(y_i) \in C$ is the PHO(SC) attribute signature, combining PHOC and Shapes to represent character presence and visual strokes [5]. Unseen labels $Y^U = \{u_1, \ldots, u_L\}$ (e.g., "være") have signatures $U = \{u_l, c(u_l)\}$, but no images during training. The goal is to learn a classifier $f_{zsl} : X \to Y^U$ for ZSL, identifying unseen words, and $f_{gzsl} : X \to Y^S \cup Y^U$ for GZSL, handling both seen and unseen words. A mapping $\phi : X \to C$ projects images to the attribute space, using cosine similarity to match predicted signatures to labels [4].

ZSL faces significant challenges in handwritten word recognition. Limited training data restricts robust mapping learning, especially for Norwegian's diacritics (æ, ø, å). Multi-writer variability introduces inconsistent stroke patterns, complicating generalization. Compositional embeddings are needed to recognize unseen words by combining sub-word features (e.g., prefixes, shapes), unlike traditional methods relying on seen examples [3,21]. Our ResNet-TPP addresses these by leveraging pre-trained features and parallel PHOC encoding for sub-word alignment.

3.2 Architecture

Our ResNet-TPP architecture as shown in fig.2 employs a pre-trained ResNet-50 backbone, replacing Pho(SC)Net's shallow CNN and ResPho(SC)Net's non-pre-trained ResNet-18, to extract a $2048 \times H \times W$ feature map [6,11]. ResNet-50's 50-layer depth, compared to ResNet-18's 18, captures richer hierarchical features, essential for distinguishing subtle diacritics (e.g., æ, ø, å) in Norwegian manuscripts [4]. The backbone comprises an initial convolutional layer (Conv1), followed by four residual blocks (Layer1âĂŞ4), each with multiple bottleneck units. As the feature map flows through these stages, the network captures increasing abstraction—from edges and corners to complex textural shapes. The final residual stage, produces the $2048 \times H \times W$ map that serves as input to PHOC and PHOS branches, ensuring spatial depth and semantic richness for handwritten text [11].

An input image, resized to 250×50 and normalized, passes through the ResNet-50 backbone, yielding a high-dimensional feature map. This shared representation splits into two streams: the PHOS branch receives a single temporal abstraction via TPP to capture coarse-to-fine shape layouts, while the PHOC branch undergoes level-wise TPP pooling, enabling hierarchical word structure modeling [4]. This modular data flow ensures each branch processes the input tailored to its representation goals, avoiding the limitations of unified architectures.

We adopt TPP [23] over Pho(SC)Net's SPP, which pools both dimensions equally, disrupting text's left-to-right structure [22,24]. TPP divides the feature map horizontally into bins (e.g., [1, 2, 5]), using max pooling to retain dominant stroke features within temporal windows. The [1, 2, 5] configuration balances coarse (1 bin) and fine (5 bins) contexts, aligning with word sequences. The PHOS branch uses a single TPP [1, 2, 5], yielding a 2048×8 vector, mapped via shallow FC layers ($4096 \rightarrow 180$ for Norwegian, 165 for English) to encode shapes (e.g., ascenders, small circle for diacritics) [5].

Our core innovation, inspired by multi-branch architectures, lies in the PHOC branch, where parallel TPP layers for each PHOC level (2, 3, 4, 5, bigram) replace ResPho(SC)Net's shared TPP [4]. Unlike end-to-end encoder-decoder architectures with a joint PHOC head, our disjoint branches prevent multi-level interference, avoiding vanishing gradients and improving interpretability. Each level learns specific sub-word decompositions (e.g., prefixes, diacritics), visualizable independently. Shared TPP compresses all levels into one vector, entangling coarse and fine patterns and causing gradient interference. Our level-specific TPP bins ([2], [3], [4], [5], [2]) and independent FC layers (e.g., $2048 \times 2 \rightarrow 4096 \rightarrow 78$ for level-2 in Norwegian) ensure clean sub-word learning, akin to multi-scale feature pyramids in object detection. Like readers parsing "norske" into "nor" and "ske," parallel branches model sub-word units, ensuring "være's diacritic (å) is distinctly captured. This enhances compositional ZSL embeddings, boosting unseen word accuracy, as validated in ablation studies (Sect. 4) [4].

Each PHOC head is shallow and narrow, comparable to ResPho(SC)Net but with finer granularity [4]. It enables transformer-style fusion, where PHOC-level embeddings are weighted via attention based on word characteristics (e.g., emphasizing diacritics in "være"), akin to transformer decoders, enhancing compositional generalization [14]. Table 1 summarizes TPP configurations.

Table 1. TPP Configurations for PHOC Levels in ResNet-TPP

PHOC Level	TPP Bins	Output Size (Norwegian)	Output Size (English)
Level 2	2	78	72
Level 3	3	117	108
Level 4	4	156	144
Level 5	5	195	180
Bigram	2	100	100
Total	—	**646**	**604**

3.3 Training and Loss

ResNet-TPP is trained using the Adam optimizer with a learning rate of 0.0001 and batch size of 32, selected for its adaptive moment estimation, which handles sparse gradients in zero-shot learning for low-resource scripts like Norwegian. The learning rate and batch size balance gradient accuracy and computational efficiency. Training runs for 100 epochs on an Nvidia V100/32GB GPU, with a scheduler reducing the rate by 0.25 if validation loss plateaus for 5 epochs, down to 1e-12, enhancing robustness [4].

The loss combines cross-entropy for PHOC (L_c) and mean squared error for PHOS (L_s), formulated as:

$$L = \sum_{i=1}^{B} \left(\lambda_c L_c(\phi_c(x_i), c_c(y_i)) + \lambda_s L_s(\phi_s(x_i), c_s(y_i))\right),$$

where $L_c = -\sum_j c_c(y_i)_j \log(\phi_c(x_i)_j)$, $L_s = \sum_j (\phi_s(x_i)_j - c_s(y_i)_j)^2$, and $\lambda_c = 4.5$, $\lambda_s = 1.5$ [4]. Higher PHOC weighting prioritizes character alignment, crucial for ZSL's compositional embeddings, while PHOS captures shape counts. Data augmentation, including perspective transformation (up to 10°), erosion (3×3 kernel), blur, and shearing (up to 5°), simulates historical degradation (e.g., ink fading, paper wear), addressing multi-writer variability [5].

Evaluation uses 5-fold cross-validation on Norwegian and GW datasets, computing top-1 accuracy via cosine similarity between predicted and true PHO(SC) signatures for ZSL (unseen classes) and GZSL (seen + unseen), tackling diacritic challenges [4].

4 Results

This section presents the experimental evaluation of our proposed ResNet-TPP architecture against ResPho(SC)Net [9] for zero-shot handwritten word recognition, addressing data scarcity in low-resource languages like Norwegian and GW dataset. We assess performance on the Norwegian and GW datasets using 5-fold and 4-fold cross-validation respectively, reporting top-1 accuracy under ZSL and GZSL protocols. These results highlight the efficacy of our pre-trained ResNet-50 backbone and independent PHOC encoders in enhancing generalization to unseen classes, supporting cultural preservation of historical scripts.

4.1 ZSL and GZSL Protocols

In Zero-Shot Learning, the model recognizes words from unseen classes (Y^U) not present in the training set, relying on attribute signatures like PHO(SC) to generalize [4]. The training set $S = \{(x_i, y_i, c(y_i))\}_{i=1}^N$ contains images x_i with seen labels $y_i \in Y^S$ and signatures $c(y_i)$, while the test set includes only unseen labels $Y^U = \{u_1, \ldots, u_L\}$ with signatures $U = \{u_l, c(u_l)\}$. The classifier $f_{zsl} : X \rightarrow Y^U$ maps images to unseen labels using cosine similarity between predicted and true PHO(SC) signatures. This tests the model's ability to identify novel words (e.g., "være" with diacritic å) absent during training, crucial for low-resource scripts like Norwegian with limited labeled data.

GZSL extends ZSL by evaluating both seen (Y^S) and unseen (Y^U) classes in the test set, requiring balanced performance [4]. The classifier $f_{gzsl} : X \rightarrow Y^S \cup Y^U$ must distinguish familiar words (e.g., "norske" seen in training) from novel ones, avoiding bias toward seen classes. Data is split such that training and validation sets contain only seen classes (Y^S), while the test set includes both seen and unseen classes in a 70%/10%/20% ratio across five folds. GZSL's challenge lies in handling multi-writer variability and degradation , where seen class familiarity can overshadow unseen class generalization. Top-1 accuracy measures correct label prediction, reflecting ResNet-TPP's robustness in low-resource ZSL and GZSL settings [3].

4.2 Experimental Setup

Each dataset is partitioned into five folds with a 70%/10%/20% train/validation/test ratio, where GZSL test sets include both seen and unseen classes. Training utilizes the Adam optimizer with a learning rate of 0.0001 and batch size of 32, optimizing a combined loss of cross-entropy for PHOC ($\lambda_c = 4.5$) and mean squared error for PHOS ($\lambda_s = 1.5$) [4]. To mimic historical degradation, we apply random perspective distortion, erosion, blur, and shearing, preprocessing images to 250×50 pixels and normalizing to $[-0.5, 0.5]$. Performance is measured as seen (S) and unseen (U) accuracies, reflecting the model's ability to handle diverse and novel word instances in ZSL and GZSL settings.

4.3 Performance on Norwegian Dataset

Table 2. Pho(SC)Net Results on Norwegian Dataset (ZSL and GZSL)

ID	ZSL-S	ZSL-U	GZSL-S	GZSL-U
0	52.03	10.00	42.23	1.50
1	63.00	31.50	52.52	6.50
2	63.00	31.50	52.52	6.50
3	66.76	33.50	54.40	6.00
4	59.54	15.75	47.08	3.75
Avg	60.87	24.45	49.75	4.85

We compare ResNet-TPP against ResPho(SC)Net and Pho(SC)Net on the Norwegian dataset, evaluating zero-shot handwritten word recognition using 5-fold cross-validation [4]. Tables 2, 3, and 4 report top-1 accuracies for ZSL and GZSL.

Table 3. ResPho(SC)Net Results on Norwegian Dataset (ZSL and GZSL)

ID	ZSL-S	ZSL-U	GZSL-S	GZSL-U
0	82.59	66.25	76.45	36.00
1	80.71	67.50	74.28	39.00
2	79.42	70.50	72.50	37.75
3	81.80	68.50	75.96	31.75
4	81.60	68.75	76.85	35.00
Avg	81.22	68.30	75.21	35.90

ResNet-TPP achieves (Table 4) average ZSL unseen accuracy of 71.20% and GZSL unseen accuracy of 39.20%, surpassing ResPho(SC)Net's (Table 3) 68.30% and 35.90%, and significantly outperforming Pho(SC)Net's (Table 2) 24.45% and 4.85%.

ZSL seen accuracy improves to 82.51% (vs. 81.22% for ResPho(SC)Net, 60.87% for Pho(SC)Net), while GZSL seen accuracy is stable at 75.33% (vs. 75.21%, 49.75%).

ResNet-TPP's gains stem from its pre-trained ResNet-50 backbone, capturing robust stroke patterns, and parallel TPP-PHOC branches, enabling sub-word feature disentanglement for diacritics (e.g., æ, ø, å), unlike ResPho(SC)Net's shared TPP and Pho(SC)Net's shallow CNN [4,11]. This enhances generalization to unseen words, supporting cultural preservation of low-resource scripts.

Table 4. ResNet-TPP Results on Norwegian Dataset (ZSL and GZSL)

ID	ZSL-S	ZSL-U	GZSL-S	GZSL-U
0	84.27	70.25	76.65	38.75
1	82.49	72.75	75.07	41.00
2	81.00	72.00	72.79	41.75
3	82.98	70.25	75.96	36.50
4	81.80	70.75	76.16	38.00
Avg	82.51	71.20	75.33	39.20

4.4 Performance on GW Dataset

We evaluate ResNet-TPP against both Pho(SC)Net and ResPho(SC)Net on the GW dataset using 4-fold cross-validation [4]. Tables 5, 6, and 7 report top-1 accuracy for ZSL and GZSL.

Table 5. Pho(SC)Net Results on GW Dataset (ZSL and GZSL)

ID	ZSL-S	ZSL-U	GZSL-S	GZSL-U
0	99.2	59.5	98.4	9.91
1	99.85	60.52	99.85	11.4
2	98.5	70.47	98.05	26.6
3	97.70	57.44	96.84	12.23
Avg	99.5	61.98	99.1	15.035

Compared to Pho(SC)Net, which achieves average ZSL seen and unseen accuracies of 99.5% and 61.98%, and GZSL unseen accuracy of 15.035%, ResNet-TPP improves ZSL unseen accuracy to 62.67% and GZSL unseen accuracy to 20.13%, showing a 5.10% absolute improvement. Notably, Pho(SC)Net's GZSL unseen accuracy on fold 0 is only 9.91%, while ResNet-TPP reaches 18.42% on the same fold—an 8.51% gain, highlighting the impact of deeper pre-trained backbones and TPP design.

ResNet-TPP also outperforms ResPho(SC)Net, which achieves lower ZSL unseen accuracy (50.39%) and GZSL unseen accuracy (14.21%). The most notable ZSL unseen gain appears in fold 0, where ResNet-TPP improves from 48.76% to 65.78% (a 17.02% increase). In GZSL, ResNet-TPP achieves a seen accuracy average of 97.36% (vs. 96.13%) and unseen accuracy of 20.13% (vs. 14.21%), resulting in a 5.92% improvement. In fold 2, ResNet-TPP achieves the best GZSL unseen accuracy of 24.76%, compared to 16.19% from ResPho(SC)Net in fold 3.

Table 6. ResPho(SC)Net Results on GW Dataset (ZSL and GZSL)

ID	ZSL-S	ZSL-U	GZSL-S	GZSL-U
0	96.97	48.76	96.17	11.57
2	97.71	56.14	97.56	15.78
3	94.75	54.28	95.80	16.19
4	95.69	42.01	94.97	13.29
Avg	96.28	50.39	96.13	14.21

Table 7. ResNet-TPP Results on GW Dataset (ZSL and GZSL)

ID	ZSL-S	ZSL-U	GZSL-S	GZSL-U
0	99.57	65.78	99.57	18.42
1	97.85	63.15	97.85	19.29
2	96.55	68.57	95.20	24.76
3	97.41	53.19	96.84	16.48
Avg	97.85	62.67	97.36	20.13

4.5 Analysis

ResNet-TPP consistently outperforms both ResPho(SC)Net and Pho(SC)Net across the Norwegian and GW datasets, as reported in Tables 8 and 9. These results are computed using 5-fold cross-validation [4].

On the Norwegian dataset, ResNet-TPP achieves a ZSL unseen accuracy of 71.20% compared to ResPho(SC)Net's 68.30% (a 2.90% improvement) and Pho(SC)Net's 24.45% (a 46.75% gain). For GZSL unseen accuracy, ResNet-TPP obtains 39.20%, outperforming ResPho(SC)Net's 35.90% (by 3.30%) and Pho(SC)Net's 4.85% (by 34.35%). ZSL seen accuracy improves to 82.51% over 81.22% and 60.87%, while GZSL seen accuracy remains stable at 75.33%, similar to ResPho(SC)Net (75.21%) and much higher than Pho(SC)Net (49.75%).

On the GW dataset, ResNet-TPP achieves a ZSL unseen accuracy of 62.67%, compared to 50.39% for ResPho(SC)Net (a 12.28% gain) and 61.98% for Pho(SC)Net (a modest 0.69% gain). GZSL unseen accuracy improves to 20.13%, surpassing ResPho(SC)Net's 14.21% (by 5.92%) and Pho(SC)Net's 15.035% (by 5.10%). Additionally, ZSL seen accuracy increases to 97.85% (vs. 96.28% and 99.5%), and GZSL seen accuracy reaches 97.36% (vs. 96.13% and 99.1%).

These improvements stem from ResNet-TPP's use of a pre-trained ResNet-50 backbone that captures robust visual features, in contrast to the shallow CNN used in Pho(SC)Net and the non-pre-trained ResNet-18 in ResPho(SC)Net [11]. Furthermore, ResNet-TPP's parallel TPP-PHOC branches promote sub-word feature disentanglement, which is essential for handling diacritics in Norwegian (e.g., æ, ø, å) and faded single-writer manuscripts in GW. The consistent gains in unseen accuracy highlight ResNet-TPP's generalizability and suitability for zero-shot word recognition in low-resource scripts.

Table 8. ResPho(SC)Net Average Results on Norwegian and GW Datasets (ZSL and GZSL)

Dataset	ZSL-S	ZSL-U	GZSL-S	GZSL-U
Norwegian	81.22	68.30	75.21	35.90
GW	96.28	50.39	96.13	14.21

Table 9. ResNet-TPP Average Results on Norwegian and GW Datasets (ZSL and GZSL)

Dataset	ZSL-S	ZSL-U	GZSL-S	GZSL-U
Norwegian	82.51	71.20	75.33	39.20
GW	97.85	62.67	97.36	20.13

5 Conclusion

This study presents a novel architecture for zero-shot handwritten word recognition, addressing the critical challenges of data scarcity and script complexity in low-resource languages like Norwegian. Leveraging a pre-trained ResNet-50 backbone and independent TPP layers for PHOS and Characters PHOC branches, our method outperforms ResPho(SC)Net and Pho(SC)Net across both Norwegian and GW datasets. Experimental results from multi-fold cross-validation demonstrate significant improvements, with ZSL unseen accuracies of 71.20% (Norwegian) and 62.67% (GW) compared to 68.30% and 50.39%, respectively, and GZSL unseen accuracies of 39.20% (Norwegian) and 20.13% (GW) versus 35.90% and 14.21%. The faster convergence and higher accuracy underscore the model's efficiency and robustness, making it a valuable tool for preserving endangered scripts and supporting cultural heritage in low-resource settings. Beyond Norwegian and GW, the architecture's scalability suggests potential for other complex-script languages, with applications in multilingual OCR for legal and medical documents. Future work will explore integrating attention mechanisms to further refine sub-word alignment and extend the framework to real-time document processing, broadening its impact on document intelligence.

Acknowledgement. We acknowledge the Simula HPC cluster for its support. The research described in this paper has utilized the Experimental Infrastructure for Exploration of Exascale Computing (eX3), funded by the Research Council of Norway under contract number 270053. This work also receives support through the Hugin-Munin project, which is financed by the Norwegian Research Council (Project number 328598).

References

1. Adamek, T., O'Connor, N., Smeaton, A.: Word matching using single closed contours for indexing handwritten historical documents. Int. J. Document Anal. Recognit. (IJDAR) **9**(2–4), 153–165 (2007)

2. Aldavert, D., Rusiñol, M., Toledo, R., Lladós, J.: A study of bag-of-visual-words representations for handwritten keyword spotting. Int. J. Document Anal. Recognit. (IJDAR) (2015). https://doi.org/10.1007/S10032-015-0245-Z
3. Almazán, J., Gordo, A., Fornés, A., Valveny, E.: Word spotting and recognition with embedded attributes. IEEE Trans. Pattern Anal. Mach. Intell. (PAMI) **36**(12), 2552–2566 (2014)
4. Bhatt, R., Rai, A., Chanda, S., et al.: PHO(SC)-CTC–a hybrid approach towards zero-shot word image recognition. Int. J. Document Anal. Recognit. (IJDAR) (2022). https://doi.org/10.1007/s10032-022-00407-6
5. Bhatt, R., Rai, A., Chanda, S., Krishnan, N.C.: Pho(SC)-CTC—a hybrid approach towards zero-shot word image recognition. Int. J. Document Anal. Recognit. (IJDAR) (2022). https://doi.org/10.1007/s10032-022-00407-6
6. Deng, J., Dong, W., Socher, R., Li, L.J., Li, K., Fei-Fei, L.: ImageNet: a large-scale hierarchical image database. In: 2009 IEEE Conference on Computer Vision and Pattern Recognition, pp. 248–255 (2009). https://api.semanticscholar.org/CorpusID:57246310
7. Ding, H., Luan, B., Gui, D., Chen, K., Huo, Q.: Improving handwritten OCR with training samples generated by glyph conditional denoising diffusion probabilistic model. ArXiv abs/2305.19543 (2023). https://api.semanticscholar.org/CorpusID:258987359
8. Gurav, A., Chanda, S., Krishnan, N.C.: HTLDiff: handwritten text line generation - a diffusion model perspective. In: 2024 IEEE Thirteenth International Conference on Image Processing Theory, Tools and Applications (IPTA), pp. 1–6 (2024). https://api.semanticscholar.org/CorpusID:274180352
9. Gurav, A., Jensen, J., Krishnan, N.C., Chanda, S.: Respho(sc)net: a zero-shot learning framework for Norwegian handwritten word image recognition. In: Pertusa, A., Gallego, A.J., Sánchez, J.A., Domingues, I. (eds.) Pattern Recognition and Image Analysis, pp. 182–196. Springer Nature Switzerland, Cham (2023). https://doi.org/10.1007/978-3-031-36616-1_15
10. Gurav, A., Krishnan, N.C., Chanda, S.: Word-diffusion: diffusion-based handwritten text word image generation. In: Antonacopoulos, A., Chaudhuri, S., Chellappa, R., Liu, C.L., Bhattacharya, S., Pal, U. (eds.) Pattern Recognition, pp. 53–72. Springer Nature Switzerland, Cham (2025). https://doi.org/10.1007/978-3-031-78495-8_4
11. He, K., Zhang, X., Ren, S., Sun, J.: Deep residual learning for image recognition. In: 2016 IEEE Conference on Computer Vision and Pattern Recognition (CVPR), pp. 770–778 (2015). https://api.semanticscholar.org/CorpusID:206594692
12. Kang, L., Toledo, J., Riba, P., Villegas, M., Fornés, A., Rusiñol, M.: Convolve, attend and spell: An attention-based sequence-to-sequence model for handwritten word recognition. In: German Conference on Pattern Recognition, pp. 459–472 (2018)
13. Kang, L., Riba, P., Wang, Y., Rusiñol, M., Forn'es, A., Villegas, M.: GANwriting: content-conditioned generation of styled handwritten word images. ArXiv abs/2003.02567 (2020). https://api.semanticscholar.org/CorpusID:212414769
14. Kass, D., Vats, E.: AttentionHTR: handwritten text recognition based on attention encoder-decoder networks. In: Uchida, S., Barney, E., Eglin, V. (eds.) Document Analysis Systems. DAS 2022, LNCS, vol. 13237. Springer, Cham (2022). https://doi.org/10.1007/978-3-031-06555-2_34

15. Krishnan, P., Dutta, K., Jawahar, C.: Deep feature embedding for accurate recognition and retrieval of handwritten text. In: 2016 15th International Conference on Frontiers in Handwriting Recognition (ICFHR), pp. 289–294 (2016). https://doi.org/10.1109/ICFHR.2016.0062
16. Krishnan, P., Jawahar, C.: HWNET V2: an efficient word image representation for handwritten documents. Int. J. Document Anal. Recognit. (IJDAR) **22**(4), 387–405 (2019). https://doi.org/10.1007/s10032-019-00336-x
17. Manmatha, R., Han, C., Riseman, E.: Word spotting: a new approach to indexing handwriting. In: 1996 Proceedings CVPR 1996, IEEE Computer Society Conference, pp. 631–637. IEEE (1996)
18. Nikolaidou, K., et al.: Wordstylist: styled verbatim handwritten text generation with latent diffusion models. In: Fink, G.A., Jain, R., Kise, K., Zanibbi, R. (eds.) Document Analysis and Recognition - ICDAR 2023, pp. 384–401. Springer Nature Switzerland, Cham (2023). https://doi.org/10.1007/978-3-031-41679-8_22
19. Retsinas, G., et al.: From seq2seq recognition to handwritten word embeddings. In: Proceedings of the British Machine Vision Conference (BMVC) (2021)
20. Rothfeder, J., Feng, S., Rath, T.: Using corner feature correspondences to rank word images by similarity. In: IEEE Computer Vision and Pattern Recognition Workshop, CVPRW 2003, vol. 3, p. 30 (2003)
21. Rusiñol, M., Aldavert, D., Toledo, R., Lladós, J.: Efficient segmentation-free keyword spotting in historical document collections. Pattern Recognit. (PR) **48**(2), 545–555 (2015)
22. Sudholt, S., Fink, G.A.: PHOCNET: a deep convolutional neural network for word spotting in handwritten documents, pp. 277–282 (2016)
23. Sudholt, S., Fink, G.A.: Attribute CNNs for word spotting in handwritten documents. Int. J. Document Anal. Recognit. (IJDAR) **21**, 199–218 (2017). https://api.semanticscholar.org/CorpusID:2120725
24. Sudholt, S., Fink, G.A.: Attribute CNNs for word spotting in handwritten documents. Int. J. Document Anal. Recognit. (IJDAR) **21**, 199–218 (2018). https://doi.org/10.1007/s10032-018-0295-0
25. Vaswani, A., et al.: Attention is all you need. In: Neural Information Processing Systems (2017). https://api.semanticscholar.org/CorpusID:13756489

Interpret, Prune and Distill Donut: Towards Lightweight VLMs for VQA on Documents

Adnan Ben Mansour[1,2,3](✉), Ayoub Karine[2], and David Naccache[1,3]

[1] DIENS, École Normale Supérieure of Paris, Paris, France
adnan.ben.mansour@ens.fr
[2] Université Paris Cité, LIPADE, Paris F-75006, France
[3] Be-ys Research, Paris, France

Abstract. Recent advances in Visually-rich Document Understanding rely on large Vision-Language Models like Donut, which perform document-level Visual Question Answering without Optical Character Recognition. Despite their effectiveness, these models are too costly for real-time or resource-constrained applications. We investigate model compression through knowledge distillation, training compact student models from a larger teacher. We leverage mechanistic interpretability to drive student architecture design within this framework. By analyzing internal computations, we identify essential subcomponents to retain, while having a clear view of which subcomponents should be approximated, skipped, or reparametrized based on their function. This approach yields Donut-MINT (Mechanistic Interpretability-based Network Trimming), a pruned Donut variant that reduces inference time and memory usage while maintaining strong performance on DocVQA, a standard benchmark for document Visual Question Answering. Our method reframes compression as circuit discovery, bridging interpretability research and practical Vision-Language Model deployment.

Keywords: Document Understanding · Mechanistic Interpretability · Structured Pruning · Knowledge-Distillation

1 Introduction

Visually-rich Document Understanding (VrDU) is a key area in multimodal artificial intelligence, with applications including invoice parsing, contract analysis, and flowchart interpretation. A central task is Visual Question Answering (VQA) on documents, where models answer questions about the content of scanned or rendered documents. Traditional approaches often rely on OCR-based pipelines [24,25,50], which are brittle and domain-specific. Recently, end-to-end Vision-Language Models (VLMs) that bypass OCR have become increasingly popular.

Several VLMs have advanced the state of the art in document VQA, including Donut [22], Pix2Struct [23], DocformerV2 [1], Qwen2.5-VL [4] and

L. Jin et al. (Eds.): ICDAR 2025 Workshops, LNCS 16225, pp. 260–278, 2026.
https://doi.org/10.1007/978-3-032-09368-4_16

InternVL3 [57]. These models are generally based on transformer architectures [12,46], combining a vision backbone with language modeling objectives to process document content holistically.

Despite strong performance, these models are computationally expensive and difficult to deploy in real-world environments with latency or resource constraints. This raises a fundamental challenge: how can we compress large VLMs while preserving task accuracy? While not the most powerful model, Donut remains a strong candidate for compression due to its OCR-free formulation and favorable trade-off between size and performance. For instance, the Donut-Hole method [42] compresses Donut via unstructured magnitude-based pruning followed by knowledge distillation [20]. In this setting, the pruned model mimics the teacher's output distributions, recovering lost performance without changing the architecture.

However, such approaches treat the teacher model as a black box. This lack of transparency limits our ability to guide compression using internal signals. In practice, VLMs used in VrDU remain poorly understood. Existing pruning and distillation methods are mostly blind to model internals, offering no principled way to distinguish essential from redundant components.

In this work, we ask whether mechanistic interpretability (MI) can help address this limitation. MI seeks to uncover the internal representations and computational mechanisms that drive model behavior. We apply MI tools to Donut's attention patterns and activations to identify functionally important components and remove others in a targeted fashion. Rather than distilling from a model we do not understand, we leverage internal insights to inform pruning and student design.

We focus our study on Donut for several reasons. First, Donut is relatively small, with approximately 250 million parameters and a shallow multimodal decoder of four transformer layers, making systematic analysis feasible. Second, Donut uses explicit cross-attention between visual features and decoder tokens, making its modality fusion interpretable and structurally disentangled. Third, it is trained to copy text spans from the image rather than generate free-form answers, which makes attribution of internal mechanisms more tractable. While our study centers on Donut, we view it as a proof of concept for a broader class of VLMs.

Our experiments show that our pruned model, Donut-MINT (Mechanistic Interpretability-guided Network Trimming), achieves significant reductions in both parameters and computation, while closely matching the teacher model's accuracy on DocVQA after distillation. An overview of our approach is provided in Fig. 1, highlighting the main stages of our method and how they compare to other baselines.

Our contributions are threefold: (1) We provide a high-level interpretation of how an end-to-end VLM solves VQA on documents, identifying the computational patterns and representations most critical to the task. (2) We demonstrate that these insights can inform the design of a student model, outperforming baseline pruning or blind distillation strategies. (3) We introduce

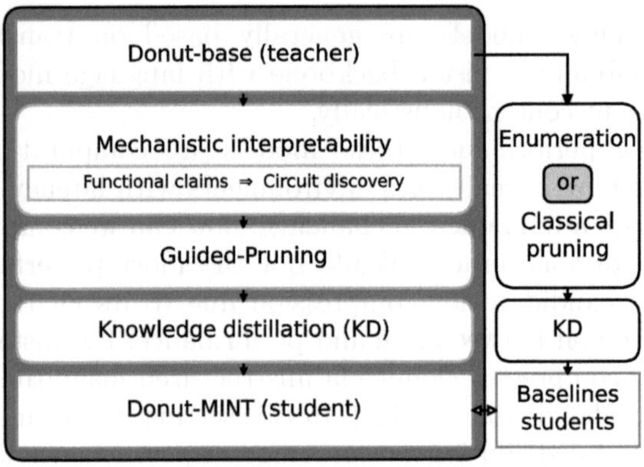

Fig. 1. Our proposed method relies on the three stages highlighted in blue: MI-based importance estimation, pruning, and knowledge distillation. We also pruned Donut-base with other SOTA techniques to allow for a fair comparison. (Color figure online)

Donut-MINT, a lightweight model that achieves competitive performance on the DocVQA dataset [31], thereby validating our methodology and illustrating how interpretability can drive principled model compression.

The rest of this paper is structured as follows. Section 2 surveys prior work on mechanistic interpretability, model pruning, and knowledge distillation. Section 3 presents our proposed framework, details the Donut architecture, and describes the tools used to derive efficient student models. Section 4 provides our interpretability claims and reports our experiments on the DocVQA dataset. Section 5 outlines implications, limitations, and future directions.

2 Related Work

2.1 Mechanistic Interpretability

Mechanistic interpretability (MI) seeks to reverse-engineer internal computations in neural networks, departing from traditional interpretability by focusing on the mechanisms themselves. As described by [39], MI is structured around three components: *features*, *circuits*, and *universality*. Features are human-interpretable properties encoded in activations; circuits are the paths connecting and processing these features; universality concerns the redundancy of such components across diverse models.

Core Methods. According to [8], MI methods fall into two categories: intervention-based and observational.

Intervention-Based Methods operate through three runs: a clean run, a corrupted run, and a patched run. The core idea is to selectively replace activations from the clean run with those from the corrupted run, or conversely (in denoising-based variants), to restore clean behavior by overwriting corrupted activations. [36] apply this to attention heads using ablation. Corruption can involve zeroing activations, but alternatives have been proposed, such as adding random noise [38] or replacing activations with their mean value over a small dataset [48]. Activation patching is closely related to causal mediation analysis [47] and is referred to as causal tracing in [32]. In some works [18,32], the corruption procedure involves resampling: the clean and corrupt runs are generated from different inputs, rather than modifying a single input. Attribution patching [34] replaces iterative patching with gradient-based estimates, accelerating discovery of circuits.

Observational Approaches require no model alteration. Vocabulary projection methods (e.g., logit lens [35], attention lens [41]) decode activations via the unembedding matrix. Probing techniques [16] and sparse autoencoders [10] train auxiliary models to recover encoded information, either via supervision or unsupervised reconstruction.

MI for VLMs. MI has progressed rapidly for LLMs, while VLMs remain underexplored. Recent efforts extend LLM techniques to the multimodal setting. [26] distinguish between VLM adaptations of LLM tools and VLM-specific methods. Causal tracing was applied to vision encoders [37] and full VLMs [5], identifying locations of stored knowledge. In VQA, [53] study visual hallucinations in LLaVA [28].

Dedicated tools include adaptations of Symmetric Token Replacement to the image modality [15], text decodings of image embeddings [14], and neuron-level analyses [3,7]. Interpretability techniques leveraging cross-attention exist in generative models [6,19], but have not been adapted to VLMs. Crucially, to the best of our knowledge, no prior work addresses interpretability in the specific context of Visual Question Answering on documents.

2.2 Compression Techniques for Transformer-Based Models

Compression of transformer-based models, which form the backbone of most VLMs, typically falls into three categories: pruning, knowledge distillation, and quantization. Since quantization alters weight precision without changing model structure, it is largely orthogonal and not considered here. We focus on methods that reshape the model or its training: structured pruning, knowledge distillation, and their combination. We conclude by reviewing how these approaches extend to vision-language models.

Structured Pruning and Knowledge-Distillation. Structured pruning removes entire components (layers, heads, neuron groups) to reduce size and computation. [56] prune feed-forward and attention blocks, [21] remove sub-matrix structures, and [27] use gradient-guided strategies. The lottery ticket hypothesis [13] suggests pruned subnetworks can perform well when reinitialized properly. However, pruning often reduces performance, necessitating retraining or fine-tuning.

Knowledge distillation [20] compresses models by aligning a compact student with a larger teacher. Techniques include output imitation, intermediate feature matching [40], attention alignment [55], and structural preservation [45]. These improve generalization while enabling smaller models.

Combining pruning and distillation enhances performance recovery after pruning. [33] use this strategy to compress LLMs without compromising downstream accuracy. The teacher is the original model; the student is the pruned one retrained to mimic the teacher's outputs.

VLM Compression for VrDU. The closest attempt at combining pruning and distillation in VLMs for VrDU tasks, although applied to document reading and key information extraction rather than VQA, is Donut-Hole [42]. However, their approach primarily focuses on reducing the parameter count through unstructured pruning [2,17,43,54], explicitly setting a subset of weights to zero based on their magnitude and forcing these weights to remain zero during retraining with KD. While this produces a model with fewer active weights, achieving actual speedups typically requires specialized hardware for sparse inference.

In another line of research, token pruning [49,51,52] has been explored, with the goal of discarding tokens in the cross-attention mechanism based on their importance or redundancy in the input image. However, this approach is inherently challenging to apply in VrDU, as most of the patches in document images are crucial for understanding and contain valuable information.

3 Method

3.1 Our Framework

Figure 1 outlines our framework. Starting from Donut-base as the teacher model, we apply mechanistic interpretability techniques to analyze its behavior on document-based questions, yielding insights formalized in Sect. 4.2. These guide a structured pruning process to obtain a distilled student model. For comparison, we also perform an exhaustive search over candidate student architectures using both structured and unstructured coarse-grained pruning strategies, following [42]. All student models are trained via knowledge distillation from the teacher. Our MI-guided pruning proceeds in two stages: first, we prune sub-layers (self-attentions, cross-attentions, and feed-forwards) from the multimodal decoder; second, we apply fine-grained pruning to individual attention heads within the remaining self and cross-attention layers.

3.2 Donut Overview

We conduct all experiments on Donut [22], used as our teacher model. It combines a Swin-B vision encoder [30] with a multimodal decoder composed of the first four layers of mBART [29]. Donut is a multilingual model pre-trained on 2M synthetic and 11M scanned document images. We use the donut-base-finetuned-docvqa checkpoint, already fine-tuned on DocVQA.

Figure 2 illustrates Donut's architecture, showing how the document image and question prompt are processed. We denote sub-layers of the four decoder layers as `S0`–`S3` (self-attentions), `C0`–`C3` (cross-attentions), and `M0`–`M3` (feed-forwards). Attention heads are indexed as in `C3.H11`, referring to the 12th head in the final cross-attention. Each sub-layer has associated activations (e.g., `C3.input`, `C3.output`), defined as needed.

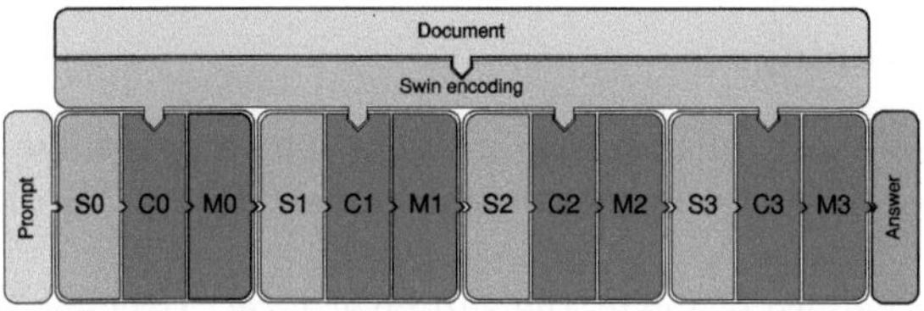

Fig. 2. Architecture overview of Donut-base. The model comprises a visual encoder and a decoder. The decoder consists of four layers, each layer $\ell \in \{0, 1, 2, 3\}$ divided into three sub-layers: self-attention $\mathtt{S}\ell$, cross-attention $\mathtt{C}\ell$ bridging modalities, and feed-forward networks $\mathtt{M}\ell$.

3.3 Mechanistic Interpretability

Mechanistic interpretability (MI) provides both an investigative lens into model internals and a foundation for structured pruning. We use MI to dissect the internal computation of Donut-base in the context of single-page document visual question answering (VQA), with the goal of identifying which components—layers, heads, or neurons—are responsible for key aspects of reasoning and representation. These insights directly inform our pruning strategy, allowing us to preserve only components essential to task performance.

To evaluate the functional relevance of model components, we employ activation patching in a teacher-forcing setting. In this setup, the model receives the full input sequence—including the question and the ground-truth answer—enabling us to compute token-level predictions for all decoding steps in a single forward pass. We monitor changes in perplexity, defined as the exponential of the cross-entropy loss, to assess the impact of patching. We also extend logit lens analysis to decoder attention heads (when dimensionality permits) for interpretability at intermediate layers.

Additionally, we apply tailored statistical measures to estimate the importance of attention heads. For self-attention (SA), we compute entropy to measure how concentrated or diffuse each head's attention is across the sequence. High entropy indicates that a token is attending to many others with similar weight, diluting the information it receives. This diffuse attention often signals that the head is noisy or uninformative. In contrast, for cross-attention (CA), which links decoder tokens to Swin-encoded image patches, we use variance. Salient cross-attention heads typically focus sharply on a few high-response patches, while low-variance heads distribute attention more evenly, often reflecting low task relevance. Due to spatial redundancy in Swin features, variance is a more robust proxy than entropy in this context.

3.4 Guided Pruning

We translate the above mechanistic insights into a two-stage guided pruning strategy, resulting in two compact student models: Donut-MINT$_{31\%}$ and Donut-MINT$_{7\%}$.

In the first stage, we prune at the sub-layer level. Mechanistic analysis identifies the initial feed-forward sub-layer (`M0`) as dispensable, and we show its function can be absorbed by adjusting the tokenizer or embedding layer. Building on this, we construct a student model retaining only 31% of the original parameters (Donut-MINT$_{31\%}$), and further reduce it to 18% by removing `M0`. Sub-layer pruning in both cases preserves the minimal circuits identified as necessary to support the full range of document VQA subtasks.

In the second stage, we apply fine-grained pruning to attention heads based on the statistical indicators described above. We remove self-attention heads with high entropy and cross-attention heads with low variance, treating them as proxies for functional irrelevance. Starting from the 18% variant, we eliminate these heads and selectively reintroduce others critical for specific sub-tasks, resulting in Donut-MINT$_{7\%}$, a model with just 7% of the original parameters but competitive performance.

3.5 Knowledge Distillation

All student models are initialized from the pretrained weights of Donut-base and retrained using knowledge distillation on the DocVQA training dataset. We follow the standard setup introduced by [20], which combines supervision from the ground-truth label with soft targets produced by the teacher model. The total loss for student training is defined as:

$$\mathcal{L}_{\text{total}} = \mathcal{L}_{\text{CE}} + \alpha\mathcal{L}_{\text{KD}} \tag{1}$$

Here, $\mathcal{L}_{\text{CE}}$ is the cross-entropy loss between the student's logits and the ground-truth labels. The $\mathcal{L}_{\text{KD}}$ term represents the softened Kullback-Leibler (KL) divergence between the teacher's and student's output distributions, computed from logits scaled by a temperature T. During training, the teacher model is kept frozen, as are the encoder and embedding layers of the student; only the student decoder is updated. The weighting factor α controls the trade-off between the ground-truth supervision and the distillation signal : it is chosen such that both loss terms contribute with similar magnitudes.

4 Results

4.1 Setup

We use the official Donut-base model pretrained on DocVQA as our teacher, avoiding costly pretraining while representing a realistic deployment target. Student models are retrained on the full DocVQA training set using our distillation pipeline. All experiments run on two NVIDIA L40S GPUs.

We evaluate on DocVQA [31], a benchmark of 50,000 QA pairs over 12,000+ visually diverse documents. We split its original validation set into three parts: one for interpretability analysis (20%), one for student model selection (20%), and one for post-pruning evaluation (60%). Since test annotations are private, final accuracy is reported via the official evaluation server[1].

Performance is measured using ANLS [9], which computes normalized edit similarity between predictions and references with a cutoff at $\tau = 0.5$. This balances tolerance to minor typos with sensitivity to semantic correctness.

To assess the efficiency of the compressed model, we adopt two metrics at a given performance level: the number of parameters and the inference cost measured in floating-point operations (FLOPs). Table 1 reports the value of these metrics for the different sub-layers of Donut-base, highlighting the primary computational bottlenecks for each metric. Parameter count is a coarse measure of model size and storage requirements, but it fails to capture actual computational cost, especially in autoregressive generation tasks where the same parameters may be reused across multiple steps. FLOPs, by contrast, estimate the number of arithmetic operations required for a full forward pass and provide a hardware-agnostic proxy for runtime and energy consumption.

While the encoder contributes most of the FLOPs in a single forward pass, we focus our analysis on the decoder for both conceptual and practical reasons. Structurally, the decoder is multimodal and must combine visual and textual information, making it more likely to implement complex and intertwined subtasks. This makes it a natural target for mechanistic interpretability and pruning. On the practical side, decoder FLOPs grow with both the context length and the number of generated tokens. Since document complexity often correlates with context size, the decoder's computational burden tends to dominate in realistic deployment scenarios.

[1] https://rrc.cvc.uab.es/?ch=17, accessed in April 2025.

4.2 Interpretability Insights

Table 1. Number of parameters and FLOPs required by Donut-base components to generate $m = 35$ new tokens given $n = 42$ prompt tokens. G and T refer to 10^9 (giga) and 10^{12} (tera) FLOPs, respectively, while M stands for 10^6 parameters.

Module	Parameters	FLOPs
Donut-base	257M	2.99T
Encoder	75M	1.30T
Embedding	59M	0
Unembedding	59M	117.17G
Decoder	64M	1.57T
Self-Attention ℓ	4M	8.58G
Cross-Attention ℓ	4M	366.11G
Feed-Forward ℓ	8M	16.69G

Proposition 1. *The task of converting pixel values into discrete tokens is entirely done in the visual encoder up to a linear transformation.* `C3` *puts attention on patches it wants to read, and heads* `C3.H4`*,* `C3.H7`*,* `C3.H8`*,* `C3.H9`*,* `C3.H10`*, and* `C3.H11` *retrieve the token at that position.*

Skipping analysis identifies `C3` as the most crucial sub-layer (Fig. 3) and Logit-lens confirms that it outputs the desired text. By directly projecting encoder embeddings through `lm_head` ∘ `C3.out_proj` ∘ `C3.v_proj`, we can fully transcribe the image patch-wise.

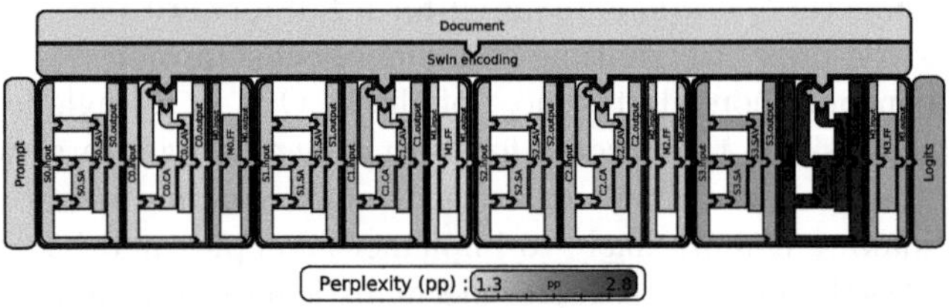

Fig. 3. Impact of skipping decoder sub-layers in Donut-base measured by perplexity. Each sub-layer of the decoder is color-coded according to the increase in perplexity observed when that sub-layer is skipped, highlighting which components are critical to model performance.

To identify the most relevant attention heads for this task, we select those with the lowest cross-attention map variance. We then project into the vocabulary space using only these heads, and we highlight image patches with low entropy in that space to better visualize which regions contribute most confidently to the transcription (Fig. 4).

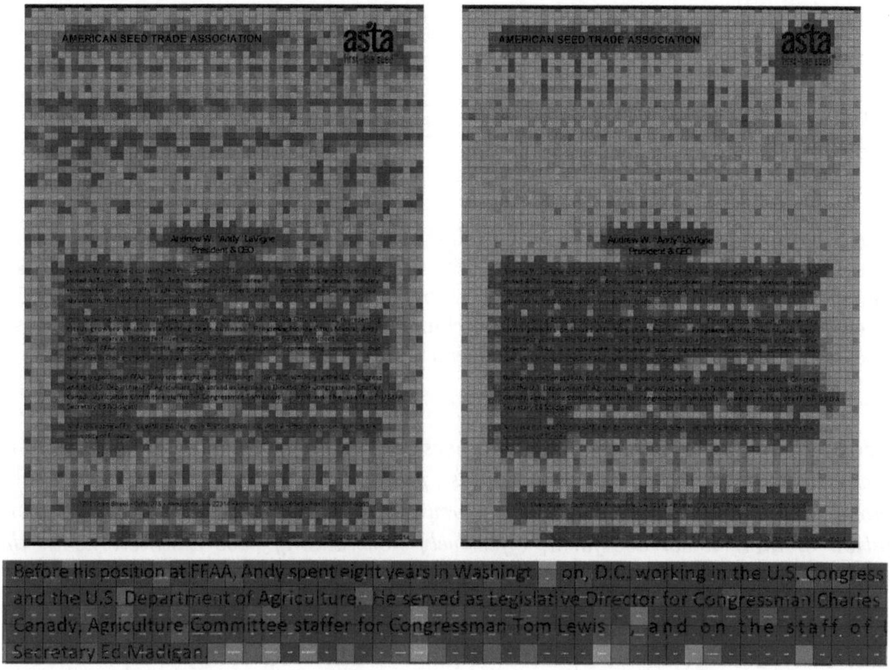

Fig. 4. (left) Projection of patch visual encoding through Swin and the linear transformation `lm_head` ∘ `C3.out_proj` ∘ `C3.v_proj` from the decoder into tokens. We colored in blue patches with a low entropy on the token distribution, and overlayed the most likely token for each. (right) We removed least important heads, leading to a clear improvement over text transcription fidelity. (bottom) This is a zoomed version to improve the lisibility of the text.

Proposition 2. *`M3` sole purpose is to remove the residual stream from `C3` which was necessary to construct the cross-attention maps.*

Since we know that `C3` already has the answer and Donut-base only outputs text present in the document, keeping `M3` becomes questionable. We found that, in fact, it removes the residual stream from `C3` thanks to logit-lens. These findings are also supported by skipping, cross-attention inspection, and activation patching. Removing `M3`, or equivalently removing `C3`'s residual, yields near-identical perplexity scores (Fig. 5), suggesting `M3` acts as a cleaner rather than a transformer.

Case-Sensitivity

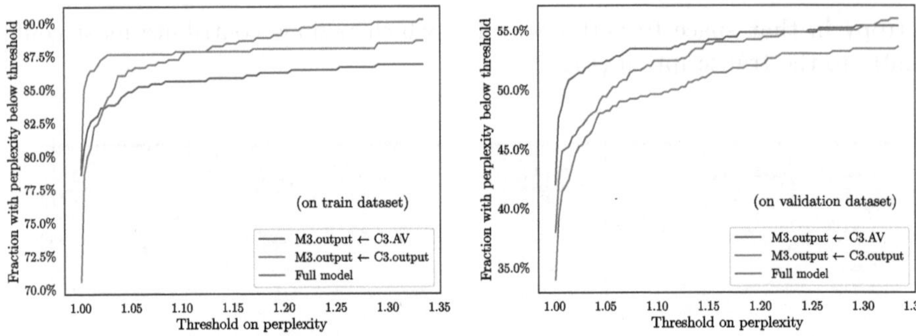

Fig. 5. Accuracy versus perplexity threshold for the first generated token of the M3 sub-layer after activation patching. Results are shown on both train and validation splits, plotting the fraction of samples with perplexity below each threshold under three conditions: skipping M3 entirely, patching M3 with activations from the last cross-attention times value (C3.AV), and the clean run.

Proposition 3. *The first two sub-layers,* S0 *and* C0*, transform token embeddings into other token-like embeddings, primarily to align special tokens with those used during pretraining. Feed-forward* M0 *maps tokens to the representation of their lowercase form, effectively neutralizing case distinctions in the embedding space.*

Using logit-lens and a novel technique we call *token reprojection*, we show that early decoder layers operate within a token embedding subspace. Given activation x, we reproject as $x' = \text{embedding}(\arg\max(\text{lm\_head}(x)))$, collapsing x to its closest interpretable token. Applying this before M0 retains output quality, suggesting that M0 enforces lowercasing while preserving semantic identity (Fig. 6).

Logit-lens is sufficient to see that M0 converts question representation into lowercase. We introduce a simple technique called *token reprojection*, which allows us to remove a contiguous block of early decoder layers without performance degradation. Given a hidden activation x at any layer, we compute $x' = \text{embedding}\,(\arg\max\,(\text{lm\_head}(x)))$, effectively projecting x back onto the vocabulary embedding manifold. This operation enforces a discrete constraint on the representation, collapsing it to the nearest interpretable token embedding. Remarkably, we find that applying token reprojection before the first feed-forward layer (M0) yields outputs nearly indistinguishable from the original model. This indicates that all intervening layers, including M0, operate within or close to the embedding subspace, and can therefore be bypassed or replaced by a simple re-embedding. By projecting activations onto discrete token embeddings, token reprojection tests whether additional information in other dimensions is necessary, enabling aggressive layer pruning without loss of performance.

Keyword-Based Questions

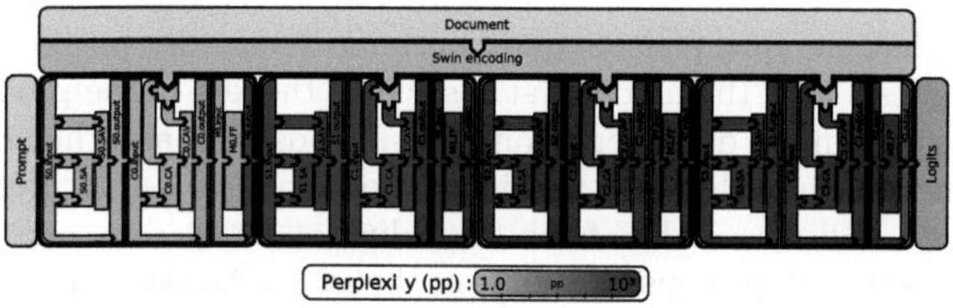

Fig. 6. Impact of *token reprojection* on decoder sub-layers in Donut-base measured by perplexity. Each sub-layer of the decoder is color-coded according to the increase in perplexity observed when that sub-layer activations are replaced with their reprojected embeddings.

Proposition 4. *Keyword-based questions refer to those targeting a generic field such as "date", "title", or "name". In these cases, sub-layers S0, S1, S2 propagate information from the keyword token to the special* `<s_answer>` *token, transforming its representation into a semantic query. This query guides the model to attend to any instance of the targeted concept within the document.*

We show the attention flow from the keyword token to <s_answer> by analyzing the self-attentions. The keyword's activation is transformed into a query vector that directs attention to all instances of the relevant concept, such as dates. Furthermore, we identify the minimal circuit capable of replicating this activation-to-query transformation. Detailed analysis can be found in ??.

We first demonstrate that the query vector constructed in `C3` is semantically grounded: for example, when prompted with "What is the date?", the corresponding cross-attention consistently targets date-related regions in the image. Visualizing the self-attention graph from the input tokens to <s_answer> reveals that information propagates from the keyword to the final token, which subsequently produces the query. Notably, this query appears to be independent of the visual input.

To further isolate the mechanism generating the query, we explored all 2^5 possible signal pathways from the embedding of the keyword token (e.g., `''date''`) to the query vector in the final cross-attention sub-layer. We considered two types of signal propagation through each sub-layer: (1) residual connections, and (2) either the self-attention value/output projections or, in the case of feed-forward layers, the linear transformations. This exhaustive search revealed that the most effective circuit consists solely of the final self-attention sub-layer.

Contextualized Questions

Proposition 5. *In structured documents such as forms or tables, some questions query values associated with visible keys (e.g., "What is the place of birth of this person?"). The model attends to the key, extracts its positional embedding, and uses this information to form a query that locates the corresponding value—typically by attending to content on the same line or visual axis.*

We show that the query at `C3.query` primarily encodes positional information. This suggests that the model first associates the key in the prompt with its visual counterpart in the document, computes its position, and then retrieves the corresponding content. This task requires at least one cross-attention sub-layer to read the key, and a second to fetch the value.

Using cross-attention maps, we confirm that `C1`, `C2`, and `C3` attend to image patches associated with the prompt. On a sanitized synthetic dataset, activation patching confirms that the query formed at `C3` retrieves content located along the same visual axis as the key.

To further analyze how `C3.query` retrieves content, we build a controlled synthetic setup: key-value pairs drawn from finite sets K and V are placed on the image in the format `key: value`. The model is prompted with "What is `key`?", and we evaluate its ability to retrieve the corresponding value. After confirming correct behavior on the clean image, we generate a patched version where values are reassigned randomly to different keys, preserving positions. This setup is illustrated in Fig. 7.

We evaluate four hypotheses regarding retrieval:

1. **Positional retrieval**: `C3.query` encodes a fixed spatial position and retrieves the content originally located there, regardless of content changes.
2. **Semantic adaptation**: `C3.query` dynamically adapts and retrieves the correct updated value for the key.
3. **Value memorization**: `C3.query` outputs a memorized token, independent of the prompt or position.
4. **Failure mode**: the model produces unrelated tokens, failing to retrieve any relevant information.

When $K = V$ are color names, the model retrieves the value at the original position in 53% of cases (hypothesis 1), correctly adapts to the new value in 31% (hypothesis 2), outputs the previously correct value in 8% (hypothesis 3), and fails in 8% (hypothesis 4). This reflects a strong positional bias, with moderate adaptability.

Replacing colors with common nouns leads to decreased performance: positional retrieval dominates at 40.4%, adaptation drops to 26.2%, memorization accounts for 3.7%, and failures rise to 29.7%. This suggests greater ambiguity or weaker priors in these tokens.

When using spelled-out numbers, we observe a more balanced behavior: 37.2% positional retrieval, 44.6% semantic adaptation, and 9.1% each for memorization and failure. This indicates that numerical tokens may support more flexible and robust grounding.

Generating All the Answer. To continue generating tokens autoregressively, the model leverages the same mechanisms used for answering contextualized questions, since the previously retrieved value now becomes part of the new prompt. Thus, the reading circuit generalizes to multi-token sequences.

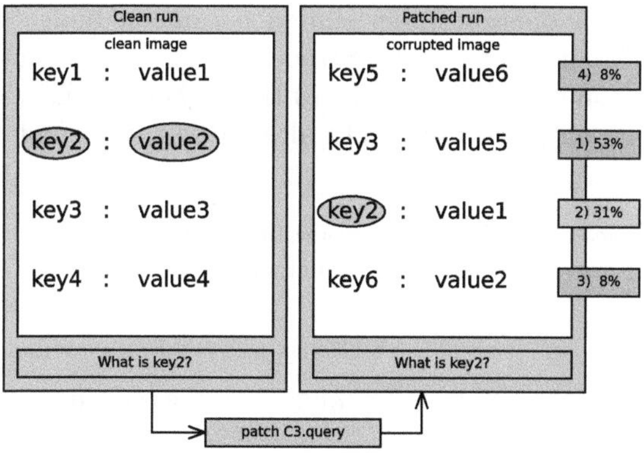

Fig. 7. Evaluating retrieval behavior of the `C3.query` vector under activation patching in a synthetic key-value setup. The figure shows two images: the original (clean) and the corrupted version where key-value pairs are shuffled. We patch `C3.query` and classify the model's output tokens according to four hypotheses: retrieving the value at the original position (positional retrieval), retrieving the updated correct value (semantic adaptation), retrieving the previously memorized value (value memorization), or producing an unrelated token (failure). Accuracy for each category is displayed in boxes, illustrating the balance between positional bias and semantic adaptation.

4.3 Comparison with State of the Art

In Table 2, we compare our mechanistic interpretability-guided pruning approach to various state-of-the-art (SOTA) techniques for reducing model size and computational cost. We consider both coarse-grained and fine-grained pruning strategies, including structured and unstructured baselines, and report their impact on ANLS, parameter count, and FLOPs.

Coarse-grained pruning removes entire sub-layers in the decoder, targeting both parameter and FLOP reductions while preserving core functionality. Our goal is to reduce the number of parameters to approximately 40% and FLOPs by half. We benchmark our MI-guided variant, Donut-MINT$_{31\%}$, against the top three students obtained via brute-force enumeration under similar constraints (24M parameters, 49% FLOPs). We name them Donut-T1, Donut-T2 and Donut-T3. All models, including Donut-MINT$_{31\%}$, are retrained through knowledge distillation to ensure fairness. As shown in Table 2, our method consistently outperforms enumeration-based students in ANLS (50% vs. 22–28%), indicating that MI can more effectively identify critical substructures.

Fine-grained pruning further reduces the model by removing individual attention heads. This yields Donut-MINT$_{7\%}$, produced by applying MI-guided structured pruning to eliminate the least functionally relevant heads, with a target of 7% of the original decoder parameters. We compare this to two strong baselines. First, the structured pruning method from [56], which removes components based on saliency metrics. Second, the unstructured pruning approach Wanda [43], which prunes weights using magnitude and activation norm heuristics. All three models operate under similar parameter budgets (7%) and comparable FLOPs (Donut-MINT: 13%, Structured: 12%, Unstructured: 100%). As seen in Table 2, Donut-MINT$_{7\%}$ achieves the highest ANLS (51%) despite the aggressive pruning, outperforming both alternatives (50% and 46%).

Table 2. Performance and efficiency comparison of student models on DocVQA under coarse-grained and fine-grained pruning constraints.

Model	↑ ANLS (on test)	↓ Parameters	↓ FLOPs
Teacher: Donut-base	**66%**	64M (100%)	1.57T (100%)
	Coarse-grained students		
Donut-T1	23%	24M (37%)	766G (49%)
Donut-T2	22%	24M (37%)	766G (49%)
Donut-T3	28%	24M (37%)	766G (49%)
Donut-MINT$_{31\%}$ (ours)	**50%**	20M (31%)	757G (48%)
	Fine-grained students		
Unstructured-Pruning [43]	46%	4.48M (7%)	1.57T (100%)
Structured-Pruning [56]	50%	4.48M (7%)	188G (12%)
Donut-MINT$_{7\%}$ (ours)	**51%**	4.48M (7%)	211G (13%)

We note that matching parameter count and FLOPs exactly across methods is infeasible, as these quantities do not scale linearly; pruning structural components such as heads or blocks impacts them differently. Still, the FLOPs remain close enough for a meaningful comparison. Notably, unstructured pruning achieves parameter reduction but no FLOPs savings, as it retains dense computation unless specialized sparse inference is used.

These results confirm that MI-guided pruning retains essential computational pathways more effectively than heuristic-based methods, especially under tight resource constraints. Our approach offers a principled alternative to trial-and-error search or hand-designed heuristics, delivering compact models with better performance preservation on Document Visual Question Answering.

5 Conclusion

In this work, we introduced a novel approach to model compression for document Visual Question Answering by leveraging mechanistic interpretability to guide pruning and architectural simplification, followed by knowledge distillation to recover performance. Rather than relying on generic or heuristic pruning, we analyzed the inner workings of a vision-language model (Donut) to identify components essential to task performance. This yielded DonutMINT, a lightweight student model that retains high accuracy while significantly reducing computational cost and parameter count. Across both coarse and fine-grained regimes, DonutMINT consistently outperformed state-of-the-art pruning baselines under comparable resource constraints.

Our analysis led to the removal of several self and cross attention heads. However, this does not mean these heads are universally redundant; their utility may depend on the task or domain. While mechanistic interpretability offers a fine-grained view of internal functions, it is not yet exploited during distillation, which uses only final logits. Future work could incorporate intermediate activations, now better understood, to define auxiliary losses and enhance distillation beyond conventional methods.

Our current pipeline relies on manual inspection and bespoke tools, but there is considerable potential in automating interpretability-driven pruning. Automation would enable scaling to larger models and diverse architectures, making principled compression more reproducible and widely applicable. The Visually rich Document Understanding (VrDU) setting poses specific challenges, due to complex multimodal interactions that resist manual probing. This motivates the need for automated corruption techniques capable of perturbing latent states in a controlled and semantically meaningful way. Promising directions include structured textual corruption [47,48] and multimodal interventions [15], though the semi-structured nature of VrDU inputs calls for more adaptive methods. Progress in automated circuit discovery [11] and efficient attribution-based patching [34,44] suggests that scalable interpretability pipelines are increasingly feasible. Beyond compression, such tools could support model editing, interpretability audits, and safety checks in high-stakes domains.

Disclosure of Interests. The authors have no competing interests to declare that are relevant to the content of this article.

References

1. Appalaraju, S., et al.: DocformerV2: local features for document understanding (2023). https://arxiv.org/abs/2306.01733
2. Back, J., Ahn, N., Kim, J.: Magnitude attention-based dynamic pruning (2023). https://arxiv.org/abs/2306.05056
3. Bai, N., Iyer, R.A., Oikarinen, T., Kulkarni, A., Weng, T.W.: Interpreting neurons in deep vision networks with language models (2025). https://arxiv.org/abs/2403.13771
4. Bai, S., et al.: Qwen2.5-VL technical report (2025). https://arxiv.org/abs/2502.13923
5. Basu, S., Grayson, M., Morrison, C., Nushi, B., Feizi, S., Massiceti, D.: Understanding information storage and transfer in multi-modal large language models (2024). https://arxiv.org/abs/2406.04236

6. Basu, S., Zhao, N., Morariu, V., Feizi, S., Manjunatha, V.: Localizing and editing knowledge in text-to-image generative models (2023). https://arxiv.org/abs/2310.13730
7. Bau, D., Zhou, B., Khosla, A., Oliva, A., Torralba, A.: Network dissection: quantifying interpretability of deep visual representations (2017). https://arxiv.org/abs/1704.05796
8. Bereska, L., Gavves, E.: Mechanistic interpretability for AI safety – a review (2024). https://arxiv.org/abs/2404.14082
9. Biten, A.F., et al.: Scene text visual question answering (2019). https://arxiv.org/abs/1905.13648
10. Burns, C., Ye, H., Klein, D., Steinhardt, J.: Discovering latent knowledge in language models without supervision (2022). https://arxiv.org/abs/2212.03827
11. Conmy, A., Mavor-Parker, A.N., Lynch, A., Heimersheim, S., Garriga-Alonso, A.: Towards automated circuit discovery for mechanistic interpretability (2023). https://arxiv.org/abs/2304.14997
12. Dosovitskiy, A., et al.: An image is worth 16x16 words: transformers for image recognition at scale (2021). https://arxiv.org/abs/2010.11929
13. Frankle, J., Carbin, M.: The lottery ticket hypothesis: finding sparse, trainable neural networks (2019). https://arxiv.org/abs/1803.03635
14. Gandelsman, Y., Efros, A.A., Steinhardt, J.: Interpreting clip's image representation via text-based decomposition (2024). https://arxiv.org/abs/2310.05916
15. Golovanevsky, M., Rudman, W., Palit, V., Singh, R., Eickhoff, C.: What do VLMS notice? A mechanistic interpretability pipeline for gaussian-noise-free text-image corruption and evaluation (2025). https://arxiv.org/abs/2406.16320
16. Gurnee, W., Nanda, N., Pauly, M., Harvey, K., Troitskii, D., Bertsimas, D.: Finding neurons in a haystack: case studies with sparse probing (2023). https://arxiv.org/abs/2305.01610
17. Han, S., Pool, J., Tran, J., Dally, W.: Learning both weights and connections for efficient neural network. In: Cortes, C., Lawrence, N., Lee, D., Sugiyama, M., Garnett, R. (eds.) Advances in Neural Information Processing Systems, vol. 28. Curran Associates, Inc. (2015). https://proceedings.neurips.cc/paper_files/paper/2015/file/ae0eb3eed39d2bcef4622b2499a05fe6-Paper.pdf
18. Hanna, M., Liu, O., Variengien, A.: How does GPT-2 compute greater-than?: Interpreting mathematical abilities in a pre-trained language model (2023). https://arxiv.org/abs/2305.00586
19. Hertz, A., Mokady, R., Tenenbaum, J., Aberman, K., Pritch, Y., Cohen-Or, D.: prompt-to-prompt image editing with cross attention control (2022). https://arxiv.org/abs/2208.01626
20. Hinton, G., Vinyals, O., Dean, J.: Distilling the knowledge in a neural network (2015). https://arxiv.org/abs/1503.02531
21. Hu, H., Zhao, P., Li, P., Zheng, Y., Wang, Z., Yuan, X.: FASP: fast and accurate structured pruning of large language models (2025). https://arxiv.org/abs/2501.09412
22. Kim, G., et al.: OCR-free document understanding transformer (2022). https://arxiv.org/abs/2111.15664
23. Lee, K., et al.: Pix2Struct: screenshot parsing as pretraining for visual language understanding (2023). https://arxiv.org/abs/2210.03347
24. Li, C., et al.: PP-OCRV3: more attempts for the improvement of ultra lightweight OCR system (2022). https://arxiv.org/abs/2206.03001
25. Li, M., et al.: TROCR: transformer-based optical character recognition with pre-trained models (2022). https://arxiv.org/abs/2109.10282

26. Lin, Z., et al.: A survey on mechanistic interpretability for multi-modal foundation models (2025). https://arxiv.org/abs/2502.17516
27. Ling, G., Wang, Z., Yan, Y., Liu, Q.: SlimGPT: layer-wise structured pruning for large language models (2024). https://arxiv.org/abs/2412.18110
28. Liu, H., Li, C., Wu, Q., Lee, Y.J.: Visual instruction tuning (2023). https://arxiv.org/abs/2304.08485
29. Liu, Y., et al.: Multilingual denoising pre-training for neural machine translation (2020). https://arxiv.org/abs/2001.08210
30. Liu, Z., et al.: Swin transformer: hierarchical vision transformer using shifted windows (2021). https://arxiv.org/abs/2103.14030
31. Mathew, M., Karatzas, D., Jawahar, C.V.: DocVQA: a dataset for VQA on document images (2020). https://arxiv.org/abs/2007.00398
32. Meng, K., Bau, D., Andonian, A., Belinkov, Y.: Locating and editing factual associations in GPT (2023). https://arxiv.org/abs/2202.05262
33. Muralidharan, S., et al.: Compact language models via pruning and knowledge distillation (2024). https://arxiv.org/abs/2407.14679
34. Nanda, N.: Attribution patching: activation patching at industrial scale (2023). https://www.neelnanda.io/mechanistic-interpretability/attribution-patching. Accessed 27 Apr 2025
35. nostalgebraist: Interpreting GPT: the logit lens (2020). https://www.lesswrong.com/posts/AcKRB8wDpdaN6v6ru/interpreting-gpt-the-logit-lens. Accessed 7 Apr 2025
36. Olsson, C., et al.: In-context learning and induction heads (2022). https://arxiv.org/abs/2209.11895
37. Palit, V., Pandey, R., Arora, A., Liang, P.P.: Towards vision-language mechanistic interpretability: a causal tracing tool for blip (2023). https://arxiv.org/abs/2308.14179
38. Rai, D., Yao, Z.: An investigation of neuron activation as a unified lens to explain chain-of-thought eliciting arithmetic reasoning of LLMS (2024). https://arxiv.org/abs/2406.12288
39. Rai, D., Zhou, Y., Feng, S., Saparov, A., Yao, Z.: A practical review of mechanistic interpretability for transformer-based language models (2025). https://arxiv.org/abs/2407.02646
40. Romero, A., Ballas, N., Kahou, S.E., Chassang, A., Gatta, C., Bengio, Y.: FitNets: hints for thin deep nets (2015). https://arxiv.org/abs/1412.6550
41. Sakarvadia, M., et al.: Attention lens: a tool for mechanistically interpreting the attention head information retrieval mechanism (2023). https://arxiv.org/abs/2310.16270
42. Shaikh, A., Cochez, M., Diachkov, D., de Rijcke, M., Yousefi, S.: Donut-hole: Donut sparsification by harnessing knowledge and optimizing learning efficiency (2023). https://arxiv.org/abs/2311.05778
43. Sun, M., Liu, Z., Bair, A., Kolter, J.Z.: A simple and effective pruning approach for large language models (2024). https://arxiv.org/abs/2306.11695
44. Syed, A., Rager, C., Conmy, A.: Attribution patching outperforms automated circuit discovery (2023). https://arxiv.org/abs/2310.10348
45. Tung, F., Mori, G.: Similarity-preserving knowledge distillation (2019). https://arxiv.org/abs/1907.09682
46. Vaswani, A., et al.: Attention is all you need (2017). https://arxiv.org/abs/1706.03762

47. Vig, J., et al.: Investigating gender bias in language models using causal mediation analysis. In: Larochelle, H., Ranzato, M., Hadsell, R., Balcan, M., Lin, H. (eds.) Advances in Neural Information Processing Systems, vol. 33, pp. 12388–12401. Curran Associates, Inc. (2020). https://proceedings.neurips.cc/paper_files/paper/2020/file/92650b2e92217715fe312e6fa7b90d82-Paper.pdf
48. Wang, K., Variengien, A., Conmy, A., Shlegeris, B., Steinhardt, J.: Interpretability in the wild: a circuit for indirect object identification in GPT-2 small (2022). https://arxiv.org/abs/2211.00593
49. Wang, Z., et al.: SmartTrim: adaptive tokens and attention pruning for efficient vision-language models (2023). https://arxiv.org/abs/2305.15033
50. Xu, Y., Li, M., Cui, L., Huang, S., Wei, F., Zhou, M.: LayoutLM: pre-training of text and layout for document image understanding. In: Proceedings of the 26th ACM SIGKDD International Conference on Knowledge Discovery & Data Mining. ACM (2019). https://doi.org/10.1145/3394486.3403172
51. Yang, D., Cao, B., Zhang, A., Gu, W., Hu, W., Chen, G.: Beyond intermediate states: Explaining visual redundancy through language (2025). https://arxiv.org/abs/2503.20540
52. Ye, X., Gan, Y., Ge, Y., Zhang, X.P., Tang, Y.: ATP-LLAVA: adaptive token pruning for large vision language models (2024). https://arxiv.org/abs/2412.00447
53. Yu, Z., Ananiadou, S.: Understanding multimodal LLMS: the mechanistic interpretability of LLAVA in visual question answering (2025). https://arxiv.org/abs/2411.10950
54. Zafrir, O., Larey, A., Boudoukh, G., Shen, H., Wasserblat, M.: Prune once for all: sparse pre-trained language models (2021). https://arxiv.org/abs/2111.05754
55. Zagoruyko, S., Komodakis, N.: Paying more attention to attention: improving the performance of convolutional neural networks via attention transfer (2017). https://arxiv.org/abs/1612.03928
56. Zhang, Y., et al.: FinerCut: finer-grained interpretable layer pruning for large language models (2024). https://arxiv.org/abs/2405.18218
57. Zhu, J., et al.: InternVL3: exploring advanced training and test-time recipes for open-source multimodal models (2025). https://arxiv.org/abs/2504.10479

Link Prediction Graph Neural Networks for Structure Recognition of Handwritten Mathematical Expressions

Cuong Tuan Nguyen(✉), Ngoc Tuan Nguyen, Triet Hoang Minh Dao, Huy Minh Nhat Nguyen, and Huy Truong Dinh

Vietnamese-German University, Ho Chi Minh City, Vietnam
{cuong.nt2,huy.td}@vgu.edu.vn,
{21024003,10423179,10423045}@student.vgu.edu.vn

Abstract. We propose a Graph Neural Network (GNN)-based approach for Handwritten Mathematical Expression (HME) recognition by modeling HMEs as graphs, where nodes represent symbols and edges capture spatial dependencies. A deep BLSTM network is used for symbol segmentation, recognition, and spatial relation classification, forming an initial primitive graph. A 2D-CFG parser then generates all possible spatial relations, while the GNN-based link prediction model refines the structure by removing unnecessary connections, ultimately forming the Symbol Label Graph. Experimental results demonstrate the effectiveness of our approach, showing promising performance in HME structure recognition.

Keywords: GNN · EGAT · math · handwriting recognition

1 Introduction

Mathematical notations are essential in scientific documents for conveying concepts in mathematics and physics. With the emergence of digital pens and tablets, there is an increasing trend toward using handwritten mathematical notations as input, leading to a demand for systems that recognize these handwritten mathematical expressions (HMEs). We refer to the problem of online handwriting recognition, where each input sample is a sequence of trajectories captured by each pen-down called handwritten strokes.

Recognizing HMEs presents a challenging problem due to the complex two-dimensional structure of mathematical expressions. Early works focus on using structural parsing methods with a context-free grammar to represent mathematical expressions [4,10,17], tree structure networks [13], or graph-based algorithms [2]. However, due to the separated tasks of symbol segmentation, symbol recognition and structural analysis, these methods have limited overall expression recognition rate.

L. Jin et al. (Eds.): ICDAR 2025 Workshops, LNCS 16225, pp. 279–291, 2026.
https://doi.org/10.1007/978-3-032-09368-4_17

With the rise of deep learning algorithms, encoder-decoder approaches solve the problem of separated tasks by an end-to-end learning model. The encoder-decoder models developed from Recurrent Neural Networks (RNNs) decoder with attention [11,12], to the transformer decoder [14,15] and multi-task learning [16]. These techniques focus on recognizing HME images and outputting one-dimensional character sequences in LaTeX format. However, they often fail to explicitly represent the hierarchical and spatial characteristics inherent to mathematical notations. Without clear symbol segmentation and two-dimensional relationships, these methods have limited effectiveness in practical scenarios that require structural clarity for computational manipulation and interactive editing.

Recent advancements have leveraged graph neural networks (GNNs) to address these limitations. Wu et al. [8] proposed a graph-to-graph (G2G) learning method using stroke-level representations, modeling handwriting strokes as nodes and spatial relationships as edges. Xie et al. [9] expanded this concept with an Edge Graph Attention Network (EGAT) focused on stroke-level interactions. Although stroke-level methods capture detailed structural information, they face the complexity of the graph due to numerous stroke primitives involved.

We propose a symbol-level graph recognition method for handwritten math expression recognition to overcome these limitations. By shifting from stroke-level to symbol-level representation, our approach reduces structural complexity of the graph. Our method integrates several key components:

First, we employ a Bidirectional Long Short-Term Memory (BLSTM) network for simultaneous symbol segmentation, recognition, and initial spatial relationship classification. Processing online handwritten data, this approach uses global contextual information to improve recognition of three tasks.

Next, we utilize a two-dimensional Cocke-Younger-Kasami (CYK) parsing algorithm to explore all possible spatial relationships among recognized symbols, constructing a symbol-level graph with recognized symbols as nodes and recognized relations as edges.

Finally, we further refine structural relationships using an Edge-featured Graph Attention Network (EGAT) on symbol-level graphs for a link prediction task. EGAT link prediction predicts whether an edge should be removed to produce the Symbol Layout Tree (SLT).

In summary, the primary contributions of this paper are:

- Formulate the problem of mathematical structure recognition to the problem of building symbol-level graph and link prediction.
- Propose a deep BLSTM and 2D CYK parser to construct symbol-level graph.
- Apply EGAT model for symbol-level link prediction to output mathematical expression as an SLT.

The remainder of this paper is structured as follows: Sect. 2 reviews related works, Sect. 3 describes our proposed methodology in detail, Sect. 4 presents experimental results and analysis, and Sect. 5 concludes with key insights and future research directions.

The source code for the experiments is available in https://github.com/ntcuong2103/math_online_egat

2 Related Works

Handwritten Mathematical Expression (HME) recognition has evolved over recent decades, progressing from grammar-driven and heuristic-based methods to sophisticated deep learning and Graph Neural Network (GNN)-based approaches. The central challenge remains accurately parsing the complex, hierarchical two-dimensional structure inherent in mathematical expressions.

2.1 Graph-Based Structural Methods

Hu and Zanibbi [2] formulate the problem of mathematical expression recognition as a Minimum Spanning Tree (MST) parsing by employing Edmonds' algorithm on directed Line-Of-Sight (LOS) graphs [1]. Their MST-based parser efficiently constructs expression trees by selecting edges that optimize spatial relationships among symbols, considerably reducing computational complexity and improving parsing accuracy compared to traditional CYK-based methods. Although MST approaches delivered promising improvements, their success heavily depended on accurate symbol segmentation and preliminary edge prediction.

Truong et al. used rule based split and merge methods for constructing the SLT [7]. The method benefits from the global context of an HME for symbol and relation classification [5].

2.2 Graph Neural Network-Based Methods

Recent advancements have shifted towards deep learning-based models, particularly those leveraging Graph Neural Networks (GNNs), due to their intrinsic capability of modeling structured data. GNNs directly encode structural relationships within a graph representation, exploiting local context through iterative message-passing mechanisms to refine node and edge predictions.

Stroke-Level GNN. Stroke-level methods treat individual pen strokes as graph primitives, bypassing explicit symbol segmentation. Wu et al. [8] introduced a pioneering Graph-to-Graph (G2G) model, conceptualizing HME recognition as graph transformation from raw stroke primitives to recognized symbol and relation graphs. Their approach employed sub-graph attention mechanisms to align input strokes precisely with their corresponding recognized symbols, simultaneously addressing segmentation and structural recognition tasks.

Building upon stroke-level concepts, Xie et al. [9] proposed the Edge-weighted Graph Attention Network (EGAT) for HME structure recognition. EGAT integrates edge features into attention computations, directly predicting symbol identities and spatial relations from stroke graphs. Although stroke-level GNN methods are advantageous in reducing error propagation from segmentation stages, they inherently produce highly dense graphs, significantly increasing computational demands.

Symbol-Level GNN. Symbol-level GNN models address stroke-level complexity by pre-segmenting strokes into symbols. Tang et al. [6] utilized a Graph Reasoning Network (GRN), structuring detected symbols from HME image into an SLT. GRN methods significantly enhance recognition accuracy by reducing the graph size and emphasizing meaningful symbol interactions, thus providing an efficient trade-off between segmentation complexity and graph representation fidelity.

Summary. Our method builds a symbol-level graph of recognized symbols and spatial relations between symbols, then applies GNN to predict whether an edge should be removed from the graph. The approach is similar to the MST, but we use GNN instead of Edmond's algorithm to remove edges. We also propose using the 2D CYK algorithm to build the symbol-level graph instead of the LOS algorithm.

3 Methodology

In this section, we propose a handwritten mathematical expression recognition method that constructs a symbol-level graph as input to the GNN for link prediction to refine the connections.

3.1 Overview of the Method

The pipeline of our method is shown in Fig. 1. A sequence of extracted features from an online HME input is fed to a deep BLSTM to predict symbols and relations. The segmented symbols are input to a symbol-level parser to build the symbol-level graph. Finally, the symbol-level graph is input to GNN for link prediction and produces SLT output.

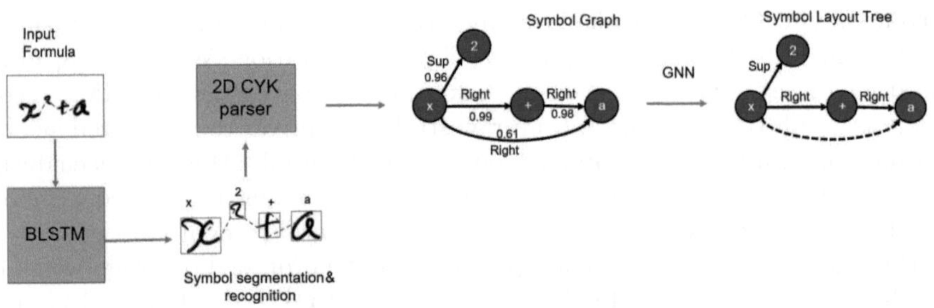

Fig. 1. Overview of the method

3.2 Build Symbol Level Graph

This consists of two main stages as shown in Fig. 2: (1) symbol segmentation, recognition, and relation detection using a temporal recognition BLSTM network, and (2) symbol-level graph construction via a 2D CYK parsing algorithm. The BLSTM outputs, at each time step corresponding to a recognized symbol or pair of symbols, a probability distribution over the relation classes. These probabilities seed our primitive graph construction.

Symbol and Relation Recognition. This stage utilizes a deep bidirectional long short-term memory (BLSTM) network to process handwritten mathematical expressions. This model leverages the global context for symbol segmentation, symbol recognition, and relation classification [5]. The method imposes a constraint to ensure the model outputs the relation classification at a precise time step, simultaneously addressing both symbol segmentation and relation classification. As shown in Level 0 of Fig. 2, the model takes the raw input sequence of strokes and segments them into symbols (e.g., 'x', '2', '+', 'a'). A key characteristic of this stage is that it primarily identifies relations between symbols that are consecutive in the writing order. We currently restrict the BLSTM's relation predictions to consecutive-in-writing-order pairs to exploit temporal context. However, writing order can vary (e.g., users draw subscripts out of sequence). In practice this can cause missing edges for non-adjacent symbols. Thus, we resolve the problem by feeding the BLSTM arbitrary symbol pairs, which discussed in the next step of symbol-level graph construction.

Symbol-Level Graph Construction. The relations identified by the BLSTM are insufficient for capturing the full 2D structure. Therefore, in the second stage, we use a parser based on a 2D Stochastic Context-Free Grammar (2D-CFG) [17] and the Cocke-Younger-Kasami (CYK) algorithm to build a more complete primitive graph. As depicted in Level 1 of Fig. 2, this parser considers all pairs of recognized symbols, not just consecutive ones. To determine the spatial relation between any two non-consecutive symbols (e.g., 'x' and 'a'), a new, temporary input sequence is constructed containing only the strokes for those two symbols. This new sequence is then fed into the same pre-trained BLSTM model to obtain a predicted spatial relation and its probability. This ensures all plausible pairwise relations are considered.

Filter Redundant Edges Using Line-of-Sight. We also applied the LOS algorithm [1] for filtering redundant edges. Let E_{CYK} be the set of edges proposed by CYK algorithm, E_{LOS} be the set of edges proposed by LOS algorithm. The output set of edges is:

$$E_{\text{CYK \& LOS}} = E_{\text{CYK}} \wedge E_{\text{LOS}} \tag{1}$$

The result of this entire process is a dense, symbol-level graph, like the one shown in the "Symbol Graph" part of 1, which serves as the input to our GNN.

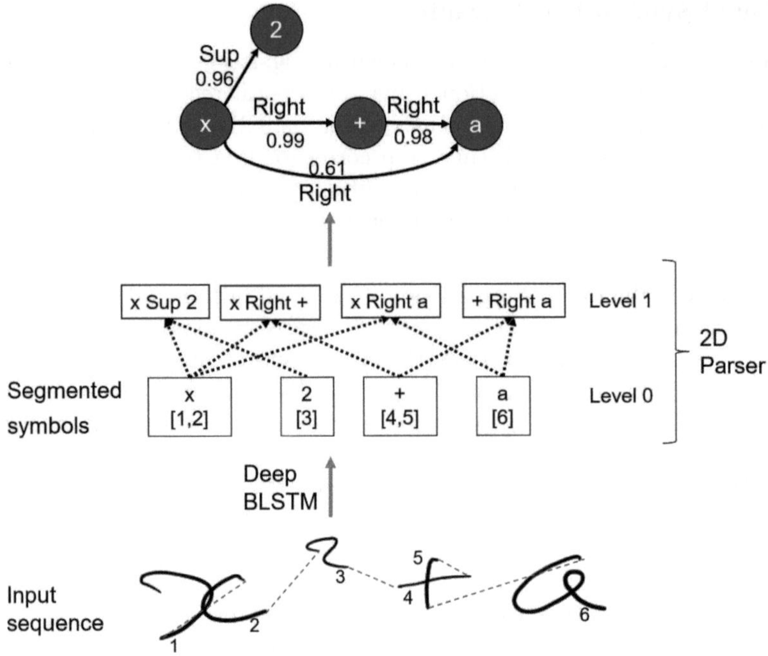

Fig. 2. Symbol-level graph constructed by deep BLSTM and 2D CYK parsing.

3.3 Edge Graph Attention Networks for Link Prediction

To further refine the symbol-level graphs generated by the CYK parsing and Line-of-Sight filtering, we utilize Edge Graph Attention Networks (EGAT). The EGATs are specifically designed to handle graph data by capturing rich structural relationships through attention mechanisms applied directly to edge features and node features. This capability makes EGAT particularly suitable for tasks involving intricate spatial relationships, such as handwritten mathematical expression recognition. Figure 3 illustrates the EGAT that refines the initial symbol-level graph to form an SLT output.

EGAT Architecture. Our EGAT model is implemented using the Deep Graph Library (DGL) on a PyTorch backend. It comprises four stacked EGATConv layers [3], each of which updates node and edge embeddings via an edge-integrated attention mechanism. The final embeddings are passed to a two-layer MLP classification head, which outputs logits over the relation classes. All layers use LeakyReLU activations and 0.2 dropout to mitigate overfitting. The attention mechanism in EGAT is defined by:

$$e_{ij} = F_{\text{att}}(f'_{ij}) \tag{2}$$

$$f'_{ij} = \text{LeakyReLU}(A[h_i||f_{ij}||h_j]) \tag{3}$$

where f'_{ij} is the updated edge features, h_i and h_j represent the embedding vectors of nodes i and j respectively, f_{ij} denotes the original edge features connecting nodes i and j. A is a learnable weight matrix that projects the concatenated features into a new embedding space suitable for attention computation. The operator $||$ indicates concatenation of two feature vectors.

Input Features. Our EGAT leverages both node and edge information:

1. **Node Features** Each node's feature is a one-hot vector encoding its recognized symbol class.
2. **Edge Features** For each candidate edge, we take its upstream predicted relation (one-hot) and multiply it by the scalar confidence score output by the BLSTM. This produces a weighted edge embedding that encodes both the relation type and its confidence.

Link Prediction. After propagation through the EGAT layers, a final MLP classifier predicts whether an edge should be included or not in the output of SLT. The prediction for each edge uses a combined feature vector, which we term the NodeEdgeFeature. It is created by concatenating the final embeddings of the source node, the edge itself, and the destination node:

$$\text{NodeEdgeFeature} = [h_i||f_{ij}||h_j] \tag{4}$$

This combined feature is then passed to the MLP classifier. This ensures the final classification is conditioned on the full context of the edge and the symbols it connects. Instead of NodeEdgeFeature, the link can also be predicted from the EdgeFeature of f'_{ij}.

Training. The loss of link prediction L is computed using Binary Cross Entropy loss. Training involves optimizing this loss across the entire symbol-level graph dataset, adjusting network parameters to minimize prediction errors. The Binary Cross-Entropy Loss:

$$L = \sum_{i=1}^{N} y_i \cdot \log p_i + (1 - y_i) \cdot \log(1 - p_i) \tag{5}$$

where y_i is the ground-truth label indicating whether an edge exists or not, and p_i is the predicted probability, N is the number of dataset samples.

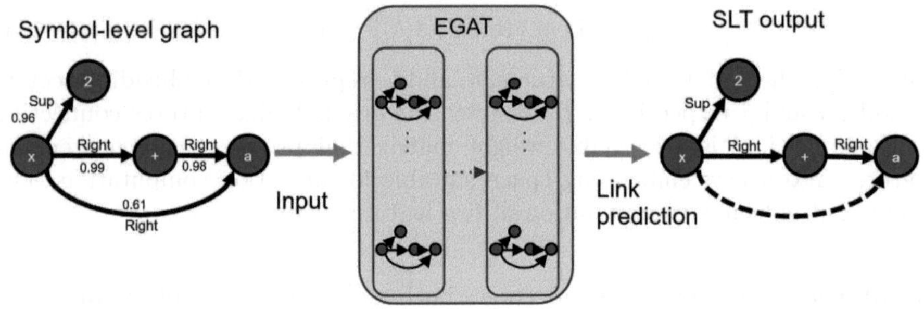

Fig. 3. The link prediction EGAT refines the initial symbol-level graph, removing edges to form an SLT.

4 Experiments

4.1 Datasets and Metrics

We evaluated our model using the CROHME (Competition on Recognition of Online Handwritten Mathematical Expressions) dataset. We used the train dataset from CROHME2019 and evaluated it on three test sets: CROHME2014, CROHME2016, and CROHME2019. The train dataset is used to train an LSTM model for symbol segmentation, symbol recognition, and relation classification, and used to train the EGAT model for link prediction.

For the deep BLSTM model, we used a stack of three BLSTM layers, each containing two LSTM layers (bidirectional) with 128 hidden units. The input sequence of coordiantes in a stroke was sampled by the Ramer method, then extracted four features: the sine and cosine of the writing direction, the normalized distance between two consecutive points and the pen-up/pen-down binary value. For the GNN, we use the stack of four EGAT layers followed by a fully connected layer for link prediction.

We measured the performance of symbol segmentation, symbol recognition, and relation classification using the BLSTM model. For symbol-level graph construction, we evaluated the coverage and redundancy of edges. Coverage measures the recall of ground-truth edges in the primitive graph:

$$\text{Coverage} = \frac{|\{e|e \in G_{\text{primitive}} \wedge e \in G_{\text{ground-truth}}\}|}{|\{e|e \in G_{\text{ground-truth}}\}|} \times 100\% \tag{6}$$

Redundancy measures the percentage of edges rather than that of ground-truth edges in the primitive graph:

$$\text{Redundant} = \left(\frac{|\{e|e \in G_{\text{primitive}}\}|}{|\{e|e \in G_{\text{ground-truth}}\}|} - 1 \right) \times 100\% \tag{7}$$

For link prediction, we evaluated link prediction accuracy. At the expression level, we evaluated the expression rate, the accuracy of recognizing entire mathematical expressions, and the structure rate, whether all the spatial relationships in a mathematical expression are correct.

4.2 Experiments and Results

In the first evaluation, we evaluated the primitive graph produced by the LSTM model and the CYK parser. The result is shown in Table 1. We also evaluated the graph produced by the LOS algorithm and that by the CYK parser filtered by the LOS graph.

Table 1. Primitive graph evaluation on CROHME 2014 dataset

Method	Seg. (%)	Seg. + Cls. (%)	Rel. (%)	Cov. (%)	Red. (%)
BLSTM-CYK	98.09 (87.84)	92.53 (60.37)	92.99 (72.01)	94.69 (79.16)	107.65
BLSTM-LOS	98.09 (87.84)	92.53 (60.37)	92.64 (69.97)	96.09 (84.78)	722.30
BLSTM CYK & LOS	98.09 (87.84)	92.53 (60.37)	92.64 (69.97)	94.33 (76.92)	87.74

The value in parentheses is measured at expression level.
Abbreviation: Seg. (Segmentation), Cls. (Classification), Rel. (Relation), Cov. (Coverage),
Red. (Redundant)

Furthermore, the values in parentheses denote the evaluation metrics at the expression level. An expression is considered correctly recognized only if both its symbols and relations are accurately classified. As a result, these metrics set an upper bound on the overall expression-level performance.

Our analysis indicates that the LOS algorithm effectively recovers missing relationships but significantly elevates graph complexity due to redundant edges. Applying LOS as a filtering step with CYK reduces redundancy substantially with minimal sacrifice in coverage, suggesting an effective compromise between coverage and complexity.

Table 2. EGAT link prediction evaluation

Input graph	Test 2014 Link acc. (%)	Test 2016 Link acc. (%)	Test 2019 Link acc. (%)
BLSTM-CYK	92.21 (57.00)	91.49 (56.22)	92.21 (53.97)
BLSTM-LOS	97.97 (53.22)	97.99 (52.67)	98.12 (54.73)
BLSTM-CYK & LOS	92.61 (60.06)	91.61 (57.09)	92.78 (60.22)

The value in parentheses is measured at expression level

Table 2 presents the link prediction accuracies on the CROHME 2014, 2016, and 2019 test sets. The LOS algorithm achieves superior link prediction performance, with accuracies of 97.97%, 97.99%, and 98.12% for each respective dataset. However, its high number of redundant edges adversely impacts the expression-level accuracy compared to alternative approaches. Notably, when CYK parsing is integrated with LOS filtering, there is a consistent improvement across all datasets with link accuracies of 92.61%, 91.61%, and 92.78%, demonstrating the robust performance of this combined approach in balancing link prediction with overall expression-level reliability.

Table 3. Math expression evaluation

Input graph	Test 2014		Test 2016		Test 2019	
	Exp. (%)	Struct. (%)	Exp. (%)	Struct. (%)	Exp. (%)	Struct. (%)
BLSTM-CYK	36.36	49.23	38.18	48.42	36.07	47.28
BLSTM-LOS	34.83	47.09	37.01	46.11	37.66	48.62
BLSTM-CYK & LOS	37.59	51.07	38.35	48.60	38.86	51.42

Table 3 illustrates the performance on full mathematical expression recognition across the three test sets. Our integrated CYK & LOS approach achieved consistently higher Expression Rates (37.59%, 38.35%, 38.86%) and Structure Recognition Rates (51.07%, 48.60%, 51.42%) compared to individual CYK or LOS methods. These results confirm the effectiveness of combining structural parsing and redundancy filtering.

Table 4. Ablation study of EGAT network

Method	Test 2014 Link acc. (%)	Test 2016 Link acc. (%)	Test 2019 Link acc. (%)
EGAT + EdgeFeature	91.50 (55.16)	91.11 (51.23)	91.82 (55.11)
EGAT + NodeEdgeFeature	92.61 (60.06)	91.61 (57.09)	92.78 (60.22)

In the ablation study summarized in Table 4, we evaluated the impact of including NodeEdgeFeatures in our EGAT model. This inclusion notably improved link accuracy, validating the critical role of comprehensive feature representation in GNN-based parsing methods.

Finally, Table 5 compares our GNN-based method with state-of-the-art models across multiple test sets. Although our approach does not surpass the top-performing systems such as MyScript or Samsung, it demonstrates competitive Expression rate and Structure Recognition Rates. Our method outperforms the graph-based method for building SLT such as the MST algorithm [2], tree BLSTM [13] by a large margin. Compare to other GNN approaches which used GNN output features for symbol and relation classification, we are behind the expression rate since our method is limited by symbol classification performance at expression-level.

Our method demonstrates competitive performance, particularly when compared to other explicit graph-based methods like MST [2], which it outperforms by a large margin. While our expression rates currently lag behind top-performing grammar-based (MyScript) and end-to-end encoder-decoder systems

Table 5. Comparison to state-of-the-art models on CROHME datasets

System	Type	Test 2014		Test 2016		Test 2019	
		Exp. (%)	Struct. (%)	Exp. (%)	Struct. (%)	Exp. (%)	Struct. (%)
MyScript	Gram	62.68	-	67.65	88.14	79.15	90.66
UPV/Wiris	Gram	37.22	-	49.61	74.28	60.13	79.15
Tokyo	Gram	25.66	-	43.94	61.55	39.95	58.22
Samsung	Gram	-	-	65.76	-	79.82	89.32
MST	Graph	26.88	-	-	-	-	-
Tree-BLSTM	Graph	29.91	-	27.03	-	-	-
Tree-construct.	Graph	44.12	58.62	41.76	49.43	-	-
TAP	EncDec	46.90	-	44.80	-	-	-
iFlyTek	EncDec	-	-	57.02	-	80.73	91.49
G2G	GNN	-	-	52.05	-	-	-
LGM-EGAT	GNN	-	-	56.41	-	58.22	83.07
GGM-EGAT	GNN	-	-	56.67	-	60.72	83.74
Our	GNN	37.59	51.07	38.35	48.60	38.86	51.42

(iFlyTek), this highlights a key trade-off. Our pipeline approach separates symbol recognition from structural analysis, meaning errors from the initial BLSTM recognizer can propagate and limit the final expression rate. However, the strength of our approach is its explicit structural output (a graph), which is more interpretable and editable than the LaTeX strings from encoder-decoder models. The strong Structure Rate (e.g., 51.42% on CROHME 2019) shows that the GNN is highly effective at its specific task: learning and predicting the correct graph structure. This indicates significant potential, and future work focused on improving the initial symbol recognizer could close the gap with end-to-end systems. Furthermore, when comparing with other GNN-based methods like LGM-EGAT [9], our performance is comparable, indicating that the current state of symbol-level GNNs for this task is still evolving. The performance gap suggests that future research should focus on more robust primitive graph construction and more powerful GNN architectures.

5 Conclusion

In this paper, we presented a novel framework for handwritten mathematical expression recognition that effectively captures structural relationships using symbol-level graph representations. Our approach addresses key limitations of existing stroke-level graph recognition methods by reducing the graph complexity. Integrating a BLSTM network for simultaneous segmentation, recognition, and spatial relation classification provided robust accuracy on three tasks. The subsequent use of a CYK parsing algorithm, coupled with a practical Line-of-Sight heuristic, generated a clear and concise initial graph structure. Finally, our customized EGAT architecture accurately predicted and refined spatial relationships, producing Symbol Layout Tree as output. Experimental evaluations on the CROHME benchmark dataset demonstrated our proposed symbol-level graph approach's competitive performance in recognition accuracy.

Future research directions include exploring improvements on symbol-level graph construction by analyzing the failure cases, improving link prediction with graph data augmentation and further optimizing computational efficiency for real-time applications.

References

1. Hu, L., Zanibbi, R.: Line-of-sight stroke graphs and parzen shape context features for handwritten math formula representation and symbol segmentation. In: Proceedings of International Conference on Frontiers in Handwriting Recognition, ICFHR **0**, 180–186 (7 2016). https://doi.org/10.1109/ICFHR.2016.0044
2. Hu, L., Zanibbi, R.: Mst-based visual parsing of online handwritten mathematical expressions. In: Proceedings of International Conference on Frontiers in Handwriting Recognition, ICFHR. vol. 0, pp. 337–342. Institute of Electrical and Electronics Engineers Inc. (7 2016). https://doi.org/10.1109/ICFHR.2016.0070
3. Kamiński, K., et al.: Rossmann-toolbox: a deep learning-based protocol for the prediction and design of cofactor specificity in rossmann fold proteins. Brief. Bioinform. **23** (1 2022). https://doi.org/10.1093/BIB/BBAB371
4. Le, A.D., Nakagawa, M.: A system for recognizing online handwritten mathematical expressions by using improved structural analysis. Int. J. Document Anal. Recogn. **19**, 305–319 (12 2016). https://doi.org/10.1007/s10032-016-0272-4
5. Nguyen, C.T., Truong, T.N., Nguyen, H.T., Nakagawa, M.: Global context for improving recognition of online handwritten mathematical expressions. Lecture Notes in Computer Science (including subseries Lecture Notes in Artificial Intelligence and Lecture Notes in Bioinformatics) **12822 LNCS**, 617–631 (2021). https://doi.org/10.1007/978-3-030-86331-9_40
6. Tang, J.M., Wu, J.W., Yin, F., Huang, L.L.: Offline handwritten mathematical expression recognition via graph reasoning network. Lecture Notes in Computer Science (including subseries Lecture Notes in Artificial Intelligence and Lecture Notes in Bioinformatics) **13188 LNCS**, 17–31 (2022). https://doi.org/10.1007/978-3-031-02375-0_2
7. Truong, T.N., Nguyen, H.T., Nguyen, C.T., Nakagawa, M.: Learning symbol relation tree for online handwritten mathematical expression recognition. Lecture Notes in Computer Science (including subseries Lecture Notes in Artificial Intelligence and Lecture Notes in Bioinformatics) **13189 LNCS**, 307–321 (2022). https://doi.org/10.1007/978-3-031-02444-3_23
8. Wu, J.W., Yin, F., Zhang, Y.M., Zhang, X.Y., Liu, C.L.: Graph-to-graph: towards accurate and interpretable online handwritten mathematical expression recognition. In: Proceedings of the AAAI Conference on Artificial Intelligence **35**, 2925–2933 (5 2021). https://doi.org/10.1609/AAAI.V35I4.16399
9. Xie, Y., Zanibbi, R., Mouchère, H.: Local and global graph modeling with edge-weighted graph attention network for handwritten mathematical expression recognition. arXiv preprint arXiv:2410.18555 (10 2024)

10. Yamamoto, R., Sako, S., Nishimoto, T., Sagayama, S.: On-line recognition of handwritten mathematical expressions based on stroke-based stochastic context-free grammar. In: tenth International Workshop on Frontiers in Handwriting Recognition. Suvisoft (10 2006)
11. Zhang, J., Du, J., Dai, L.: Track, attend, and parse (tap): An end-to-end framework for online handwritten mathematical expression recognition. IEEE Trans. Multimed. **21**, 221–233 (1 2019). https://doi.org/10.1109/TMM.2018.2844689
12. Zhang, J., et al.: Watch, attend and parse: an end-to-end neural network based approach to handwritten mathematical expression recognition. Pattern Recogn. **71**, 196–206 (11 2017). https://doi.org/10.1016/J.PATCOG.2017.06.017
13. Zhang, T., Mouchère, H., Viard-Gaudin, C.: A tree-blstm-based recognition system for online handwritten mathematical expressions. Neural Comput. Appl. **32**, 4689–4708 (5 2020). https://doi.org/10.1007/s00521-018-3817-2
14. Zhao, W., Gao, L.: Comer: Modeling coverage for transformer-based handwritten mathematical expression recognition. Lecture Notes in Computer Science (including subseries Lecture Notes in Artificial Intelligence and Lecture Notes in Bioinformatics) **13688 LNCS**, 392–408 (2022). https://doi.org/10.1007/978-3-031-19815-1_23
15. Zhao, W., Gao, L., Yan, Z., Peng, S., Du, L., Zhang, Z.: Handwritten mathematical expression recognition with bidirectionally trained transformer. Lecture Notes in Computer Science (including subseries Lecture Notes in Artificial Intelligence and Lecture Notes in Bioinformatics) **12822 LNCS**, 570–584 (2021). https://doi.org/10.1007/978-3-030-86331-9_37
16. Zhu, J., Gao, L., Zhao, W.: Ical: Implicit character-aided learning for enhanced handwritten mathematical expression recognition. Lecture Notes in Computer Science (including subseries Lecture Notes in Artificial Intelligence and Lecture Notes in Bioinformatics) pp. 21–37 (2024). https://doi.org/10.1007/978-3-031-70549-6_2
17. Álvaro, F., Sánchez, J.A., Benedí, J.M.: Recognition of on-line handwritten mathematical expressions using 2d stochastic context-free grammars and hidden Markov models. Pattern Recogn. Lett. **35**, 58–67 (2014). https://doi.org/10.1016/J.PATREC.2012.09.023

Rule-Based Reinforcement Learning for Document Image Classification with Vision Language Models

Michael Jungo[1,2](✉) and Andreas Fischer[1,2]

[1] University of Applied Sciences and Arts Western Switzerland, Delémont, Switzerland
{michael.jungo,andreas.fischer}@hefr.ch
[2] University of Fribourg, Fribourg, Switzerland

Abstract. Rule-based reinforcement learning has been gaining popularity ever since DeepSeek-R1 has demonstrated its success through simple verifiable rewards. In the domain of document analysis, reinforcement learning is not as prevalent, even though many downstream tasks may benefit from the emerging properties of reinforcement learning, particularly the enhanced reason capabilities. We study the effects of rule-based reinforcement learning with the task of Document Image Classification which is one of the most commonly studied downstream tasks in document analysis. We find that reinforcement learning tends to have better generalisation capabilities to out-of-distritbution data, which we examine in three different scenarios, namely out-of-distribution images, unseen classes and different modalities. Our code is available at https://github.com/jungomi/vision-finetune.

Keywords: Document Image Classification · Vision Language Models · Large Language Models · Reinforcement Learning

1 Introduction

Document Image Classification remains an important task as a first step for many document analysis approaches. It can be as simple as deciding to which department it needs to be sent to, up to more complex systems that perform automated analyses to extract and process the specific contents of various types of documents.

With the rise of popularity in Large Language Models (LLMs) and their steadily improving capabilities across multiple modalities, they have become of interest for a wide variety of tasks. A key reason for this, is their ability of in-context learning [5], which makes them adaptable to any new task with only a few examples. Many different Vision Language Models (VLMs) [20,43] are available, with commercial models like GPT-4-Vision [1] and Gemini [34], but also open-source models such as Llama 3.2-Vision [10], Qwen-2.5-VL [3] and Gemma 3 [35].

L. Jin et al. (Eds.): ICDAR 2025 Workshops, LNCS 16225, pp. 292–309, 2026.
https://doi.org/10.1007/978-3-032-09368-4_18

Even though the training data of most models is rarely fully disclosed, they are expected to also have been pre-trained on document understanding, including publicly available datasets for document visual question answering [23–26,33], as well as synthetic datasets [8,17,41]. These models are therefore equipped with a good understanding of documents and provide an excellent starting point for Document Image Classification.

The RVL-CDIP [12] dataset is the most commonly used benchmark for Document Image Classification. It consist of 400 000 images of scanned documents across 16 classes, which is a labelled subset of the IIT-CDIP collection of tobacco litigation documents [18]. Over the years, many different models have been evaluated on this dataset, but LLMs, including VLMs, have not been explored in much detail.

The contributions of this paper can be summarised as follows:

- We show that reinforcement learning (RL) can be used as an alternative to supervised fine-tuning (SFT) for document image classification.
- We compare the generalisation capabilities of RL and SFT in three scenarios: out-of-distribution images, unseen classes and different modalities.
- We examine the effect of the reasoning ability that is induced by RL.

2 Related Work

2.1 Rule-Based Reinforcement Learning for Vision Language Models

After the success of rule-based reinforcement learning achieved by DeepSeek-R1 [11], naturally, the desire of applying it to other models emerged and researchers started applying it to other domains and modalities. R1-onevision [40], Vision-R1 [42] and VLM-R1 [31] successfully applied the R1 style reinforcement learning to vision language models shortly after the release of DeepSeek-R1 with the general consensus that rule-based RL on its own can achieve competitive performance compared to SFT or even surpass it. Zhou et al. [44] managed to reproduce the "aha-moment" from DeepSeek-R1 in the form of visual reasoning in a 2B VLM without any prior SFT. Jigsaw-R1 [36] applies RL to jigsaw puzzles, under the premise that the reassembling of shuffled patches provide visual understanding that is transferable to downstream tasks, and found that the model is able to generalise to other visual tasks that require spatial reasoning.

Rule-based RL has found its way to many other visual task, for example in the medical domain, Med-R1 [14] and MedVLM-R1 [28] concurrently investigated rule-based reinforcement learning, where they observed that it improves the generalisation and reliability of the VLM across eight distinct medical imaging modalities, such as MRI, CT or X-ray. To the best of our knowledge, downstream tasks involving document images have not yet been studied, which compels us to explore it in the context of document image classification.

2.2 Document Classification with Large Language Models

Scius-Bertrand et al. [29] utilised LLMs to classify a subset of the RVl-CDIP with zero-shot and one-shot prompting as well as fine-tuning them. At the time, VLMs were a novelty and only GPT-4-Vision was available, hence they focused primarily on classifying the documents based on their textual content that was extracted with an OCR engine.

The zero-shot and one-shot already achieved good performances, with the highest being GPT-4 at 61.8% accuracy when only given the OCR as input and 69.9% when the images are included in the input. When they fine-tuned Mistral on the 1 600 training samples from that subset, its accuracy increased from 45.4% up to 83.4%. They showed that fine-tuning with a relatively small subset, compared to the full dataset, improves the results considerably.

Given the fact that the image based classification has achieved better results compared to the OCR, it would be expected that fine-tuning a VLM has even more potential. We use this opportunity to study the fine-tuning of VLMs on the subset they created, which ensures that we have a dataset that is sufficiently large to see the impact of the training while not being too large to the point of not being able to finish the experiments in a reasonable time.

3 Fine-Tuning Methods

There are multiple stages in the pre-training of LLMs. After being trained on large unsupervised data, where the goal is a simple next token prediction in order to learn fundamental text understanding and intricacies of the language, there are two primary stages to shape the LLMs into the adaptable form that is commonly used for all sorts of tasks. Firstly, supervised fine-tuning (SFT) is used to learn specific answers to a given question, this also includes instruction following, which teaches the model to respond to a query rather than just completing the input by adding additional text that is most likely to follow, as it was done during the initial pre-training. Afterward, reinforcement learning (RL) is employed to steer the model's responses into a more preferential form, which may be stylistic choices, but also for safety [4,7,21] by avoiding inappropriate phrases or by refusing to answer unconscionable queries.

While those are well established practices, most downstream tasks, e.g. document classification, only apply additional SFT to adapt the model to their specific tasks. One of the most prevalent reasons is the fact that RL requires a reward and value model to be trained alongside the LLM itself. This not only makes it much more demanding in terms of resources, but also adds a training complexity that made it unsuitable for downstream fine-tuning. With the release of DeepSeek-R1 [11], reinforcement learning with verifiable rewards (RLVR) [15,22,38] has gained a lot of attention, as they have shown that their Group Relative Policy Optimisation (GRPO) [30] training method can achieve excellent results by replacing the reward and value models with simple verifiable reward functions. For downstream tasks such as classification, it is straightforward to define a reward function, making it much more accessible and a viable option.

3.1 Supervised Fine-Tuning (SFT)

Supervised fine-tuning has been the de facto standard fine-tuning method for classical tasks with a clear cut answer, such as classification, where the answer is restricted to one of the known classes. In the context of LLMs, SFT not only serves to adapt the model to the given task but also to enforce an expected output format, which makes it easy to parse the expected class. In the simplest form, the model should reply purely with the predicted class without any additional explanation. While that can be achieved through more restrictive prompts, the fine-tuning bakes it into the model, making it more reliable. This is achieved with a simple cross-entropy loss, which is applied to the tokens of the response in order to alternate the model's parameters such that the likelihood of producing the desired class is increased. The loss is applied exclusively to the response, meaning that only the tokens in the response are learned.

3.2 Reinforcement Learning (RL)

While reinforcement learning has been included in the training of LLMs, it is primarily used to align the responses with human preferences, where annotators are presented with multiple responses which they have to rank in order of preference. As having a human in the loop would be prohibitively expensive, reinforcement learning from human feedback (RLHF) [27] trains a reward model that estimates the human preferences based on the collected preference annotations for a dataset with curated responses. Needing an additional reward model, as well as a dataset with annotated preferences, makes it unappealing for most downstream tasks, particularly in a low-resource scenario.

Group Relative Policy Optimisation (GRPO). Group Relative Policy Optimisation (GRPO) removed the reward model entirely and replaced it with simple verifiable rewards, which can be implemented with any deterministic function that can verify the quality of a response and assign it a reward value. This opens up a lot of possibilities for downstream tasks with a verifiable outcome, for example, the classification can easily be verified by checking whether the output corresponds to the expected class.

To get the same effect as the human preference ranking, for each query q a group of responses $\{o_1, \ldots, o_G\}$ are sampled from the old policy $\pi_{\theta_{old}}$ and the relative advantages within the group G of responses is calculated. GRPO optimises the policy π_θ to maximise the following objective:

$$\mathcal{J}_{GRPO}(\theta) = \mathbb{E}_{q \sim P(Q), \{o_i\}_{i=1}^{G} \sim \pi_{\theta_{old}}(\cdot \mid q)} \left[\frac{1}{G} \sum_{i=1}^{G} \frac{1}{|o_i|} \sum_{t=1}^{|o_i|} \left\{ min \left(\frac{\pi_\theta(o_i \mid q)}{\pi_{\theta_{old}}(o_i \mid q)} A_i, clip \left(\frac{\pi_\theta(o_i \mid q)}{\pi_{\theta_{old}}(o_i \mid q)}, 1 - \epsilon, 1 + \epsilon \right) A_i \right) - \beta \mathbb{D}_{KL} \left(\pi_\theta(\cdot \mid q) \| \pi_{ref}(\cdot \mid q) \right) \right\} \right] \tag{1}$$

$$A_i = \frac{r_i - mean(\{r_1, r_2, \ldots, r_G\})}{std(\{r_1, r_2, \ldots, r_G\})} \tag{2}$$

where A_i is the group-normalised advantage of the sample i, which is calculated by comparing the reward r_i to the entire group. Concretely, this means that a response that is better than the average response from that group will produce a positive advantage, which encourages the model to choose this particular response, as the objective becomes larger. For the opposite case, when the response is worse than the average response of the group, it will result in a negative advantage, which discourages the model by reducing the likelihood of producing that response. The $\mathbb{D}_{KL}$ term refers to the KL divergence between the distribution of the current policy π_θ and the reference policy π_{ref}, which ensures that the model does not deviate too far from the original model. β and ϵ are hyperparamters, which are the coefficient for the KL penalty and the clipping threshold, respectively.

Reinforcement Learning with Verifiable Rewards (RLVR). In RLVR the rewards are calculated with rule-based reward functions. These are for the most part simple checks, which evaluate the quality of the response. In the case of classification, it can be a binary check for whether it is correct, in which case the reward would be 1.0 if it is correct and otherwise 0.0. As the reward is any real valued number, a reward can span an entire spectrum, where negative values would be a penalty. An example for continuous values would be length reward, where each additional token increases the reward. If such a reward function is chosen, the magnitude of the value needs to be managed, particularly to not overpower other rewards which might be fixed.

We chose two reward functions, namely a format reward to ensure that the final classification can be easily extracted, while also enforcing the inclusion of a reasoning trace, and one for the classification accuracy, which is a simple check of whether the predicted class was correct.

- **Format**: To ensure that the predicted class can be easily extracted from the response, the class must be given inside an <answer></answer> tag. To make the format a little more strict and promote providing the reasoning steps for the decision, the response also needs to include a <reasoning></reasoning> tag. Each of the tags being present gives a reward of 0.5 while an additional 0.5 is awarded if they are given in the exact order of reasoning followed by answer. Any superfluous occurrence of the tags will induce a reward penalty, in the form of a negative penalty of -0.5 for every additional occurrence.

- **Classification:** Once the predicted class has been successfully extracted from the response, a reward of 1.0 is given if the classification was correct, otherwise it is 0.0.

3.3 Parameter Efficient Fine-Tuning (PEFT)

Due to the large number of parameters in LLMs, full fine-tuning is prohibitively costly as it demands a lot of GPU resources. To address this issue, parameter efficient fine-tuning (PEFT) methods have been introduced, where only a small number of extra parameters are fine-tuned while keeping the original model's parameters untouched. Since only the newly added parameters are fine-tuned, it requires a lot less memory for the gradients, as well as reducing the storage requirements, as only these parameters need to be stored, rather than having to store the whole model again due to minor adjustments of the base parameters.

One of the most used method is Low-Rank Adaptation (LoRA) [13], which injects a set of low-rank weight matrices, called adapters, to already existing weights. For any adapted weight, the input is passed through the original weight as well as the adapter, whose outputs are then combined by taking their sum. Instead of updating the original weight in-place, the weight update is entirely reflected by the adapter. As the weight updates supposedly have a low intrinsic rank [2,19], the adapters are decomposed into two low-rank matrices to further reduce the number of added parameters. Formally, given a weight matrix $W_0 \in \mathbb{R}^{d \times k}$, the weight update ΔW is decomposed into BA, where $B \in \mathbb{R}^{d \times r}, A \in \mathbb{R}^{r \times k}$ with the rank $r \ll min(d, k)$. The output of this layer is calculated as follows:

$$h(x) = W_0 x + \Delta W x = W_0 x + BAx \tag{3}$$

Even though this adds a slight latency during the training, the overall benefits are much greater and the latency becomes negligible. Furthermore, the latency can be removed entirely for inference once the model has finished training, by merging the adapters into the weight, due to the mathematical equivalence of $W_0 x + BAx = (W_0 + BA)x$.

The memory requirements for training can be additionally reduced by quantising the pre-trained model to 4-bit. QLoRA [9] showed that weight quantisation to 4-bit precision did not sacrifice performance compared to the full 16-bit precision when LoRA adapters are fine-tuned, but substantially reduces the memory footprint of the model. We therefore decided to use QLoRA, as it established itself as the commonly preferred PEFT method.

4 Experiments

To investigate the difference between SFT and RL and their ability to generalise to out-of-distribution data, we consider three scenarios, that represent different types of previously unseen data. The first, more commonly studied scenario is the evaluation on a dataset with the same classes but with images from a completely different source. Whereas for the second scenario, we remove a fraction of the available document classes for the training and examine whether the model can classify previously unseen classes. As LLMs are rather flexible in what classes they should predict, due to the variable prompts, it raises the question whether the fine-tuning helps or hinders the adaptation to unseen classes, considering the same task. The third scenario focuses on different modalities, specifically images and text, where the same documents are given to the model either as an image or as their textual content, that was extracted through an OCR system. Lastly, the reasoning traces that have been enabled by the RL and their effect on the classification are examined.

4.1 Experimental Setup

All our experiments are based on LLama-3.2-11B-Vision-Instruct[1] as the model that is fine-tuned using QLoRA [9], either with SFT or RL. For both fine-tuning methods the LoRA adapters are exclusively added to the LLM, therefore the vision encoder is untouched. We use a group size of $G = 8$ for the RL, meaning that 8 responses are sampled for every input and the advantages are calculated based on these generated samples. The KL divergence $\mathbb{D}_{KL}$ in Eq. 1 requires the reference model π_{ref}, which is the base model before fine-tuning. Since the base model is unaltered due to the use of LoRA adapters, we can access it by temporarily disabling the LoRA adapters, which does not incur any additional memory requirement to store the reference model.

4.2 Out-Of-Distribution Images

The images from RVL-CDIP have a very particular look, which is typical for documents from the late 20th century. Additionally, since they have been scanned at the time, it resulted in fairly low quality images in today's standards, approximately 100dpi, as well as introducing noise and other artefacts.

In contrast, Larson et al. [16] created a dataset containing more modern documents, where a large proportion are born-digital instead of being scanned versions of physical documents. They followed the same annotation strategy as RVL-CDIP with the same 16 classes and named it RVL-CDIP-N, where the *N* stands for *new distribution*. The dataset contains 1 002 images of documents that were collected from either web searches or the public DocumentCloud[2] repository.

It is worth noting that since these documents are publicly available on the internet, there is a chance some of them may also have been included in the pre-training data of available models. However, it is much less likely that the classification task was part of the pre-training. The same could also be said for the original RVl-CDIP dataset, but born-digital documents are more likely

[1] https://huggingface.co/meta-llama/Llama-3.2-11B-Vision-Instruct.
[2] https://documentcloud.org/.

to end up in the large pre-training datasets, as the PDF files provide readily accessible information about the contents and structure of the documents.

Table 1. Out-of-Distribution Images. LLama-3.2-11B-Vision-Instruct trained on the 1 600 images of the original RVL-CDIP training set with either SFT or RL and evaluated the classification accuracy on the RVL-CDIP test set (in-distribution) and the RVL-CDIP-N test set (out-of-distribution).

Training Method	RVL-CDIP *ID*	RVL-CDIP-N *OOD*
SFT	85.62	94.23
RL	77.50	96.21

We trained the model on the RVL-CDIP training set, once with a supervised fine-tuning (SFT) and once with reinforcement learning (RL), and evaluated it on the accompanying RVL-CDIP test set, which contains in-distribution images, as well as on out-of-distribution images from the RVl-CDIP-N test set. The results are shown in Table 1.

While the model trained with SFT has a better accuracy on the in-distribution images, the model trained with RL narrowly surpasses it on the out-of-distribution images. Even though the difference between the two models is quite small in absolute terms, the relative differences going from in-distribution to out-of-distribution is considerably larger in favour of RL. This indicates that RL generalises better to new images that stray away from the characteristics found in the training data.

The lower accuracy of the in-distribution result from RL compared to SFT might be due to some training instability, see Sect. 5.1 for the details about the encountered training instabilities.

4.3 Unseen Classes

In order to evaluate the generalisation capabilities of the model to previously unseen classes, we withhold some classes from the training. This is achieved by splitting the dataset into two subsets, where only 10 out of the 16 classes in RVl-CDIP are kept for the training by excluding the remaining 6 classes. The test set follows the split of classes listed in Table 2, which allows evaluating the models on unseen classes, while still keeping the images from the same distribution. Since the dataset has an equal number of images for every class, the exact choice of classes has no effect on the size of the two subsets and were therefore randomly chosen.

During the training, only the images from the 10 classes are shown to the model, while also only asking for these classes in the prompt. As a consequence, the model also needs to be able to adapt to a new prompt with new, previously unseen, classes, rather than always answering with the classes it may have

Table 2. Classes Split. The 16 classes of the RVL-CDIP dataset split into two subsets, the 10 in-distribution classes used for training, and the 6 remaining out-of-distribution classes, that are exclusively used for the evaluation.

10 classes *ID*	6 classes *OOD*
letter	email
form	handwritten
advertisement	news article
scientific report	invoice
scientific publication	presentation
specification	questionnaire
file folder	
budget	
resume	
memo	

memorised during the training. To assess the adaptability of the models, including their instruction following capabilities, every test set is evaluated with three prompts, which contain all 16 classes, only the 10 classes used during training or the 6 unseen classes, respectively. This means, that some combinations of prompts and test sets have no overlap in the classes, therefore if the model is following the instructions correctly, it should always predict one of the provided classes. Specifically, when asked for the 6 unseen classes in the prompt, but evaluated on the test set of the 10 classes it was trained on, the model should not predict one of the classes it might have memorised during training.

The results in Table 3 show the different combinations of prompts and test sets for the 10 in-distribution classes, the 6 out-of-distribution classes and also all classes combined. The models trained with SFT and RL are very close on the test set for the 10 classes they were trained on, indicating that both methods can achieve equally good results. However, they start to deviate quite strongly when looking at the out-of-distribution classes, where the SFT model drops to 43.23% accuracy, i.e. less than half of the in-distribution accuracy, when evaluated on the 6 unseen classes, compared to the 78.65% of the RL model. A slight drop is always expected when moving to previously unseen data, but the RL certainly managed to maintain much more of the performance than the SFT.

Another interesting aspect is the instruction following capabilities of the models. When the models are evaluated on the test set with the 10 classes they were trained, but the prompt asks them to only choose from one of the 6 classes, they should in theory achieve an accuracy of 0%, since there is no overlap in classes between the prompt and the actual data. Unfortunately, this is not the case, and they still occasionally respond with one of the classes they were trained on, ignoring the ones provided in the prompt. While the RL is only doing it roughly

Table 3. Unseen Classes. LLama-3.2-11B-Vision-Instruct trained on the 1 000 images of the 10 selected classes from the original RVL-CDIP training set, where the prompt contains only the 10 available classes. Each test set is evaluated with three variations of the prompt containing either all classes, the 10 classes seen during training or the 6 unseen classes, respectively. Values ingrey indicate that the classes in the prompt differ from the actual classes in the test data.

		Test Data		
Training Method	Prompt	*10 classes*	*6 classes*	*All classes*
SFT on all classes†	All classes	87.50	80.73	85.62
RL on all classes†	All classes	80.36	74.48	77.50
SFT	10 classes	89.29	0.00	56.25
	6 classes	76.79	43.23	61.87
	All classes	89.29	18.23	63.13
RL	10 classes	90.18	0.00	55.00
	6 classes	9.82	78.65	32.50
	All classes	88.40	58.33	78.75
RL after SFT	10 classes	88.39	0.00	55.00
	6 classes	54.46	48.44	53.75
	All classes	91.07	33.85	66.87

† Models that were trained on all classes as a reference.

1 out of 10 cases, which achieves an accuracy of 9.82%, the SFT does it much more frequently, resulting in an accuracy of 76.79%. The unexpected accuracy points to the issue of the model having memorised the classes it was trained, and in fact, the SFT model only responds with one of the asked classes in less than 2% of the cases. The expected results would be 0%, which can be clearly observed in the inverse case, where the model is evaluated on the test set with the 6 unseen classes, but the prompt contains only the 10 classes it was trained on, granted that the new classes were never presented to the model at any point. This exposes a broader problem of SFT overfitting on the training data, which leads to memorisation instead of generalisation, which is inline with the findings of Chu et al. [6] but for document image classification.

Since it is common practice to apply SFT first and then RL afterwards, we also examined the effects of following this strategy in the context of downstream tasks. RL after SFT alleviates the issue of memorisation to a certain degree, where the previous 76.79% accuracy of the mismatched classes is reduced to 54.46%. There is a caveat to this change, as the model responds in only roughly double the number of cases with the asked classes compared to before, but instead, starts mangling classes, such as *scientific journal article*, or completely deviating from any of the available classes, e.g. *appendix* or *membership investment notice*. This means that the damaged caused by SFT is not as easily reversible.

4.4 New Modality

Another form of generalisation is the change of modality. All models have been trained to classify document images, however the same task could also be performed with the extracted text of the documents. Therefore, a model that has learned the fundamental task of document classification, should also be able to perform that task on a different modality to a certain extent, granted that the base model supports other modalities. In the case of Llama-3.2 Vision, it is and extension of Llama-3.2 by adding a vision encoder that projects the images into embedding space of the base LLM, therefore the underlying text understanding is still present, making it a great option to evaluate the two different modalities.

Table 4. New Modality. LLama-3.2-11B-Vision-Instruct trained on the 1 600 images or their text contents, extracted through an OCR model, of the original RVL-CDIP training set. The classification accuracy is evaluated for each combination of modalities. Values ingrey indicate the test data having a different modality than the model was trained on.

Training		Test Data	
Method	Data	Image	OCR
SFT	Image	85.62	60.62
	OCR	43.75	81.88
RL	Image	77.50	52.50
	OCR	25.00	71.25

Contrary to the previous results, this time the model trained with SFT generalises better to the other modality than the one with RL, as is evident by Table 4. In either case, the drop-off is more severe when going from OCR to images. The very steep drop-off from 71.25% to 25% of the RL model can be partially explained by the training instabilities that are discussed in Sect. 5.1, because the responses contain formatting issues such as missing the closing tag of the answer or not having any tags at all. The same cannot be said for the other direction, going from images to OCR, which has only a few rare cases where the format was not respected, hence it cannot be the sole reason.

4.5 Reasoning

Increased test-time compute has been shown to improve the outputs of LLMs [32]. Chain-of-Thought (CoT) [37] is a technique to include a series of reasoning steps into the desired output of the LLMs, which is intended to help the model get to the solution by breaking it down into smaller intermediate steps. Using CoT in a supervised setting would require curating a dataset of high quality reasoning traces for the given task, which is time and labour intensive. On the other hand, GRPO can integrate reasoning traces into the training by including them in the generations of the samples and only enforcing the format, rather than the exact reasons the model is supposed to give. There is no guarantee that the reasons are always correct, even if the answers are correct, but it still allows the model to generate extra tokens for increased test-time compute (colloquially referred to as "thinking time").

To judge the impact of the reasoning at test-time, we compare the model trained with GRPO with and without the reasoning in the answer. Inspired by Qwen3's [39] "non-thinking" mode, we use the same model that was trained to include reasoning, but prefilling the output with an empty reasoning tag, i.e. <reasoning></reasoning>, which forces the model to go straight to the answer, as the reasoning was already completed.

Table 5. Impact of Reasoning. The same models that were trained on the 10 classes as in Table 3, but comparing the models after RL with reasoning (✓) and without (✗), where the response is prefilled with an empty reasoning tag, i.e. <reasoning></reasoning>. Values in grey indicate that the classes in the prompt differ from the actual classes in the test data.

			Test Data		
Training Method	Reasoning	Prompt	*10 classes*	*6 classes*	*All classes*
RL	✓	10 classes	90.18	0.00	55.00
	✓	6 classes	9.82	78.65	32.50
	✓	All classes	88.40	58.33	78.75
	✗	10 classes	75.89	0.00	46.25
	✗	6 classes	0.02	58.33	23.75
	✗	All classes	71.43	48.44	67.50
RL after SFT	✓	10 classes	88.39	0.00	55.00
	✓	6 classes	54.46	48.44	53.75
	✓	All classes	91.07	33.85	66.87
	✗	10 classes	91.07	0.00	55.00
	✗	6 classes	59.82	50.00	53.13
	✗	All classes	90.20	32.30	66.25

In Table 5 we can see that the accuracy of the model trained with pure RL deteriorates across the board when the reasoning is disabled. This would suggest that the increased test-time compute is indeed helpful for the model to make the correct classification, but we observed that the model starts to disregard the imposed format and just puts everything inside the answer tag. This makes it impossible to extract the final classification in a reliable manner and therefore undermines the fine-tuning efforts. The model is still able to adhere to the format and responds with a singular answer tag containing only the predicted class, which may not necessarily always be the correct class, but in roughly a third of the cases, that is no longer the case. In the most extreme scenario, namely when the model is evaluated on the 10 classes but the prompt contains the other 6 classes, over half of the responses completely ignore the format. It seems that the model is more likely to forego the format in scenarios where it struggles to find the correct solution.

This pattern cannot be observed for the model that was first trained with SFT and then followed by RL. Removing the reasoning has very little effect in comparison to the pure RL and more of often than not it is slightly improving the accuracy. As we observed in Sect. 4.3, the underlying behaviour of this model is predominantly attributed to SFT, therefore the reasoning is an after-effect and removing it has no discernible impact, which arguably brings it closer to the original SFT output structure.

The image appears to be a file folder, as indicated by the rectangular shape with a tab on the bottom right corner and the presence of a label with a number, which is a common feature of file folders.

GT: file folder
Pred: file folder

(a)

The image depicts a news article, which is a type of document that reports on current events or news stories. The content of the article includes a headline, text, and a photo of a person, which are all typical features of a news article. The layout and design of the page, including the use of columns and headlines, further support this classification.

GT: news article
Pred: news article

(b)

The image depicts a document that appears to be a spreadsheet or table, with rows and columns, and handwritten notes and numbers. The top of the page has a header with the words "Lorillard" and "Transit", and the page is filled with rows of numbers and handwritten notes. The document appears to be a financial or accounting document, possibly a ledger or a budget.

GT: budget
Pred: form

(c)

The image depicts a document with a formal tone, featuring a header with the sender's and recipient's names, a subject line, and a brief summary of the content. The document is addressed to a specific individual and includes a timestamp and a CC field, which are common elements found in email messages. The presence of these features suggests that this document is an email.

GT: letter
Pred: email

(d)

Fig. 1. Reasoning Examples. Predictions of LLama-3.2-11B-Vision-Instruct trained with reinforcement learning (RL) with the reasoning the model provided in the response before giving its final answer.

A few examples of the reasoning given by the model are depicted in Fig. 1. The reasons often describe what attributes were found to be relevant for the prediction, giving the user a better understanding as to why a class was potentially chosen. For example, the news article (b) is identified by the very common layout of news papers. Sometimes the provided reasoning makes perfect sense, but the model gets the classification wrong. This can be seen in (c) where the reasoning ends with *"[...] possibly a ledger or a budget"* but the model then predicts *form* instead of the actual class *budget*. It was able to identify it correctly, but when it came to making the final decision, it most likely put more emphasis on the other aspects that could equally be attributed to a form. Lastly, there are some cases like (d), where not only the reasoning makes perfect sense, but also the chosen class, yet the classification is incorrect. These stem from annotation errors, as the RVL-CDIP is known to have inaccuracies in the annotations, particularly since there are overlapping classes that led to ambiguities, which also made it harder for the annotators to choose the correct class.

5 Discussions

The primary discussion points are related to the question of whether RL is worth considering for more traditional tasks such as classification. Based on the results presented in this paper, we think that there is merit to at least try RL for downstream tasks that have a verifiable answer, as it seems to improve the generalisation capabilities of the model to previously unseen data with more flexibility. There are however a few points that need to be taken into account, which are not purely result oriented.

5.1 Training Instability of RL

RL relies heavily on the base model being able to produce at least one correct answer so that not all rewards are zero. Thankfully, this was the case for our training, but zero advantages, which arrive when there is no variation in quality of responses, good or bad, may occur at any point during the training, which makes the model stagnate. Without any intervention, this could make the training unstable or even impossible.

An instability that we encountered during the training of the model on the 1 600 images of the RVL-CDIP dataset, was the odd behaviour of the model producing the answer before the response. This goes against the idea of increasing test-time compute, as the final answer was already given and everything that follows does not help improve the answer itself. Even after forcing the order with a strict format reward, it still occurred and seems to have had a negative effect on the model, to the point where it even produced a reasoning outside of the prescribed tags and sometimes copying the same reasoning after the answer to have it the reasoning tag. This did not occur when training on the subset with only 10 classes, which is probably why the results in Table 3 are better for the model that was not trained on all classes.

Since RLVR has only recently been in the spotlight for LLMs, this is expected to be improved with future research, and hopefully methods will be developed to alleviate the instabilities in order to get the full potential out of it.

5.2 Efficiency

Another negative aspect of RL compared to SFT, is the fact that the training is much less efficient, since for each batch a group of samples needs to be generated at every iteration. The generation is usually much slower because of the autoregressive nature of LLMs, where one token is generated at a time, resulting in many more forward passes than during SFT. To make matters worse, the responses are generally much longer, since it also needs to generate the reasoning. The combination of all that makes the training much less efficient. Similarly, during inference, the responses are also longer due to the reasoning.

In terms of memory requirements, due to the use of LoRA and having access to the base model, there is no additional memory required for any other model. The only difference is the memory required for the batches. Since the length of the reasoning has no effective limit, unless a length reward is integrated, the batches might get much larger, which in turn require more memory, particularly for the attention layers. This means that the batch size needs to be reduced compared to SFT. And since the generated samples do not have a uniform length, some batches might leave a lot of head space, which affects the hardware utilisation efficiency.

5.3 Explainability

The goal of having the model include the reasoning in the answer is primarily for the increased test-time compute, but it also offers the benefit of having better explainability in order to understand why a certain decision was made. Although we have shown some examples of the reasoning, which seem to be coherent and helpful, in particular in the cases where the prediction was incorrect, they have not been scrutinised and evaluated in detail. As most RLVR methods do not impose any particular demands or restrictions on the reasoning but only the format, the provided reasoning is never guided and may not be as helpful for other tasks. This is a first step in the direction of explainability that can be included into the training without needing an annotated reasoning dataset, but additional research is need in this area to get the most out of it.

6 Conclusion

With the experiments conducted in this paper we showed that RL tends to have a better generalisation capability in the context of document image classification, where the model adapts more easily to out-of-distribution images that come from an entirely different era, as well as previously unseen classes. This indicates that RL is more akin to learning the underlying fundamentals of the document classification task rather than being overly focused on the specific classes at hand. However, it is not the case for different modalities, i.e. going from images to text as input or vice-versa, where the model trained with RL struggles much more, which may be explained by the encountered training instabilities. On the other hand, SFT is much simpler to train and generally performs strongly on in-distribution data, while suffering more on out-of-distribution data. The biggest drawback of SFT is the memorisation aspect, which worsens its instruction following ability as it tends to be referring back to the classes it was trained on even if they were no longer a viable option given by the prompt.

Based on these findings, RL will hopefully be taken into consideration as a viable option for more downstream tasks, with the understanding that the training is more challenging, which will take some tinkering to mitigate the possible training instabilities that may arise. Additional research is necessary to arrive to a more consistent and stable training to make the effort of RL worthwhile for a wide range of applications. Besides the improved generalisation capabilities, a compelling aspect that emerges from RL is the reasoning that the model provides, which not only improves its result through increased test-time compute, but also allows to get an insight into the choices that were made.

References

1. Achiam, J., et al.: GPT-4 technical report. arXiv preprint arXiv:2303.08774 (2023)
2. Aghajanyan, A., Gupta, S., Zettlemoyer, L.: Intrinsic dimensionality explains the effectiveness of language model fine-tuning. In: Zong, C., Xia, F., Li, W., Navigli, R. (eds.) Proceedings of the 59th Annual Meeting of the Association for Computational Linguistics and the 11th International Joint Conference on Natural Language Processing (vol. 1: Long Papers), pp. 7319–7328. Association for Computational Linguistics (2021). https://doi.org/10.18653/v1/2021.acl-long.568
3. Bai, S., et al.: Qwen2.5-VL Technical Report. arXiv preprint arXiv:2502.13923 (2025)
4. Bai, Y., et al.: Training a helpful and harmless assistant with reinforcement learning from human feedback. arXiv preprint arXiv:2204.05862 (2022)
5. Brown, T., et al.: Language models are few-shot learners. Adv. Neural. Inf. Process. Syst. **33**, 1877–1901 (2020)
6. Chu, T., Zhai, Y., Yang, J., Tong, S., Xie, S., Levine, S., Ma, Y.: SFT memorizes, RL generalizes: a comparative study of foundation model post-training. In: The Second Conference on Parsimony and Learning (Recent Spotlight Track) (2025). https://openreview.net/forum?id=d3E3LWmTar
7. Dai, J., et al.: Safe RLHF: safe reinforcement learning from human feedback. In: The Twelfth International Conference on Learning Representations (2024). https://openreview.net/forum?id=TyFrPOKYXw
8. Deitke, M., et al.: Molmo and PixMo: open weights and open data for state-of-the-art multimodal models. arXiv preprint arXiv:2409.17146 (2024)
9. Dettmers, T., Pagnoni, A., Holtzman, A., Zettlemoyer, L.: QLoRA: efficient fine-tuning of quantized LLMs. Adv. Neural. Inf. Process. Syst. **36**, 10088–10115 (2023)
10. Grattafiori, A., et al.: The llama 3 herd of models. arXiv preprint arXiv:2407.21783 (2024)
11. Guo, D., et al.: DeepSeek-R1: incentivizing reasoning capability in LLMs via reinforcement learning. arXiv preprint arXiv:2501.12948 (2025)
12. Harley, A.W., Ufkes, A., Derpanis, K.G.: Evaluation of deep convolutional nets for document image classification and retrieval. In: Proc. Int. Conf. on Document Analysis and Recognition (ICDAR), pp. 991–995 (2015)
13. Hu, E.J., et al.: LoRa: low-rank adaptation of large language models. ICLR **1**(2), 3 (2022)
14. Lai, Y., Zhong, J., Li, M., Zhao, S., Yang, X.: Med-R1: reinforcement learning for generalizable medical reasoning in vision-language models. arXiv preprint arXiv:2503.13939 (2025)

15. Lambert, N., et al.: T\" ulu 3: pushing frontiers in open language model post-training. arXiv preprint arXiv:2411.15124 (2024)
16. Larson, S., Lim, Y.Y.G., Ai, Y., Kuang, D., Leach, K.: Evaluating out-of-distribution performance on document image classifiers. Adv. Neural. Inf. Process. Syst. **35**, 11673–11685 (2022)
17. Laurençon, H., Marafioti, A., Sanh, V., Tronchon, L.: Building and better understanding vision-language models: insights and future directions. In: Workshop on Responsibly Building the Next Generation of Multimodal Foundational Models (2024)
18. Lewis, D., Agam, G., Argamon, S., Frieder, O., Grossman, D., Heard, J.: Building a test collection for complex document information processing. In: Proceedings of the 29th Annual International ACM SIGIR Conference on Research and Development in Information Retrieval, pp. 665–666 (2006)
19. Li, C., Farkhoor, H., Liu, R., Yosinski, J.: Measuring the intrinsic dimension of objective landscapes. In: International Conference on Learning Representations (2018)
20. Li, Z., Wu, X., Du, H., Nghiem, H., Shi, G.: Benchmark evaluations, applications, and challenges of large vision language models: a survey. arXiv preprint arXiv:2501.02189 **1** (2025)
21. Liu, Y., et al.: Trustworthy LLMs: a survey and guideline for evaluating large language models' alignment. In: Socially Responsible Language Modelling Research (2023). https://openreview.net/forum?id=oss9uaPFfB
22. Liu, Z., et al.: Visual-RFT: visual reinforcement fine-tuning. CoRR **abs/2503.01785** (2025). https://doi.org/10.48550/arXiv.2503.01785
23. Lu, P., et al.: Learn to explain: multimodal reasoning via thought chains for science question answering. Adv. Neural. Inf. Process. Syst. **35**, 2507–2521 (2022)
24. Mathew, M., Bagal, V., Tito, R., Karatzas, D., Valveny, E., Jawahar, C.: InfographicVQA. In: Proceedings of the IEEE/CVF Winter Conference on Applications of Computer Vision, pp. 1697–1706 (2022)
25. Mathew, M., Karatzas, D., Jawahar, C.: DocVQA: a dataset for VQA on document images. In: Proceedings of the IEEE/CVF Winter Conference on Applications of Computer Vision, pp. 2200–2209 (2021)
26. Mishra, A., Shekhar, S., Singh, A.K., Chakraborty, A.: OCR-VQA: visual question answering by reading text in images. In: 2019 International Conference on Document Analysis and Recognition (ICDAR), pp. 947–952. IEEE (2019)
27. Ouyang, L., et al.: Training language models to follow instructions with human feedback. Adv. Neural. Inf. Process. Syst. **35**, 27730–27744 (2022)
28. Pan, J., et al.: MedVLM-R1: incentivizing medical reasoning capability of vision-language models (VLMs) via reinforcement learning. CoRR **abs/2502.19634** (2025). https://doi.org/10.48550/arXiv.2502.19634
29. Scius-Bertrand, A., Jungo, M., Vögtlin, L., Spat, J.M., Fischer, A.: Zero-shot prompting and few-shot fine-tuning: revisiting document image classification using large language models. In: International Conference on Pattern Recognition, pp. 152–166. Springer (2025)
30. Shao, Z., et al.: DeepSeekMath: pushing the limits of mathematical reasoning in open language models. arXiv preprint arXiv:2402.03300 (2024)
31. Shen, H., et al.: VLM-R1: a stable and generalizable r1-style large vision-language model. arXiv preprint arXiv:2504.07615 (2025)
32. Snell, C., Lee, J., Xu, K., Kumar, A.: Scaling LLM test-time compute optimally can be more effective than scaling model parameters. arXiv preprint arXiv:2408.03314 (2024)

33. Tanaka, R., Nishida, K., Yoshida, S.: VisualMRC: machine reading comprehension on document images. In: Proceedings of the AAAI Conference on Artificial Intelligence, vol. 35, pp. 13878–13888 (2021)
34. Team, G., et al.: Gemini: a family of highly capable multimodal models. arXiv preprint arXiv:2312.11805 (2023)
35. Team, G., et al.: Gemma 3 technical report. arXiv preprint arXiv:2503.19786 (2025)
36. Wang, Z., et al.: Jigsaw-R1: a study of rule-based visual reinforcement learning with jigsaw puzzles. arXiv preprint arXiv:2505.23590 (2025)
37. Wei, J., et al.: Chain-of-thought prompting elicits reasoning in large language models. Adv. Neural. Inf. Process. Syst. **35**, 24824–24837 (2022)
38. Wen, X., et al.: Reinforcement learning with verifiable rewards implicitly incentivizes correct reasoning in base LLMs. arXiv preprint arXiv:2506.14245 (2025)
39. Yang, A., et al.: Qwen3 technical report. arXiv preprint arXiv:2505.09388 (2025)
40. Yang, Y., et al.: R1-OneVision: advancing generalized multimodal reasoning through cross-modal formalization. arXiv preprint arXiv:2503.10615 (2025)
41. Yang, Y., et al.: Scaling text-rich image understanding via code-guided synthetic multimodal data generation. arXiv preprint arXiv:2502.14846 (2025)
42. Zhan, Y., et al.: Vision-R1: evolving human-free alignment in large vision-language models via vision-guided reinforcement learning. CoRR **abs/2503.18013** (2025). https://doi.org/10.48550/arXiv.2503.18013
43. Zhang, J., Huang, J., Jin, S., Lu, S.: Vision-language models for vision tasks: a survey. IEEE Trans. Pattern Anal. Mach. Intell. (2024)
44. Zhou, H., Li, X., Wang, R., Cheng, M., Zhou, T., Hsieh, C.J.: R1-zero's "Aha Moment" in visual reasoning on a 2B Non-SFT model. CoRR **abs/2503.05132** (2025). https://doi.org/10.48550/arXiv.2503.05132

ICDAR 2025 Workshop on Multi-modal Mathematical Reasoning in Documents (M3RD 2025)

Boosting Handwritten Mathematical Expression Recognition Through Contextual Reasoning with Vision Large Language Models (vLLMs)

Merouane Zouaid, Yejing Xie(✉), and Harold Mouchère

Laboratoire des Sciences du Numérique de Nantes (LS2N), Nantes Université, École Centrale Nantes, CNRS, LS2N, UMR 6004, Nantes 44000, France
{yejing.xie,harold.mouchere}@univ-nantes.fr

Abstract. Handwritten Mathematical Expression Recognition (HMER) has traditionally relied on dedicated models specifically designed for this task. With the emergence of Vision Large Language Models (vLLMs), we investigate their potential for HMER tasks. This paper presents a comprehensive benchmark study evaluating various state-of-the-art vLLMs on the CROHME dataset. We develop an experimental pipeline that includes language model inference, cleaning and conversion to structural graphs, and evaluation using graph-based metrics. Our experiments explore two prompting strategies: without context (image-to-text only) and with context (image-to-text with surrounding contextual information). Results show that while vLLMs demonstrate promising capabilities, dedicated models still outperform general-purpose vLLMs on HMER tasks. We analyze the performance gaps and discuss future directions for improving vLLM-based mathematical expression recognition.

Keywords: Handwritten Mathematical Expression Recognition · Vision Language Models · Mathematical Contextual Reasoning

1 Introduction

Handwritten Mathematical Expression Recognition (HMER) is a research challenge that combines aspects of computer vision and natural language processing. The goal is to convert a handwritten mathematical expression image into a structured, machine-readable format such as LaTeX. Typically, an encoder-decoder architecture is employed [17], where a visual encoder extracts spatial features from the image, and a decoder generates the corresponding mathematical sequence.

Despite recent progress, HMER remains a difficult problem. This is largely due to the inherent ambiguity of handwritten symbols, the two-dimensional spatial structure of mathematical notation, and the complex grammar of mathematical language. Even for well-educated humans, interpreting handwritten mathematics can be challenging without sufficient context.

L. Jin et al. (Eds.): ICDAR 2025 Workshops, LNCS 16225, pp. 313–326, 2026.
https://doi.org/10.1007/978-3-032-09368-4_19

Meanwhile, large language models (LLMs) have demonstrated remarkable performance in language understanding, reasoning, and generation tasks. Their capabilities suggest a new opportunity for enhancing mathematical expression recognition by introducing models that not only see but also reason about the content. However, most existing benchmarks using LLMs in mathematical tasks are limited to elementary-level reasoning, such as arithmetic, basic algebra and basic geometry. These tasks do not fully reflect the challenges of parsing handwritten expressions in more advanced or diverse contexts.

In this work, we investigate the potential of Vision-Language Large Models (vLLMs) for handwritten mathematical expression recognition. Unlike traditional approaches that rely solely on visual features, vLLMs are capable of jointly processing images and text, offering the opportunity to integrate contextual reasoning into the recognition process. We explore whether these models can effectively interpret handwritten mathematical input and whether taking advantage of the surrounding context, such as neighboring expressions, accompanying text, or layout information, can enhance the accuracy of recognition in realistic settings.

The study could be applied as a benchmark for evaluating vLLM on HMER, which is built from real-world mathematical content and emphasizes advanced, context-rich equations. The goal is to assess whether vLLMs can move beyond surface-level transcription and demonstrate a deeper understanding of complex mathematical expressions in realistic settings.

2 Related Work

2.1 Handwritten Mathematical Expression Recognition

Early approaches to Handwritten Mathematical Expression Recognition (HMER) relied on structural recognition methods that represented mathematical expressions as trees or graphs. These models typically decomposed the task into sequential stages—symbol segmentation, classification, relation prediction, and structural analysis—which often led to error propagation. Grammar-based methods used predefined rules like context-free grammars (CFGs) to define expression syntax [1,3,7,27], while graph-based approaches such as Stroke Label Graph (SLG) applied parsing strategies like minimum spanning trees to infer structure [5,6,13,21].

With the rise of deep learning, end-to-end models such as encoder-decoder architectures have become dominant, reducing reliance on handcrafted features and grammar rules. CNN-based encoders (e.g., DenseNet) are widely used for offline image input, while RNNs and Graph Neural Networks (GNNs) are applied for online stroke data [18–20,23,28,29,32]. Most models generate linear LaTeX sequences using GRU or Transformer-based decoders [9,30,33,34], but often overlook the inherent hierarchical structure of mathematical expressions. To address this, recent works have explored tree-based decoders [8,18,22,31,35] and graph-based representations [11,23,26], incorporating structural constraints into deep models.

However, most HMER methods still treat expressions in isolation, ignoring the surrounding textual or contextual cues. In contrast, LLMs offer new potential for contextual understanding and symbolic reasoning, but existing benchmarks focus primarily on elementary math problems. To bridge this gap, we propose a new benchmark that evaluates vLLMs in the context of real-world handwritten expressions, emphasizing structural ambiguity and the role of surrounding context in improving recognition.

2.2 Vision Large Language Models

Recent advances in large language models have led to powerful vision-language models capable of understanding and reasoning about visual content. Table 1 presents a comparison between dedicated models and vision large language models across key characteristics.

Table 1. Comparison between Dedicated Models and Vision Large Language Models

Characteristic	Dedicated Models	Vision LLMs
Purpose	Task-specific (HMER)	General-purpose
Model Size	$\sim$10M parameters	Billions-Trillions
Knowledge Scope	Narrow, focused	Broad, cross-domain
Training Data	math-specific	Web, books, code
Architecture	Encoder-Decoder	Transformer-based

Leading vision-language models demonstrate varying capabilities for mathematical tasks. Gemini 2.5 Pro [15] shows strong performance on mathematical reasoning benchmarks with particular strengths in LaTeX generation and step-by-step explanations. InternVL 2.5 [2] represents the leading open-source alternative, excelling in visual-mathematical reasoning on benchmarks like MathVista [10] and OlympiadBench [4].

2.3 Mathematical Reasoning Benchmarks

The evaluation of language models on mathematical tasks encompasses several specialized benchmarks. Table 2 summarizes the key mathematical reasoning benchmarks and their characteristics.

These benchmark tests cover different aspects of mathematical ability, but mainly focus on elementary mathematical problem-solving skills. The content rarely involves more complex advanced mathematics, physics, computer science, and other areas of in-depth research. Therefore, this study uses HMER problems to measure the understanding ability of vLLMs in high-difficulty research fields, filling the gap in mathematical expression reasoning and questioning.

Table 2. Mathematical Reasoning Benchmarks Overview

Benchmark	Focus	Size	Modality
MATH	Competitive problem-solving	12,500 problems	Text
MathVista	Visual mathematical reasoning	5,141 examples	Visual
MathArena	Comprehensive evaluation	Various	Mixed
AIME	Advanced mathematics	Competition problems	Text

3 Pipeline

In this session, we present the complete pipeline developed for our task, from data processing to model interaction. We begin with Dataset Preparation, outlining how raw inputs are curated, annotated, and structured to ensure quality and consistency. Next, we detail the Language Model Inference stage, where pre-trained models are applied to extract predictions or generate outputs relevant to our objective. Finally, we focus on Prompt Engineering, describing how carefully crafted prompts are used to guide the behavior of large language models and improve task-specific performance (Fig. 1).

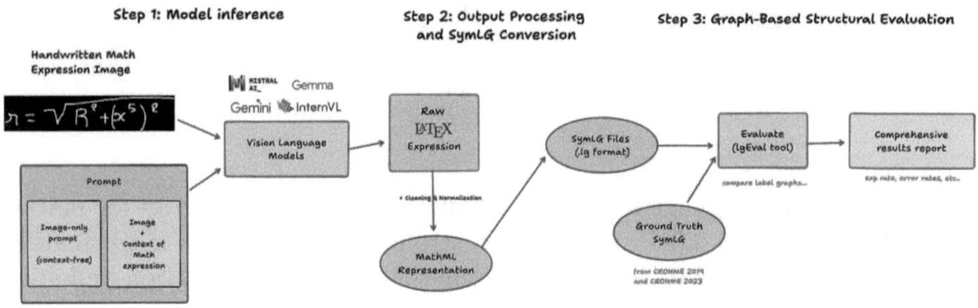

Fig. 1. Pipeline

3.1 Datasets Preparation

The CROHME 2023 dataset [25] serves as the foundation of our study. It contains a wide range of complex handwritten mathematical expressions collected from real-world sources such as Wikipedia and arXiv. As a result, the expressions are highly diverse and closely tied to research-related contexts, covering various scientific and mathematical domains.

While CROHME has been widely used for evaluating handwriting recognition models, previous works typically focus on isolated expressions. However, in real-world scenarios—especially in academic documents—mathematical expressions are rarely standalone. They are often embedded in surrounding text that provides additional semantic context. This context can help disambiguate visually similar expressions and guide more accurate recognition.

To explore this idea, we enhance the CROHME dataset by collecting contextual information around each expression, an example as shown in Fig. 2. Specifically, for each original, non-normalized expression in the dataset, we extract at around 200 characters before and after the expression from its source document. All visible characters, including mathematical symbols, punctuation, and text, are retained. This results in a paired dataset where each handwritten expression is accompanied by its surrounding textual context, enabling us to investigate the role of natural language in mathematical expression recognition.

Example of Context and Math Expression

As we chose the orientations according to the sign of the charges, this is exactly the same as saying that the whole system is BPS if and only if the two strings are parallel. But Q and $\tilde{Q}$ need not be equal. It suffices that they have same sign. But if $Q = 2\pi N$ for some $N \neq \pm 1$, then we think that we can view this as N copies of a string with $N = \pm 1$ which are on top of each other, and which then could be separated from each other without any energy cost. So we will in the sequel assume that $N = \pm 1$.

$$r = \sqrt{R^2 + (x^5)^2}$$

If string number one moves with a velocity v_1 which is much less than one (the speed of light), and we parameterize Σ_1 as $X^0 = \tau\ X^I = {v_1}^I \tau\ X^5 = \sigma$ and the associated Dirac brane D as $Y^0 = \tau\ Y^{1,2,3} = {v_1}^{1,2,3}\tau\ Y^4 = \tilde{\sigma} + {v_1}^4 \tau\ Y^5 = \sigma$, then the scalar field produced by Σ_1 becomes $\phi = D_1 - \sqrt{1 - \boldsymbol{v_1}^2}\frac{|Q_1|}{2V_3 S^2} + Ordo((v_1)^3)$ where $S = R(1 + Ordo(v_1))$. Similarly, the retarded potential becomes $B_{05} = -\frac{Q_1}{2V_3 S^2} + Ordo((v_1)^3)\ B_{IJ} = \omega_{IJ} + Ordo(v_1)\ B_{0J} = v^I \omega_{IJ} + Ordo((v_1)^2)\ B_{I5} = v_{1I} B_{05}$. Here ω_{MN} is the potential generated by a static magnetically charged string with Dirac-brane singularity along D. We assume that we can neglect the accelaration. It is then exactly the *same* quantity S that enters in all these expressions. This is a crucial observation.

Fig. 2. An example of a mathematical expression with its surrounding context, as used in our dataset. The context marked in blue will assist vLLM in identifying the handwritten manuscript of the yellow math expression. (Color figure online)

3.2 Language Model Inference

Our experimental pipeline consists of three main stages: language model inference, cleaning and conversion to structural graphs, and evaluation.

We evaluate multiple vLLMs including Gemini variants (Gemini-3-4b-it, Gemini-3-12b-it, Gemini-3-27b-it), Mistral-Small, InternVL2.5, and Gemini 2.5 Pro. For each model, we use carefully designed prompts that instruct the model to act as a Handwritten Mathematical Recognition Assistant.

We explore two prompting strategies:

- **Without context**: Direct image-to-text conversion where only the mathematical expression image is provided
- **With context**: Enhanced prompting that includes contextual information before and after the expression to help the model better understand the mathematical context

Prompt Engineering. Our system instruction defines the model's role as a mathematical recognition assistant with specific guidelines for LaTeX output format. The prompt is structured as shown in Fig. 3. The dictionary of the CROHME dataset is provided as a token list in the system part. The user part includes the image, the question text and also the context extracted from source data.

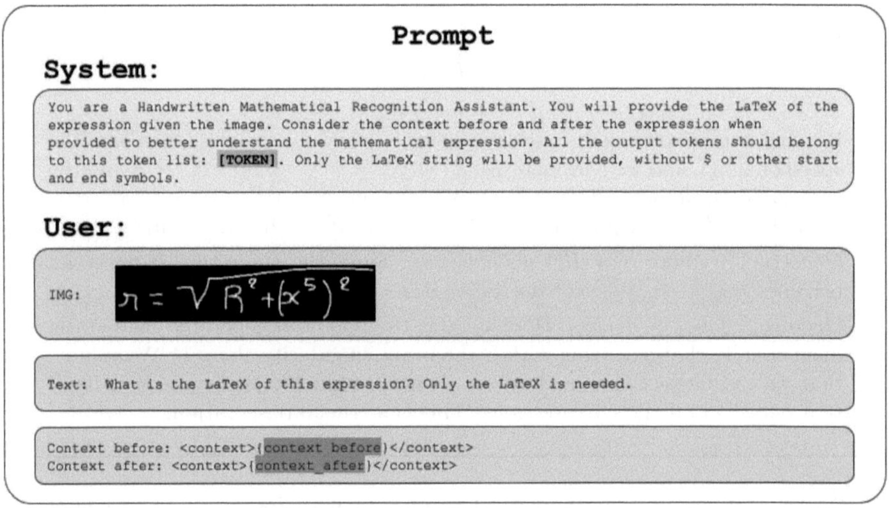

Fig. 3. Prompt Engineering

3.3 Graph Structure Evaluation

In mathematical expression recognition, evaluating results based solely on linear representations, such as LaTeX strings, often introduces ambiguity. The same expression can have multiple equivalent representations depending on formatting preferences, symbol ordering, or rule-based normalization. This ambiguity complicates both model training and evaluation, requiring numerous handcrafted rules and normalization heuristics.

To address these limitations, we adopt the Symbol Layout Graph (SLG) as a more robust and unique representation of mathematical expressions. In an SLG, each node corresponds to a visible symbol in the expression, and each edge captures a spatial relation between a pair of symbols. Specifically, six types of spatial relations (Right, Superscript, Subscript, Above, Below, Inside) are defined to cover all the situations.

This graph-based representation naturally reflects the two-dimensional structure of mathematical notation, making it significantly more expressive and less ambiguous than one-dimensional linear strings, especially useful for evaluation.

With the help of LgEval[1], the evaluation metrics include **Expression Correct Rate (ExpRate)**, which measures the percentage of expressions with an exact graph matching. Less than one ($\leq \mathbf{1}$) and two ($\leq \mathbf{2}$) edit distance, indicating the percentage of predictions within one or two graph edit operations from the ground truth. The **Structure Correct Rate (Struc.)**, which indicates whether the spatial layout (i.e., edge connectivity) is correct, even if some symbol (node) labels types are misclassified.

4 Experimental Results

4.1 Benchmark Performance

Table 3 and Fig. 2 present the performance comparison between vLLMs and dedicated models on CROHME 2019 [12] and 2023 datasets. The results demonstrate a clear performance gap between dedicated models and general-purpose vLLMs.

According to the results, the best-performing vLLM (Gemini 2.5 Pro) achieves 62.22% expression rate on CROHME 2019, while dedicated models like Sonia reach 77.73%. This represents a significant performance gap of approximately 15% points. However, compared with the dedicated models trained on the specific datasets, the performance of models is much better than the vLLM. Despite the huge difference in the number of parameters (Fig. 4).

4.2 Impact of Context

Table 4 and Fig. 5 shows the effect of providing contextual information to vLLMs during recognition.

The results show that providing context consistently improves performance across all tested models, with InternVL2.5 showing the most significant improvement (15.1% points on CROHME 2019).

[1] https://univ-nantes.io/crohme/lgeval.

Table 3. Performance comparison on CROHME datasets *without context.* Exp. is Expression Rate, ≤ 1 and ≤ 2 is the error tolerance levels, Struc. is the structural accuracy. '-' indicates the private close-source models.

Systems	Params	CROHME 2019				CROHME 2023			
		Exp.↑	1↑	2↑	Struc.↑	Exp.↑	1↑	2↑	Struc.↑
Vision Large Language Models									
Gemma-3-4b-it [16]	4B	12.93	20.52	24.94	27.11	10.09	17.09	21.40	23.53
Gemma-3-12b-it [16]	12B	23.52	33.86	40.70	42.29	23.23	35.66	43.41	45.06
Gemma-3-27b-it [16]	27B	35.11	49.04	57.05	61.80	29.88	45.72	54.46	57.76
Mistral-Small [14]	24B	17.60	23.44	27.44	28.19	16.53	23.18	27.36	30.23
InternVL2.5 [2]	78B	36.28	45.12	51.04	50.96	49.65	59.62	66.24	65.14
Gemini 2.5 Pro [15]	-	62.22	73.48	78.57	79.82	60.68	71.90	77.51	79.08
Dedicated Models									
CoMER [33]	6M	60.72	72.89	75.56	76.23	59.82	69.23	74.12	74.16
TST [24]	12M	70.81	80.57	83.15	83.65	67.25	78.68	82.73	83.04
Sonia [25]	-	77.73	87.41	90.08	90.58	70.81	81.74	86.13	86.95
YP-OCR [25]	-	73.39	85.74	88.49	89.16	67.86	80.86	85.13	85.99

Table 4. Performance comparison with and without context. w/o = without context, w/ = with context.

Systems	Context	CROHME 2019				CROHME 2023			
		Exp.↑	1↑	2↑	Struc.↑	Exp.↑	1↑	2↑	Struc.↑
Mistral-Small	w/o	17.60	23.44	27.44	28.19	16.53	23.18	27.36	30.23
	w/	18.77	24.02	28.44	30.86	17.44	24.53	29.32	31.49
InternVL2.5	w/o	36.28	45.12	51.04	50.96	49.65	59.62	66.24	65.14
	w/	51.38	59.22	63.80	62.72	54.94	62.85	68.94	67.25
Gemini 2.5 Pro	w/o	62.22	73.48	78.57	79.82	60.68	71.90	77.51	79.08
	w/	67.64	76.73	80.98	81.65	67.47	76.73	80.98	81.65

5 Discussion

5.1 Expression Difficulty Analysis

In this analysis, the *difficulty* of a mathematical expression refers to how challenging it is for recognition models to correctly identify and interpret the expression. This is quantified based on the recognition rates achieved by various models: expressions that are consistently recognized correctly are considered "easy," while those that are frequently misrecognized are deemed "hard." Thus, difficulty here reflects the capability of current models to detect and accurately process different types of mathematical expressions (Fig. 6).

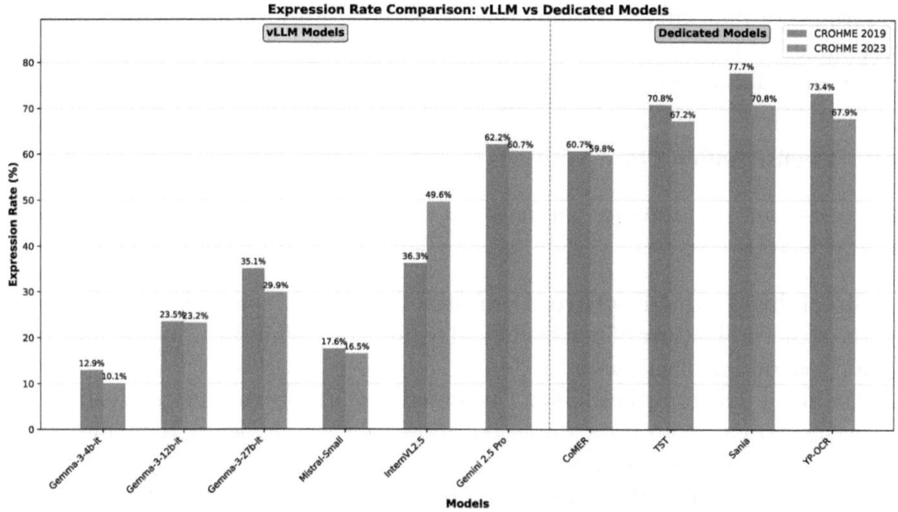

Fig. 4. Expression Rate Comparison of dedicated models and vLLMs on CROHME 2019 and 2023.

Our statistical analysis categorizes expressions by difficulty based on recognition rates across all models. The distribution reveals the inherent complexity of the CROHME dataset, as shown in Fig. 7. The chart demonstrates that the majority of expressions fall into challenging categories, with very hard expressions constituting the largest portion of the dataset. This skewed distribution

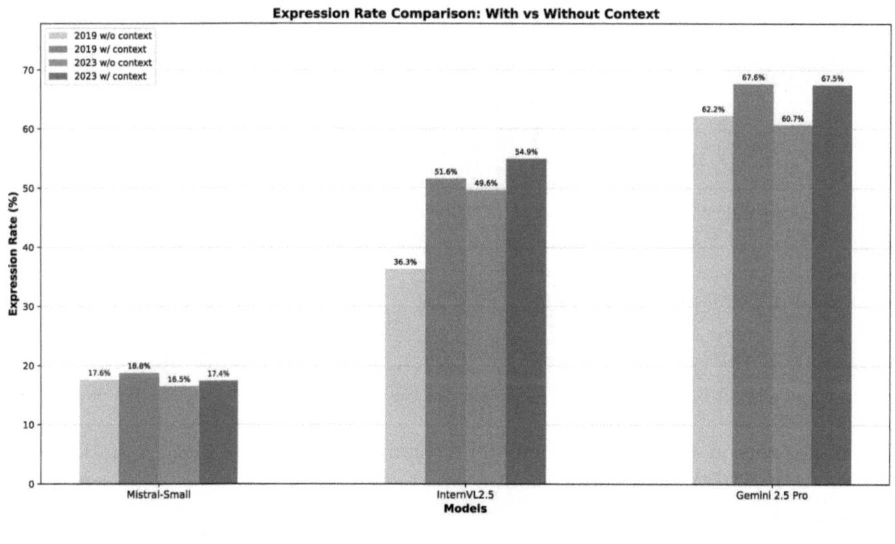

Fig. 5. Expression Rate Comparison across models and years with/without context.

toward more difficult expressions reflects the real-world nature of handwritten mathematical content, where complex structures and diverse notational styles create substantial recognition challenges for current systems.

5.2 Performance Gap Analysis

The consistent performance gap between dedicated models and vLLMs suggests several factors:

1. **Specialized Architecture**: Dedicated models are specifically designed for the structural nature of mathematical expressions, while vLLMs are optimized for general language tasks.
2. **Training Data Specificity**: Dedicated models are trained exclusively on mathematical expression datasets, allowing them to learn domain-specific patterns more effectively.
3. **Evaluation Methodology**: The structural evaluation used in CROHME may favor models designed specifically for mathematical layout understanding.

5.3 Context Utilization

The consistent improvement from contextual information indicates that vLLMs can leverage surrounding information to better understand mathematical expressions. However, it remains unclear whether models truly understand the mathematical context or simply benefit from additional textual cues.

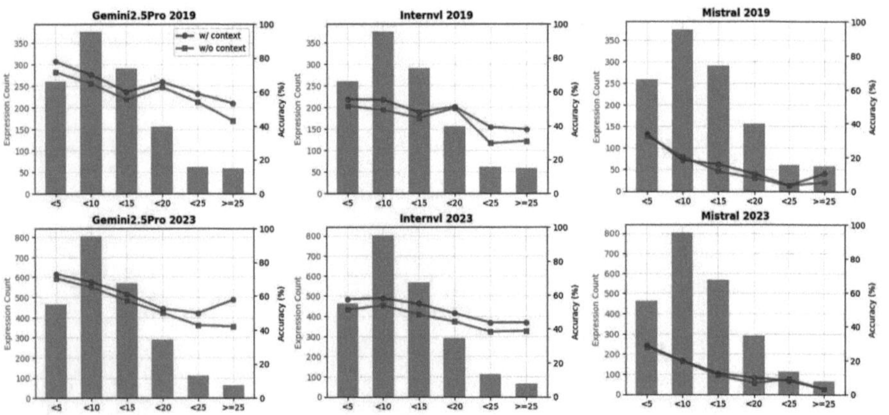

Fig. 6. Expression count distribution and recognition accuracy by node count for three vLLMs (2019/2023), with and without context.

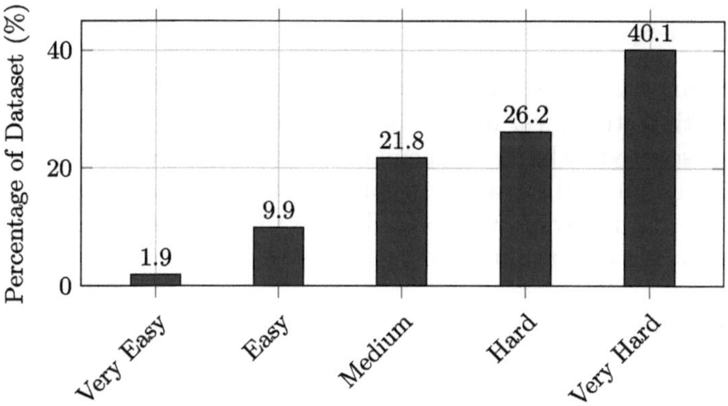

Fig. 7. Distribution of expression difficulty in the CROHME dataset, categorized by recognition rate.

5.4 Limitations

Our study identifies several limitations inherent in applying general-purpose vision-language models to handwritten mathematical expression recognition. The most significant finding is the substantial performance gap between vLLMs and smaller, specialized HMER models, where even the best-performing vLLM (Gemini 2.5 Pro) achieved 62.22% expression rate compared to 77.73% for dedicated models on CROHME 2019. While providing contextual information consistently improved performance across all tested models, the mechanism underlying this improvement remains unclear—whether models perform deep contextual reasoning or simply benefit from surface-level textual cues. Additionally, our analysis reveals that 40.1% of expressions in the CROHME dataset fall into the "very hard" category with recognition rates below 29%, indicating fundamental challenges in parsing complex, multi-level mathematical structures that affect both vLLMs and dedicated models alike.

6 Future Work

Based on our experimental findings, several research directions emerge as particularly promising for advancing vLLM-based mathematical expression recognition.

6.1 Enhanced Context Integration

The consistent improvements observed with contextual prompting suggest potential for more sophisticated multimodal fusion approaches. Future work could explore architectural modifications that create shared representational spaces for visual and textual modalities, moving beyond simple prompt concatenation to enable deeper integration of contextual understanding with visual recognition processes.

6.2 Domain-Specific Adaptation

The performance gap with dedicated models points toward the need for targeted adaptation strategies. Rather than relying on zero-shot capabilities, fine-tuning large-scale vLLMs on mathematical expression datasets could leverage their broad foundational knowledge while developing domain-specific competencies. This approach may prove more effective than training specialized models from scratch while maintaining the versatility of general-purpose architectures.

6.3 Structural Reasoning Enhancement

The challenges with complex expressions highlight limitations in current models' understanding of mathematical structure. Incorporating explicit structural supervision during training, potentially through graph-based loss functions or layout-aware objectives, could help models develop better representations of the hierarchical and spatial relationships fundamental to mathematical notation. Such approaches might address the current brittleness when encountering novel or complex expression layouts.

7 Conclusion

We presented a comprehensive benchmark study of Vision Large Language Models applied to Handwritten Mathematical Expression Recognition. Our experimental pipeline evaluated multiple state-of-the-art vLLMs using both context-free and context-aware prompting strategies, with evaluation based on structural graph comparison rather than simple text matching.

While vLLMs demonstrate promising capabilities for HMER, our results show that dedicated models still significantly outperform general-purpose vision-language models. The best vLLM (Gemini 2.5 Pro) achieved 62.22% expression rate compared to 77.73% for the leading dedicated model (Sonia). However, the consistent improvement from contextual information suggests potential for enhancement through better prompt engineering and context utilization.

Our findings highlight both the promise and current limitations of vLLMs for specialized tasks like mathematical expression recognition. Future work should focus on bridging the performance gap through fine-tuning approaches, improved context integration, and enhanced structural understanding capabilities.

References

1. Awal, A.M., Mouchère, H., Viard-Gaudin, C.: A global learning approach for an online handwritten mathematical expression recognition system. Patt. Recogn. Lett. **35**, 68–77 (2014), frontiers in Handwriting Processing
2. Chen, Z., et al.: Internvl: scaling up vision foundation models and aligning for generic visual-linguistic tasks. In: Proceedings of the IEEE/CVF Conference on Computer Vision and Pattern Recognition, pp. 24185–24198 (2024)
3. Chou, P.A.: Recognition of equations using a two-dimensional stochastic context-free grammar. In: Visual Communications and Image Processing IV, vol. 1199, pp. 852–865. SPIE (1989)

4. He, C., et al.: Olympiadbench: A challenging benchmark for promoting agi with olympiad-level bilingual multimodal scientific problems. arXiv preprint arXiv:2402.14008 (2024)
5. Hu, L., Zanibbi, R.: Mst-based visual parsing of online handwritten mathematical expressions. In: 2016 15th International Conference on Frontiers in Handwriting Recognition (ICFHR), pp. 337–342. IEEE (2016)
6. Koschinski, M., Winkler, H.J., Lang, M.: Segmentation and recognition of symbols within handwritten mathematical expressions. In: 1995 International Conference on Acoustics, Speech, and Signal Processing, vol. 4, pp. 2439–2442. IEEE (1995)
7. Lavirotte, S., Pottier, L.: Mathematical formula recognition using graph grammar. In: Document Recognition V, vol. 3305, pp. 44–52. SPIE (1998)
8. Li, Z., Yang, W., Qi, H., Jin, L., Huang, Y., Ding, K.: A tree-based model with branch parallel decoding for handwritten mathematical expression recognition. Pattern Recogn. **149**, 110220 (2024)
9. Lin, Q., Huang, X., Bi, N., Suen, C.Y., Tan, J.: Cclsl: combination of contrastive learning and supervised learning for handwritten mathematical expression recognition. In: Proceedings of the Asian Conference on Computer Vision, pp. 3724–3739 (2022)
10. Lu, P., et al.: Mathvista: evaluating mathematical reasoning of foundation models in visual contexts. arXiv preprint arXiv:2310.02255 (2023)
11. Mahdavi, M., Zanibbi, R.: Visual parsing with query-driven global graph attention (qd-gga): preliminary results for handwritten math formula recognition. In: Proceedings of the IEEE/CVF Conference on Computer Vision and Pattern Recognition Workshops, pp. 570–571 (2020)
12. Mahdavi, M., Zanibbi, R., Mouchère, H., Viard-Gaudin, C., Garain, U.: Icdar 2019 crohme+ tfd: Competition on recognition of handwritten mathematical expressions and typeset formula detection. In: 2019 International Conference on Document Analysis and Recognition (ICDAR), pp. 1533–1538. IEEE (2019)
13. Matsakis, N.E.: Recognition of handwritten mathematical expressions. Ph.D. thesis, Massachusetts Institute of Technology (1999)
14. Mistral AI: Mistral-small-3.1-24b-instruct-2503. https://huggingface.co/mistralai/Mistral-Small-3.1-24B-Instruct-2503 (2024). Accessed 27 June 2025
15. Team, G., et al.: Gemini: a family of highly capable multimodal models. arXiv preprint arXiv:2312.11805 (2023)
16. Team, G., et al.: Gemma: Open models based on gemini research and technology. arXiv preprint arXiv:2403.08295 (2024)
17. Truong, T.N., Nguyen, C.T., Zanibbi, R., Mouchère, H., Nakagawa, M.: A survey on handwritten mathematical expression recognition: The rise of encoder-decoder and GNN models. Patt. Recogn. 110531 (2024)
18. Truong, T.N., Nguyen, H.T., Nguyen, C.T., Nakagawa, M.: Learning symbol relation tree for online mathematical expression recognition. arXiv preprint arXiv:2105.06084 (2021)
19. Wang, J., Du, J., Zhang, J., Wang, B., Ren, B.: Stroke constrained attention network for online handwritten mathematical expression recognition. Pattern Recogn. **119**, 108047 (2021)
20. Wang, J., Wang, Q., Du, J., Zhang, J., Wang, B., Ren, B.: MRD: a memory relation decoder for online handwritten mathematical expression recognition. In: Document Analysis and Recognition–ICDAR 2021: 16th International Conference, Lausanne, Switzerland, September 5–10, 2021, Proceedings, Part III 16, pp. 39–54. Springer (2021)

21. Winkler, H.J., Lang, M.: Online symbol segmentation and recognition in handwritten mathematical expressions. In: 1997 IEEE International Conference on Acoustics, Speech, and Signal Processing. vol. 4, pp. 3377–3380. IEEE (1997)
22. Wu, C., et al.: Tdv2: A novel tree-structured decoder for offline mathematical expression recognition. In: Proceedings of the AAAI Conference on Artificial Intelligence, pp. 2694–2702 (2022)
23. Wu, J.W., Yin, F., Zhang, Y.M., Zhang, X.Y., Liu, C.L.: Graph-to-graph: towards accurate and interpretable online handwritten mathematical expression recognition. In: Proceedings of the AAAI Conference on Artificial Intelligence, pp. 2925–2933 (2021)
24. Xie, Y., Mouchère, H.: TST: Tree structured transformer for handwritten mathematical expression recognition. In: ICDAR 2025 (2025)
25. Xie, Y., et al.: Icdar 2023 crohme: Competition on recognition of handwritten mathematical expressions. In: International Conference on Document Analysis and Recognition, pp. 553–565. Springer (2023)
26. Xie, Y., Zanibbi, R., Mouchère, H.: Local and global graph modeling with edge-weighted graph attention network for handwritten mathematical expression recognition. arXiv preprint arXiv:2410.18555 (2024)
27. Yamamoto, R., Sako, S., Nishimoto, T., Sagayama, S.: On-line recognition of handwritten mathematical expressions based on stroke-based stochastic context-free grammar. In: Tenth international workshop on frontiers in handwriting recognition. Suvisoft (2006)
28. Zhang, J., Du, J., Dai, L.: Track, attend, and parse (tap): An end-to-end framework for online handwritten mathematical expression recognition. IEEE Trans. Multimedia **21**(1), 221–233 (2018)
29. Zhang, J., Du, J., Yang, Y., Song, Y.Z., Dai, L.: SRD: a tree structure based decoder for online handwritten mathematical expression recognition. IEEE Trans. Multimedia **23**, 2471–2480 (2020)
30. Zhang, J., et al.: Watch, attend and parse: An end-to-end neural network based approach to handwritten mathematical expression recognition. Pattern Recogn. **71**, 196–206 (2017)
31. Zhang, T., Mouchere, H., Viard-Gaudin, C.: Tree-based blstm for mathematical expression recognition. In: 2017 14th IAPR International Conference on Document Analysis and Recognition (ICDAR). vol. 1, pp. 914–919. IEEE (2017)
32. Zhang, T., Mouchère, H., Viard-Gaudin, C.: A tree-BLSTM-based recognition system for online handwritten mathematical expressions. Neural Comput. Appl. **32**, 4689–4708 (2020)
33. Zhao, W., Gao, L.: Comer: modeling coverage for transformer-based handwritten mathematical expression recognition. In: European Conference on Computer Vision, pp. 392–408. Springer (2022)
34. Zhao, W., Gao, L., Yan, Z., Peng, S., Du, L., Zhang, Z.: Handwritten mathematical expression recognition with bidirectionally trained transformer. In: Lladós, J., Lopresti, D., Uchida, S. (eds.) ICDAR 2021. LNCS, vol. 12822, pp. 570–584. Springer, Cham (2021). https://doi.org/10.1007/978-3-030-86331-9_37
35. Zhu, J., Zhao, W., Li, Y., Hu, X., Gao, L.: Tamer: tree-aware transformer for handwritten mathematical expression recognition. arXiv preprint arXiv:2408.08578 (2024)

SCANS: An Efficient Geometric Problem Solver with Content-Aware Attention and Adaptive Fusion

Zhi Chen[1,2,3(✉)], Yuhan Yang[1,2,3], Xiangdong Su[1,2,3], Haoran Zhang[1,2,3], Xinxiang Zhou[1,2,3], Wei Chen[1,2,3], and Guanglai Gao[1,2,3]

[1] College of Computer Science, Inner Mongolia University, Hohhot, China
cssxd@imu.edu.cn

[2] National & Local Joint Engineering Research Center of Intelligent Information Processing Technology for Mongolian, Hohhot, China

[3] Inner Mongolia Key Laboratory of Multilingual Artificial Intelligence Technology, Hohhot, China

Abstract. Geometric problem solving (GPS) involves inferring unknown quantities or relations from a diagram and its related description. It is a visual reasoning task that is challenged by two difficulties. One is to first locate the key geometric elements in the diagram, and then merge them with the relevant description. Existing neural solvers often fail to exploit the inherent structural information in geometric diagrams, since they do not distinguish between content-rich and blank regions during feature extraction. Meanwhile, these methods typically neglect the varying importance of features in the multi-modal fusion process. We propose a salient content-aware neural solver (SCANS) to address the above two challenges. Specifically, a content-aware structural attention module first constructs a structural prior, which enables it to focus on the informative regions in the diagram while suppressing the empty background. Then, a hierarchical adaptive fusion module is applied to merge visual and textual features across spatial, channel, and modality dimensions and learn task-specific fusion weights. Such a design facilitates fine-grained perception of geometric relations and balances the contributions from two modalities. Extensive experiments on the Geometry3K and PGPS9K datasets validate the effectiveness of the proposed SCANS.

Keywords: Geometry problem solving · Content-aware structural attention · Multi-modal fusion

1 Introduction

Geometry Problem Solving (GPS) is a long-standing challenge in artificial intelligence that attracts significant interest from both the computer vision and natural language processing communities [10,18,20]. The task is inherently difficult as

L. Jin et al. (Eds.): ICDAR 2025 Workshops, LNCS 16225, pp. 327–343, 2026.
https://doi.org/10.1007/978-3-032-09368-4_20

it requires complex mathematical reasoning across multiple modalities, such as visual diagrams and textual descriptions. A typical geometry problem consists of a textual description and a corresponding geometric figure. The text provides the given information and the goal of the solution, while the image complements this with latent geometric relationships that require fine-grained visual understanding. Solving these problems requires the effective integration of textual and visual information.

However, achieving such integration is hampered by the inherent sparsity of geometric diagrams. As illustrated in Fig. 1, most pixels form a uniform background, while crucial lines, points, and labels cover only a small area. When the image is divided into standard patches, many patches contain background alone. These redundant patches waste computation and weaken the attention assigned to informative regions, which in turn limits the modeling of essential geometric relations. Recent studies [12,15] have explored richer geometric representations, yet they rarely tackle the problem of isolating salient structures from empty regions.

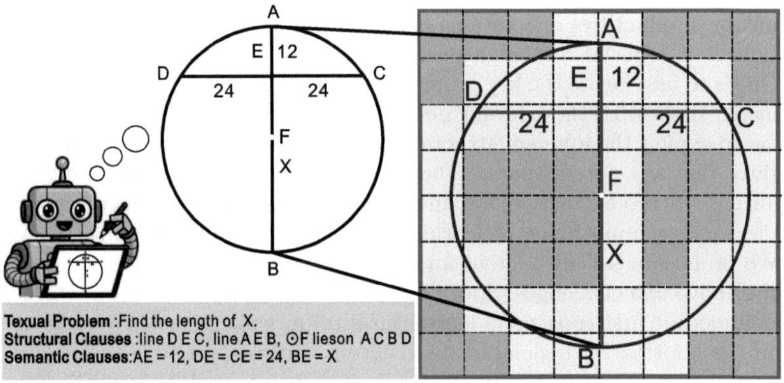

Fig. 1. Example of a plane geometry problem that includes two modalities: visual diagrams and textual descriptions. When the diagram is divided into patches, the red areas, which are completely blank, distract the model' s attention. (Color figure online)

To contend with this gap, current neural solvers follow two main routes. The first route keeps the diagram in pixel space and encodes it with visual backbones that are often adapted from natural-image models [6,7,12,15]. Because these backbones are not tuned for sparse line drawings, researchers introduce auxiliary tasks such as position restoration or geometric element prediction to improve spatial awareness [7,12] and masked image modeling to cope with the sparse foreground [15]. These strategies raise accuracy, but they fail to distinguish between regions containing geometric structures and blank background regions within the encoder. As a result, the learned features remain vulnerable to noise from these irrelevant regions. The second route converts the diagram into a symbolic or linguistic form before reasoning [10,16,24,32,35]. Typical choices include geometry logic graphs that link tuples of points, lines, and angles [16], or full textual descriptions that can be processed by large language models [32]. This abstraction removes redundant pixels and simplifies the downstream inference, yet it hinges on a precise parser. Once the parser mislabels a line or angle, the error propagates through the pipeline and the solver cannot easily recover, which constrains the upper bound of performance.

Beyond these two routes, the solver must still bridge a semantic gap between diagram and text. The textual description states facts and goals in a precise logical form, while the diagram conveys spatial relations that lack explicit structure. This mismatch pushes their feature embeddings apart and weakens cross modal alignment. The weight of each modality also shifts with the problem because some questions hinge on the exact arrangement of lines and angles whereas others rely mainly on textual clues. Most current fusion modules [15,26] apply a fixed strategy that cannot adapt to this variation, and they often miss key cues or create conflicting evidence when the task demands a different balance.

To tackle the background noise in diagrams, we introduce content-aware structural attention, inspired by research in semantic segmentation [30]. This mechanism first scans low-level visual cues such as edge density and pixel variance and assigns a content score to every image patch. The score produces three complementary masks that encode local content, absolute position, and global structure. Feeding these masks into the self-attention of the vision branch guides the model toward lines, points, and labels while ignoring blank areas. With cleaner visual tokens in place, we next build a hierarchical adaptive fusion module to address the shifting balance between the two modalities. A cross-attention layer [31] creates an initial link between the refined visual sequence and the textual tokens, then a spatial-attention map further highlights the most informative image regions, and finally a coordinated attention block revisits the fused sequence to adjust spatial, channel, and modality weights in a unified pass. This three-step process lets the solver emphasize the cues that matter most for the final geometric reasoning task.

The contributions of this paper are summarized as follows:

- We present a salient content-aware neural solver (SCANS), tailored to the sparse structure of geometric diagrams, which integrates multi-modal cues with high fidelity and achieves superior performance on the Geometry3K and PGPS9K datasets.
- We introduce content-aware structural attention, which derives structural priors from low level visual statistics so the network concentrates on informative lines, points, and labels while ignoring blank background, thereby overcoming a critical weakness of general vision backbones in geometry tasks.
- We develop a hierarchical adaptive fusion module that aligns and merges modalities in three successive stages and ends with coordinated attention over spatial, channel, and modality dimensions, which enables problem-specific balancing of textual and visual evidence.

2 Related Work

2.1 Geometry Problem Solving

GPS can be divided into two categories: geometric theorem proving and geometric numerical calculation. Early research in GPS primarily focused on the automatic proving of geometric theorems [14,23]. Geometric numerical calculation [20] has garnered increasing attention in recent years with the emergence of new datasets and methods. This paper focuses on geometric numerical calculation and reviews two main classes of methods: symbolic solvers and neural solvers.

Symbolic solvers typically rely on rule-based inferences to solve geometry problems [10,16,18,20,24]. GEOS [20] parses geometric diagrams and applies a heuristic search algorithm to match and apply rules to find a solution. To address GEOS's inability to utilize external knowledge, [18] proposed integrating geometric theorem knowledge from math textbooks into the model, enabling it to perform reasoning based on rigorous theorems. Inter-GPS [10] utilizes a formal language to unify diagrams and text and then processes them with a symbolic engine. However, the efficiency of traditional heuristic search is limited. To address this, GeoDRL [16] models the solving process as a Markov decision process and employs deep reinforcement learning to autonomously optimize the reasoning strategy. E-GPS [24] is a top-down geometric problem-solving model that starts reasoning from the solution goal and works backward to generate both the final answer and interpretable solution steps. Although symbolic solvers are logically rigorous, their main challenge lies in the need for experts to perform complex rule design and manual construction, and they are difficult to extend to more diverse or complex geometry problems.

Neural solvers treat geometric problem solving as a unique visual language task. They typically employ an encoder-decoder architecture that treats the problem-solving process as the generation of a series of operations [5–7,15]. NGS [7] and GEOformer [6] use self-supervised auxiliary tasks such as geometric element prediction to refine diagram representations. DPE-NGS [5] improves on the NGS model by integrating a more efficient text encoder that encodes long text problems more efficiently. SCA-GPS [15] attempts to align common characters in text and images to enhance the understanding of diagrams through pre-training in multi-label classification and masked image modeling. Although neural solvers show great potential in processing large-scale datasets, they are still inadequate in processing and fusing geometric image information. In this paper, we propose a content-aware neural solver that improves geometric problem solving by precisely attending to and fusing key geometric elements.

2.2 Multi-modal Reasoning

Multi-modal reasoning tasks aim to combine and understand information from different modalities, such as text, images, and tables, and are typically presented in a question-answering format. Examples include Visual Question Answering

[3,8], Document Question Answering [22,27], and Table Question Answering [11]. To accomplish multi-modal reasoning, it is essential to first understand and integrate multi-modal content. Some prior works [17,19] on Visual Question Answering adopt implicit reasoning frameworks, fusing multi-modal information through joint encoding. In contrast, other works [13,29] utilize domain-specific languages to perform explicit symbolic reasoning. This process involves converting visual features into a symbolic language and then performing logical deduction.

However, the programmatic languages used in these methods primarily consider basic operations, such as object counting [1], and are difficult to apply directly to geometry problems that require multi-step numerical calculations and knowledge of geometric theorems. Therefore, GPS is considered a particularly distinct and complex form of multi-modal reasoning. To this end, we propose SCANS, a novel architecture that employs an adaptive fusion module to dynamically integrate multi-modal content for the task of GPS.

3 Methodology

3.1 Overall Framework

This paper proposes a salient content-aware neural solver (SCANS) for the task of GPS. Figure 2 illustrates the overview of our proposed architecture. Given a geometry problem P comprising a geometric diagram D and a textual problem description T_{prob}, such that $P = \{D, T_{\text{prob}}\}$, we first employ two independent encoders to extract features from the text and the diagram, respectively.

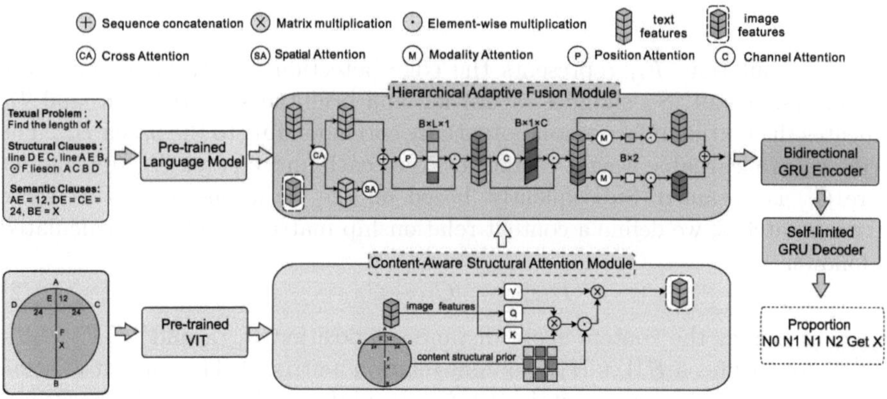

Fig. 2. Overview of SCANS model. The SCANS model processes an input image and text by first refining visual features with the Content-Aware Structural Attention module, then fusing them with text features via the Hierarchical Adaptive Fusion Module, and finally generating the solution program with an encoder-decoder.

Following this, we introduce a content-aware structural attention module. This module generates prior masks based on the structural properties of the geometric image, guiding the model to focus on key regions within the diagram. Subsequently, the hierarchical adaptive fusion module employs a multi-stage fusion strategy and adaptively weights the multi-modal features. This allows the model to dynamically allocate its attentional focus between textual descriptions and geometric shapes based on problem-specific characteristics, emphasizing salient information crucial for the solution process. The fused features are then fed into a bidirectional GRU encoder, and finally, a self-constraining GRU decoder sequentially generates the solution program in an auto-regressive manner.

3.2 Content-Aware Structural Attention Module

Our approach leverages a frozen, pre-trained Vision Transformer (ViT) as a powerful feature extractor. Initially, an image is partitioned into $N = HW$ non-overlapping patches, where H and W are the number of patches in the vertical and horizontal dimensions. Each patch is denoted by $B_{i,j}$, with i and j as its row and column indices. We calculate a content score for each image patch based on its information density, which is then used to explicitly model the relationships among all patches. Finally, this relational information is embedded into the model's self-attention layers, consequently forming a content-aware structural attention.

Specifically, by performing operations such as edge detection, line detection, analysis of saturation and brightness channels, and extraction of textual content from the image, the content features of each image patch are obtained. These features are then fused using a weighted sum:

$$C_{ij} = w_1 E_{ij} + w_2 L_{ij} + w_3 S_{ij} + w_4 T_{ij} \tag{1}$$

in this formulation, E_{ij} represents the edge detection result, L_{ij} denotes the line analysis result, S_{ij} signifies saturation and brightness information, and T_{ij} indicates the textual information potentially corresponding to the patch B_{ij}. The terms w_x are learnable weight coefficients. For each image patch B_{ij}, its content score C_{ij} is obtained. Subsequently, based on the variations in content score between patches, we define a content relationship matrix D, which is calculated as follows:

$$D_{ij,i'j'} = |C_{ij} - C_{i'j'}| \tag{2}$$

$D_{ij,i'j'}$ indicates the content score distance at position (i,j), and (i',j') , ultimately constitutes a $HW \times HW$ content relation matrix D. The content relation matrix, however, does not contain the distance information of spatial positions, so we generate a spatial relation matrix by calculating the spatial distances between all image blocks. This can be defined as :

$$S_{ij,i'j'} = |i - i'| + |j - j'| \tag{3}$$

$S_{ij,i'j'}$ represents the spatial Manhattan distance between the positions (i,j) and (i',j'). These values constitute a spatial prior matrix S, of shape $HW \times HW$.

To enhance the model's sensitivity to structural and shape similarities, we assume that similar geometric structures and shape features should receive comparable attention. We introduce a hyperparameter threshold τ to quantify structural and shape similarity and construct a binary structural consistency mask R, defined as:

$$R_{ij,i'j'} = \begin{cases} 1 & \text{if } |C_{ij} - C_{i'j'}| < \tau \\ 0 & \text{otherwise} \end{cases} \tag{4}$$

$R_{ij,i'j'}$ represents whether the patches at positions (i, j) and (i', j') are structurally similar. τ is a predefined similarity threshold. When the content density difference $|C_{ij} - C_{i'j'}|$ is below the threshold, the mask R assigns the patch pair a value of 1, marking them as structurally and shape-consistent.

These three prior matrices are fused to generate a comprehensive structural prior matrix P. The process begins by scaling the distance matrices D and S with a learnable negative log-decay value decay. This transforms them from simple distance metrics into penalty matrices, where a greater distance imposes a stronger negative bias. These decay-adjusted matrices are then combined via a weighted sum with learnable weights α and β. Finally, the structural consistency mask R is added to form the comprehensive structural prior matrix P. The formulation is as follows:

$$P = (\alpha \cdot (S \cdot \text{decay}) + \beta \cdot (D \cdot \text{decay})) + R \tag{5}$$

After deriving several content structural priors, the input image features are initially transformed via three linear transformations to generate query, key, and value vectors and then query and key vectors are first enhanced using rotary position embeddings [21] to provide spatial information for each token and enhance spatial awareness, yielding Q_r and K_r. The transformed query and key vectors are used to obtain the initial attention scores via scaled dot-product. Subsequently, the comprehensive structural priors P are injected as an additive bias directly into the attention score computation before the softmax operation. This realizes the content-aware structural attention, formulated as:

$$Attention(Q, K, V) = \text{softmax}\left(\frac{Q_r K_r^T}{\sqrt{d}} + P\right) V \tag{6}$$

where the prior term P adjusts the attention scores based on patch characteristics: contributions from patches with plain backgrounds or sparse content are suppressed, while content-rich patches receive greater emphasis. Furthermore, the prior explicitly bolsters attention towards spatially adjacent and structurally similar regions. This design, in turn, strengthens local feature modeling and facilitates the learning of cross-region coupling.

3.3 Hierarchical Adaptive Fusion Module

To achieve a robust and adaptive fusion of multi-modal information, we design a hierarchical adaptive fusion module. Our approach follows a progressive, three-stage strategy: it first establishes a coarse cross-modal alignment using a cross-attention network, then enhances modality-specific features with specialized attention mechanisms, and finally performs a global optimization via a sequence-level coordinated attention mechanism. Formally, let the text features be $T \in \mathbb{R}^{N_t \times d}$ and the visual features be $V \in \mathbb{R}^{N_v \times d}$, where N_t and N_v denote the sequence lengths of the text and visual modalities respectively, and d is the feature dimension.

At the initial stage, we utilize a multi-modal cross-attention network to establish deep cross-modal alignment. The network is composed of both self-attention and cross-attention modules, enabling the visual modality to extract complementary information from the textual modality. Specifically, the text features are first processed via a self-attention mechanism to construct intra-modal contextual representations:

$$T' = \text{SelfAttention}(T, T, T) \tag{7}$$

Subsequently, visual features attend to the text features via cross-attention to obtain guided information:

$$V' = \text{CrossAttention}(V, T', T') \tag{8}$$

To account for the spatial structural characteristics of geometric images, we design a dedicated spatial-attention module to enhance the visual features obtained after cross-modal attention. This module first pools the feature representation along the channel dimension to generate two descriptive feature maps: an average-pooling feature map and a max-pooling feature map. These two maps are then concatenated and processed through a 2D convolutional layer to generate the final spatial weighting map. The process is formulated as:

$$A_{\text{spatial}} = \sigma\left(\text{Conv}_{2D}\left(\text{AvgPool}_{\text{ch}}(V') \oplus \text{MaxPool}_{\text{ch}}(V')\right)\right) \tag{9}$$

where $\text{AvgPool}_{\text{ch}}$ and $\text{MaxPool}_{\text{ch}}$ denote average and max pooling operations along the channel dimension of the feature sequence V', $\oplus$ denotes channel-wise concatenation, and σ is the Sigmoid activation function. The spatially enhanced visual features are:

$$V_{\text{enhanced}} = V' \odot A_{\text{spatial}} \tag{10}$$

where $\odot$ denotes element-wise multiplication. This mechanism enables the model to adaptively focus on geometrically salient regions while suppressing the influence of background noise. After obtaining the spatially enhanced visual features, we concatenate them with the text features T to form a complete multi-modal sequence $S = [V_{\text{enhanced}}; T] \in \mathbb{R}^{(N_v+N_t)\times d}$.

To perform global optimization over this fused sequence, we introduce a sequence-level coordinated attention Mechanism. This mechanism imposes attention constraints jointly across spatial position, feature channels, and modality dimensions, ensuring that the most critical feature combinations for geometric reasoning are prioritized.

Position Attention. To capture the varying importance across different positions in the concatenated sequence, position attention dynamically computes positional weights for each element in the sequence. The weight for each position i in the sequence is computed as:

$$w_{\text{pos}}^{(i)} = \sigma\left(W_{\text{pos2}} \cdot \text{ReLU}\left(W_{\text{pos1}} \cdot S^{(i)} + b_{\text{pos1}}\right) + b_{\text{pos2}}\right) \tag{11}$$

where $W_{\text{pos1}} \in \mathbb{R}^{(d/r)\times d}$, $W_{\text{pos2}} \in \mathbb{R}^{1\times(d/r)}$, and r is the dimension reduction ratio, $b_{\text{pos1}} \in \mathbb{R}^{d/r}$ and $b_{\text{pos2}} \in \mathbb{R}$ are learnable bias terms that provide adaptive thresholding for position-wise feature activation. This enables the model to adaptively attend to key positions within the sequence based on task-specific characteristics, regardless of whether they correspond to important textual tokens or critical geometric patches in the image.

Channel Attention. To identify the channels within the feature dimensions that are most critical for geometric understanding, channel attention first extracts global channel statistics using adaptive average pooling. Specifically, we transpose the input sequence S from shape (B, L, d) to (B, d, L) to align with the standard 1D convolutional input format, and then apply pooling:

$$g_c = \text{AdaptiveAvgPool1d}(S^T) \in \mathbb{R}^{B\times d\times 1} \tag{12}$$

where S^T denotes the transposed sequence. Then, we apply a 1D convolutional layer to compute channel-wise attention weights:

$$W_{\text{channel}} = \sigma\left(\text{Conv1d}\left(\text{ReLU}\left(\text{Conv1d}(g_c)\right)\right)\right) \tag{13}$$

where the first convolution reduces channels from d to $d/16$, and the second convolution restores them to d. This squeeze-and-excitation design suppresses redundant channels while enhancing those most relevant to the geometric semantics of the task.

Modality Attention. Considering that the relative importance of textual and visual information may vary across tasks, we introduce a modality attention mechanism that dynamically adjusts weights for each modality based on their global semantic representations. Specifically, we first compute the global representations of text and visual modalities as:

$$r_{\text{text}} = \frac{1}{N_t}\sum_{i=1}^{N_t} S^{(i)}, \quad r_{\text{visual}} = \frac{1}{N_v}\sum_{i=N_t+1}^{N_t+N_v} S^{(i)} \tag{14}$$

The final modality weights are computed via a two-layer MLP with dimension reduction:

$$[\alpha_{\text{text}}, \alpha_{\text{visual}}] = \text{Softmax}\left(W_{\text{mod2}} \cdot \text{ReLU}\left(W_{\text{mod1}} \cdot \frac{r_{\text{text}} + r_{\text{visual}}}{2}\right)\right) \tag{15}$$

where $W_{\text{mod1}} \in \mathbb{R}^{(d/r)\times d}$ and $W_{\text{mod2}} \in \mathbb{R}^{2\times(d/r)}$ with $r = 4$. This two-layer design with ReLU activation provides enhanced nonlinear modeling capability compared to direct linear projection.

Final Fusion Representation. The attention weights from the three dimensions are applied in a hierarchical manner through multiplicative integration. First, position and channel attention are jointly applied to the entire sequence:

$$S_{\text{enhanced}}^{(i)} = S^{(i)} \odot w_{\text{pos}}^{(i)} \odot W_{\text{channel}} \tag{16}$$

Then, modality-specific weights are selectively applied based on the sequence position:

$$S_{\text{final}}^{(i)} = \begin{cases} S_{\text{enhanced}}^{(i)} \odot \alpha_{\text{text}} & \text{if } i \leq N_t \\ S_{\text{enhanced}}^{(i)} \odot \alpha_{\text{visual}} & \text{if } i > N_t \end{cases} \tag{17}$$

This hierarchical fusion strategy ensures that position and channel attention operate globally across the entire sequence, while modality attention provides specialized weighting for textual and visual components, respectively. Through the hierarchical and progressive processing of the HAFM module, the resulting multi-modal representation S_{final} possesses enhanced cross-modal semantic alignment capability and an adaptive modality-specific weighting mechanism. This representation is subsequently fed into a GRU encoder for sequence encoding.

4 Experiments

4.1 Setup

Implementation Details. We employ a pre-trained Vision Transformer [15] as the visual feature extractor, which consists of 12 Transformer layers, 12 attention heads, and a hidden dimension of 768. For preprocessing, input images are resized by scaling their longest edge to 224 pixels and are then center-padded to a 224×224 canvas. The image is subsequently partitioned into an 8×8 grid of 64 patches, with each patch being 28×28 pixels. The parameters of the ViT backbone are kept frozen during training. Our Hierarchical Adaptive Fusion Module (HAFM) is composed of a 6-layer multi-modal cross-attention network, with each layer having 8 attention heads and a hidden dimension of 256. The bidirectional GRU (Bi-GRU) encoder is a 2-layer network with an input embedding dimension of 256 and a hidden state size of 512. Similarly, the self-constraining GRU decoder also has 2 layers, with both its input embedding and hidden state dimensions set to 512. Our model is implemented in PyTorch and trained on a single NVIDIA P100 GPU.

For training, we use the AdamW optimizer with a weight decay of 0.01. We employ a learning rate schedule with a warmup phase spanning the first 40 epochs. The initial learning rate is set to 1e-3 and is decreased via a step decay schedule at epochs 160, 280, 360, 440, and 500, with a decay rate of 0.5. The model is trained for a total of 540 epochs with a batch size of 90. The maximum input sequence length is capped at 400 tokens, and the maximum output sequence length is 40. For inference, we use a beam search with a beam size of 10.

Datasets. We use two widely used plane geometry datasets, Geometry3K [10] and PGPS9K [34], to evaluate the performance of the proposed SCANS. These datasets contain geometry problems sourced from U.S. grades 5-12, encompassing 30 distinct problem types. Geometry3K consists of 8,433 training and 589 test samples, while PGPS9K contains 8,022 training and 1,000 test samples. The datasets provide ground-truth solutions in the form of programs. Each solution program is composed of multiple steps, where every step includes an operator and its associated operands. The operator corresponds to a geometric theorem, and the operands are arranged according to the theorem's formal structure. A Python-based program executor is used to calculate the final numerical results from these solution programs.

Metrics. To ensure a fair comparison, the SCANS performance evaluation utilized three metrics consistent with previous work [34]: completion, selection, and top-3. In completion, the neural solver takes the first executable solution program as the result of completion. The choice is to select a correct solution from the four candidates, or a random one if no answer is available. In top3, a correct solution is considered correct if it is one of the top three solutions with the highest confidence.

4.2 Comparison with State-of-the-Art Methods

We compare SCANS with state-of-the-art models. As shown in Table 1, for a more intuitive and fair comparative analysis, we classify the reference methods into three groups: symbolic solvers, neural solvers, and the current popular multimodal large models. The experimental results show that our proposed SCANS exhibits excellent performance in all evaluation metrics.

For Symbolic Solvers, InterGPS [10] is a rule-based symbolic solver. Depending on the input source of the formal language, InterGPS exhibits three result types: "Predict" means that all formal languages are predicted by its parser, "Diagram GT" means that the formal clauses of diagrams use truth-valued annotations, and "All GT" means that the formal clauses of both diagrams and textual problems use truth-valued annotations. Deep Reinforcement Learning improves the search strategy of InterGPS. Building on Inter-GPS, GeoDRL [16] enhances geometric deductive reasoning by combining logistic graph inference and deep reinforcement learning to improve the theorem selection and search strategy. Experimental results show that our method is generally superior to mainstream symbolic solvers. While the InterGPS (All GT) setting achieves the highest score in completion mode, it relies on fully ground-truth formal annotations for both diagrams and text and is not an end-to-end predictive model. In all other metrics, our model surpasses the existing symbolic solvers. Furthermore, our method achieves a significant 4.9% performance improvement over InterGPS (All GT) in the choice mode on the Geometry3K dataset.

For neural solvers, NGS [7] is a pioneering neural-geometric solver that relies heavily on textual problems for solving, while Geoformer [6] jointly handles geometric proofs and computational problems. PGPSNet [34] has designed

Table 1. Comparison of state-of-the-art GPS solvers on the Geometry3K and PGPS9K datasets in terms of Completion, Choice, and Top-3 accuracy.

Method	Geometry3K			PGPS9K		
	Completion	Choice	Top-3	Completion	Choice	Top-3
Human Expert [10]	-	90.9	-	-	-	-
interGPS (predict) [10]	44.6	56.9	-	-	-	-
InterGPS (DiagramGT) [10]	64.2	71.7	-	59.8	68.0	-
InterGPS (All GT) [10]	**69.0**	75.9	-	-	-	-
GeoDRL [16]	-	68.4	-	-	-	-
NGS [7]	35.3	58.8	62.0	34.1	46.1	60.9
Geoformer [6]	36.8	59.3	62.5	35.6	47.3	62.3
PGPSNet [34]	65.0	77.9	80.7	62.7	70.4	79.5
PGPSNet-v2-S [33]	68.2	79.8	81.6	63.4	71.7	81.2
DenseGPS [12]	66.7	77.4	80.2	60.7	70.1	79.9
GeoX [25]	58.6	72.5	69.4	52.7	63.3	65.4
GPT-4V [2]	38.6	42.3	-	31.8	40.3	-
Qwen-VL [4]	22.1	26.7	-	20.1	23.2	-
LLaVA-v1.5 [9]	7.6	11.2	-	6.3	9.1	-
mPLUG-Owl2 [28]	12.1	17.4	-	10.1	13.1	-
Ours	67.7	**80.8**	**82.5**	**65.7**	**75.5**	**83.1**

unique linguistic pre-training tasks to fully exploit high-level semantic features. PGPSNet-v2 [33] continues to build on this by adding a validator for the solution program. DenseGPS [12] adds the auxiliary tasks of geometric element prediction and graph position recovery to PGPSnet's graph encoder. In contrast, our model SCANS outperforms PGPSNet in choice mode, improving by 2.9% on Geometry3K and 5.1% on PGPS9K datasets.

We also compared our approach with generalized multi-modal large models and showed that GPT-4V [2] and Qwen-VL [4] perform poorly in GPS scenarios. This poor performance may be attributed to the fact that the general-purpose multi-modal models are mainly trained for natural scene images and lack the ability to specifically model the spatial structure and geometric relationships specific to geometric images.

4.3 Ablation Study

We conduct ablation experiments on the PGPS9K dataset to examine the effectiveness of the modules proposed in our SCANS. As shown in Table 2, we select PGPSNet [34] as our baseline for these experiments. In contrast to the PGPSNet, our Content-Aware Structural Attention (CASA) module introduces geometric priors to guide the model's focus towards salient elements while suppressing background noise. The results show that incorporating the CASA module alone outperforms the baseline, improving performance in the completion, choice, and top-3 metrics by 0.5%, 3.1%, and 1.8%, respectively.

Table 2. Ablation study of modules on PGPS9K.

Module	PGPS9k		
	Completion	Choice	Top-3
Baseline	62.7	70.4	79.5
+ CASA	63.2	73.5	81.3
+ HAFM	63.3	73.2	82.6
+ CASA + HAFM	65.7	75.5	83.1

Additionally, our Hierarchical Adaptive Fusion Module (HAFM) further enhances GPS performance by employing a coordinated attention mechanism to dynamically balance the contributions from textual and visual modalities. When added to the baseline, this module alone improves the completion, choice, and top-3 scores by 0.6%, 2.8%, and 3.1%, respectively. Ultimately, our full model, SCANS, which integrates both CASA and HAFM, achieves a final accuracy of 75.5% on the choice metric, marking a 5.1% total improvement over the baseline.

4.4 Qualitative Analysis

Attention Visualization. To qualitatively analyze our model's visual understanding capabilities, we visualized the attention heatmap distributions generated by our proposed SCANS and the PGPSNet baseline's visual encoder when

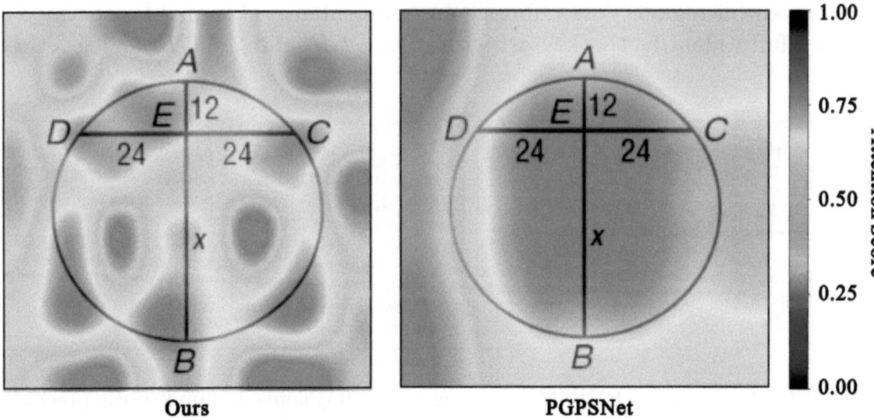

Fig. 3. A qualitative comparison of spatial-attention heatmaps from our model (left) and the PGPSNet baseline (right).

processing geometric diagrams, as shown in Fig. 3. The visualization reveals a significant difference in the attention allocation strategies between the two models.

As depicted in Fig. 3 (right), the attention mechanism of PGPSNet, while generally covering the diagram area, produces a heatmap that is diffuse across the central region, failing to precisely attend to the key geometric elements. In contrast, our model (left) exhibits a far more precise and semantically aware attention distribution. The model accurately localizes and assigns high attention scores (indicated by yellow and red regions) to the core geometric and numerical elements essential for problem-solving, such as the vertices (A, B, C, D), the key intersection point (E), and the numerical labels (12, 24). Furthermore, our model effectively suppresses attention on irrelevant background regions, demonstrating its superior ability to distinguish between key foreground entities and background noise. This targeted attention mechanism validates that our CASA module successfully guides the model to focus on the most informative parts of the diagram, providing precise visual guidance for automated geometric problem-solving.

Case Analysis. As shown in Fig. 4, we conduct a detailed case study to further explore the strengths and weaknesses of different solvers. In case (a), the correct solution path requires the model to understand the textual information that "PQRS is a rhombus" and integrate it with the visual structure of its intersecting diagonals. It means $\angle QTR$ must be 90°. However PGPSNet fails to correctly combine these two pieces of information. It is instead misled by textual details such as $\angle SPR$, causing it to apply an irrelevant theorem and ultimately fail. In contrast, our SCANS model accurately maps the textual conditions to the visual elements, correctly identifying that $\triangle QTR$ is a right-angled triangle, which leads to the correct solution. The challenge in case (b) lies in its visual complexity. Our SCANS model leverages its content-aware capability to concentrate processing resources on regions containing geometric content. This enables the model to successfully identify the key structures: $\triangle CFH$ and $\triangle CGH$ and leverage

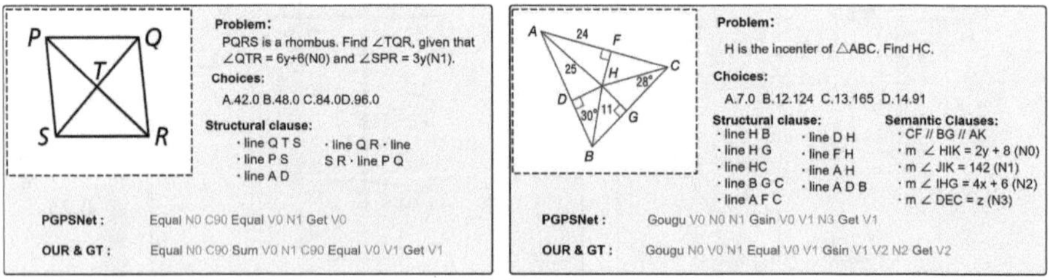

Fig. 4. Case analysis on PGPS9K. Solving these problems requires both precisely attending to key visual elements and accurately mapping textual conditions to the diagram.

complementary information from the geometric diagram and text to derive the critical insight that $\triangle CFH \cong \triangle CGH$, ultimately solving the problem.

5 Conclusion

We propose a salient content-aware neural solver (SCANS) to address the unique challenges of geometric problem solving by effectively processing geometric images and fusing multi-modal information. Benefiting from the content-aware structural attention mechanism, SCANS is endowed with the ability to focus on salient geometric regions while suppressing interference from blank background patches. Thanks to the hierarchical adaptive fusion module, SCANS further improves cross-modal integration through progressive text-guided visual enhancement and coordinated attention across multiple dimensions. The experimental results demonstrate the superiority of SCANS enhanced with geometry-specific content awareness and adaptive multi-modal fusion capabilities.

References

1. Acharya, M., Kafle, K., Kanan, C.: Tallyqa: answering complex counting questions. In: Proceedings of the Thirty-Third AAAI Conference on Artificial Intelligence and Thirty-First Innovative Applications of Artificial Intelligence Conference and Ninth AAAI Symposium on Educational Advances in Artificial Intelligence. AAAI Press (2019). https://doi.org/10.1609/aaai.v33i01.33018076
2. Achiam, J., et al.: Gpt-4 technical report. arXiv preprint arXiv:2303.08774 (2023)
3. Anderson, P., et al.: Bottom-up and top-down attention for image captioning and visual question answering (2017)
4. Bai, J., et al.: Qwen-vl: A versatile vision-language model for understanding, localization, text reading, and beyond (2023)
5. Cao, J., Xiao, J.: An augmented benchmark dataset for geometric question answering through dual parallel text encoding. In: Calzolari, N., et al., (eds.) Proceedings of the 29th International Conference on Computational Linguistics, pp. 1511–1520. International Committee on Computational Linguistics, Gyeongju, Republic of Korea (Oct 2022)
6. Chen, J., Li, T., Qin, J., Lu, P., Lin, L., Chen, C., Liang, X.: Unigeo: unifying geometry logical reasoning via reformulating mathematical expression. arXiv preprint arXiv:2212.02746 (2022)
7. Chen, J., et al.: GeoQA: A geometric question answering benchmark towards multimodal numerical reasoning. In: Zong, C., Xia, F., Li, W., Navigli, R. (eds.) Findings of the Association for Computational Linguistics: ACL-IJCNLP 2021, pp. 513–523. Association for Computational Linguistics, Online (Aug 2021). https://doi.org/10.18653/v1/2021.findings-acl.46
8. Kim, W., Son, B., Kim, I.: Vilt: vision-and-language transformer without convolution or region supervision. In: Meila, M., Zhang, T. (eds.) Proceedings of the 38th International Conference on Machine Learning. Proceedings of Machine Learning Research, vol. 139, pp. 5583–5594. PMLR (18–24 Jul 2021)
9. Liu, H., Li, C., Wu, Q., Lee, Y.J.: Visual instruction tuning. Adv. Neural. Inf. Process. Syst. **36**, 34892–34916 (2023)

10. Lu, P., et al.: Inter-GPS: Interpretable geometry problem solving with formal language and symbolic reasoning. In: Zong, C., Xia, F., Li, W., Navigli, R. (eds.) Proceedings of the 59th Annual Meeting of the Association for Computational Linguistics and the 11th International Joint Conference on Natural Language Processing (Volume 1: Long Papers), pp. 6774–6786. Association for Computational Linguistics, Online (Aug 2021). https://doi.org/10.18653/v1/2021.acl-long.528
11. Lu, P., et al.: Dynamic prompt learning via policy gradient for semi-structured mathematical reasoning. arXiv preprint arXiv:2209.14610 (2022)
12. Ma, B., Jian, P., Pan, C., Wang, Y., Ma, W.: A geometric neural solving method based on a diagram text information fusion analysis. Sci. Rep. **14**(1), 31906 (2024)
13. Mao, J., Gan, C., Kohli, P., Tenenbaum, J.B., Wu, J.: The neuro-symbolic concept learner: Interpreting scenes, words, and sentences from natural supervision (2019)
14. Nevins, A.J.: Plane geometry theorem proving using forward chaining. Artif. Intell. **6**(1), 1–23 (1975)
15. Ning, M., Wang, Q.F., Huang, K., Huang, X.: A symbolic characters aware model for solving geometry problems. In: Proceedings of the 31st ACM International Conference on Multimedia, pp. 7767–7775. MM '23, Association for Computing Machinery, New York, NY, USA (2023). https://doi.org/10.1145/3581783.3612570
16. Peng, S., Fu, D., Liang, Y., Gao, L., Tang, Z.: GeoDRL: a self-learning framework for geometry problem solving using reinforcement learning in deductive reasoning. In: Rogers, A., Boyd-Graber, J., Okazaki, N. (eds.) Findings of the Association for Computational Linguistics: ACL 2023, pp. 13468–13480. Association for Computational Linguistics, Toronto, Canada (Jul 2023). https://doi.org/10.18653/v1/2023.findings-acl.850
17. Perez, E., Strub, F., De Vries, H., Dumoulin, V., Courville, A.: Film: Visual reasoning with a general conditioning layer. In: Proceedings of the AAAI conference on artificial intelligence, vol. 32 (2018)
18. Sachan, M., Dubey, K., Xing, E.: From textbooks to knowledge: a case study in harvesting axiomatic knowledge from textbooks to solve geometry problems. In: Proceedings of the 2017 Conference on Empirical Methods in Natural Language Processing, pp. 773–784 (2017)
19. Santoro, A., et al.: A simple neural network module for relational reasoning. In: Advances in Neural Information Processing Systems, vol. 30 (2017)
20. Seo, M., Hajishirzi, H., Farhadi, A., Etzioni, O., Malcolm, C.: Solving geometry problems: combining text and diagram interpretation. In: Màrquez, L., Callison-Burch, C., Su, J. (eds.) Proceedings of the 2015 Conference on Empirical Methods in Natural Language Processing, pp. 1466–1476. Association for Computational Linguistics, Lisbon, Portugal (Sept 2015). https://doi.org/10.18653/v1/D15-1171
21. Su, J., Ahmed, M., Lu, Y., Pan, S., Bo, W., Liu, Y.: Roformer: enhanced transformer with rotary position embedding. Neurocomputing **568**, 127063 (2024)
22. Tito, R., Karatzas, D., Valveny, E.: Document collection visual question answering. In: International Conference on Document Analysis and Recognition, pp. 778–792. Springer (2021)
23. Wong, W.K., et al.: A computer-assisted environment for understanding geometry theorem proving problems and making conjectures. Int. J. Intell. Inf. Database Syst. **3**(3), 231–245 (2009)
24. Wu, W., et al.: E-gps: Explainable geometry problem solving via top-down solver and bottom-up generator. In: Proceedings of the IEEE/CVF Conference on Computer Vision and Pattern Recognition, pp. 13828–13837 (2024)
25. Xia, R., et al.: Geox: geometric problem solving through unified formalized vision-language pre-training. arXiv preprint arXiv:2412.11863 (2024)

26. Xiao, T., et al.: Learning to solve geometry problems via simulating human dual-reasoning process. arXiv preprint arXiv:2405.06232 (2024)
27. Xu, Y., Li, M., Cui, L., Huang, S., Wei, F., Zhou, M.: Layoutlm: pre-training of text and layout for document image understanding. In: Proceedings of the 26th ACM SIGKDD International Conference on Knowledge Discovery & Data Mining, pp. 1192–1200. KDD '20, Association for Computing Machinery, New York, NY, USA (2020). https://doi.org/10.1145/3394486.3403172
28. Ye, Q., et al.: mplug-owl2: Revolutionizing multi-modal large language model with modality collaboration. In: Proceedings of the IEEE/CVF Conference on Computer Vision and Pattern Recognition, pp. 13040–13051 (2024)
29. Yi, K., Wu, J., Gan, C., Torralba, A., Kohli, P., Tenenbaum, J.: Neural-symbolic vqa: Disentangling reasoning from vision and language understanding. In: Bengio, S., Wallach, H., Larochelle, H., Grauman, K., Cesa-Bianchi, N., Garnett, R. (eds.) Advances in Neural Information Processing Systems. vol. 31. Curran Associates, Inc. (2018)
30. Yin, B.W., Cao, J.L., Cheng, M.M., Hou, Q.: Dformerv2: geometry self-attention for RGBD semantic segmentation. In: Proceedings of the Computer Vision and Pattern Recognition Conference, pp. 19345–19355 (2025)
31. Yu, Z., Yu, J., Cui, Y., Tao, D., Tian, Q.: Deep modular co-attention networks for visual question answering. In: Proceedings of the IEEE/CVF Conference on Computer Vision and Pattern Recognition, pp. 6281–6290 (2019)
32. Zhang, J., Moshfeghi, Y.: GOLD: Geometry problem solver with natural language description. In: Findings of the Association for Computational Linguistics: NAACL 2024, pp. 263–278. Association for Computational Linguistics, Mexico City, Mexico (Jun 2024). https://doi.org/10.18653/v1/2024.findings-naacl.19
33. Zhang, M.L., Li, Z.Z., Yin, F., Lin, L., Liu, C.L.: Fuse, reason and verify: Geometry problem solving with parsed clauses from diagram. arXiv preprint arXiv:2407.07327 (2024)
34. Zhang, M.L., Yin, F., Liu, C.L.: A multi-modal neural geometric solver with textual clauses parsed from diagram. In: IJCAI (2023)
35. Zhao, J., Zhang, T., Sun, J., Tian, M., Huang, H.: Pi-GPS: Enhancing geometry problem solving by unleashing the power of diagrammatic information. arXiv preprint arXiv:2503.05543 (2025)

GeoGRPO: Investigating the Stepwise-GRPO Enhancement in RLHF Framework

Kecheng Liang, Xinyu Li, Weixing Chen, and Yang Liu(✉)

Sun Yat-sen University, Guangzhou, China
{liangkch,lixy589,chenwx228}@mail2.sysu.edu.cn, liuy856@mail.sysu.edu.cn

Abstract. Solving geometry problems remains a major challenge for multi-modal large language models (MLLMs), primarily due to their reliance on extensive, high-quality datasets and their limited capacity to generalize intermediate reasoning steps. To overcome these issues, we introduce GeoGRPO (Geometry-aware Group Relative Policy Optimization), a two-phase training approach designed to explicitly guide and optimize each stage of the reasoning process. In the first phase, we conduct supervised fine-tuning of a Qwen2.5-VL-7B policy model using our step-structured Geometry170K dataset, enhancing its capability to generate organized geometric proofs. In the second phase, we propose a novel Process Reward Model that assesses each generated reasoning step against expert solutions, providing fine-grained rewards. These detailed rewards are integrated into a newly designed GRPO objective, which steers the policy training towards step-level accuracy and logical consistency. When evaluated on datasets including GeoQA, Geometry3K, MathVista, and PGPS9K, GeoGRPO-Qw-7B achieves top-1 accuracies of 80.6%, 53.2%, 73.4%, and 50.6%, respectively, significantly outperforming baseline models. Furthermore, among open-source LLMs/MLLMs of similar scale, GeoGRPO-Qw-7B consistently leads in performance across all benchmarks, highlighting the effectiveness of fine-grained process supervision in enhancing geometric reasoning capabilities.

Keywords: mathematical geometry reasoning · Multimodal LLM · GRPO

1 Introduction

Prior to the application of large language models (LLMs) to geometry reasoning tasks, several researchers explored classical approaches that combine natural language processing (NLP) and computer vision (CV) techniques to tackle geometric problem solving. For instance, Chen et al. [6] proposed a framework that separately encodes the problem text and diagram using pretrained text and diagram encoders. These representations are then passed through a joint reasoning

L. Jin et al. (Eds.): ICDAR 2025 Workshops, LNCS 16225, pp. 344–361, 2026.
https://doi.org/10.1007/978-3-032-09368-4_21

and decoding module to convert the input into a formal representation, upon which symbolic solvers operate. Building upon this idea, Cao et al. [4] introduced a dual-text encoder architecture to alleviate limitations in text-diagram alignment, improving the semantic coherence between multimodal inputs. Similarly, Murphy et al. [24] also employed formalization-based methods for mathematical problem solving.

While these approaches leverage formal representation and symbolic search to solve geometry problems, they fundamentally diverge from the core philosophy of mathematical reasoning. Instead of exhaustively searching over predefined axioms and theorems, we often expect a solver to identify an intelligent, reasoning-driven path toward the solution. This mismatch highlights the limitations of traditional formalization-based pipelines and points to a promising direction for multimodal LLMs: enabling models to reason naturally over both textual and visual inputs, rather than relying on brute-force symbolic search. [54] Consequently, this marks a significant step forward in the development of more human-like and efficient reasoning systems for geometry problem solving.

Over the past six months, following the release of OpenAI's o1 model, reinforcement learning-based training paradigms for large language models (LLMs) have garnered increasing attention. A new wave of models—demonstrating significantly stronger capabilities in mathematical reasoning and programming—has emerged, outperforming those trained solely with supervised fine-tuning. The core idea of this training approach is to incorporate the chain-of-thought reasoning process directly into training, enabling the model to internalize step-by-step reasoning patterns during optimization. Furthermore, reinforcement learning offers a broader exploration space, enhancing the model's ability to generate diverse and creative reasoning paths.

Among these models, DeepSeek-R1 represents a notable example. It leverages a result-based reward GRPO algorithm, which assigns binary rewards (1 or 0) based on the correctness of the model's final answer. GRPO is an enhanced variant of Proximal Policy Optimization (PPO) that replaces traditional neural network-based value estimation with a more efficient estimation strategy, significantly improving both computational and memory efficiency. Following the release of o1, GRPO-based approaches like DeepSeek-R1 made a strong impact in the field, demonstrating competitive performance with relatively low training cost.

However, while result-based reinforcement learning suffices for general tasks, it falls short in more complex reasoning especially in multimodal mathematical domains such as geometry problem solving. [55] In these cases, giving reward based solely on the final answer can be misleading: the model may arrive at the correct answer despite flawed or incomplete reasoning. This undermines the effectiveness of reward signals and limits the model's capacity to learn robust reasoning patterns.

To address this limitation, our work focusing on LLM-based geometry problem solving, proposes a novel GRPO training method with process-level supervision. Unlike DeepSeek-R1, which relies on rule-based binary rewards, our

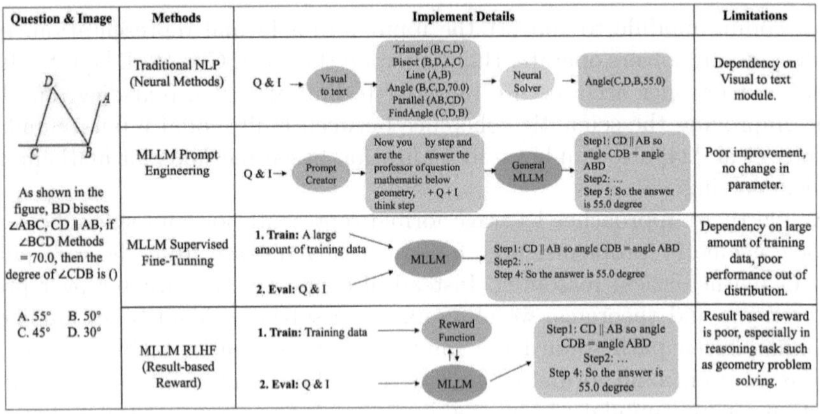

Fig. 1. Limitations of existing methods in Math Geometry Problem Reasoning.

approach draws inspiration from the openR project [34], employing an expert model to assign stepwise scores to LLM-generated solutions. This enables fine-grained supervision across the entire reasoning trajectory. Additionally, rather than training from scratch, we adopt a cold-start training paradigm where supervised fine-tuning (SFT) precedes GRPO optimization. This hybrid strategy helps stabilize the reinforcement learning phase and guides the model toward more reliable convergence. We further refine the GRPO objective function to better reflect the current policy's performance, ensuring more focused and effective optimization.

In summary, our work performed an process-level supervised, step-level GRPO method on multimodal LLM. In system design, We introduced in detail the reasons for the method design based on the previous GRPO training methods. Through experiments, we demonstrated the traing process to show the effectiveness of prm and our method. Also, we tested our method on three datasets: MathVista testmini(GPS), Geometry3K, GeoQA, PGPS9K, which show our training method does improve the LLM performing. At last, We analyzed this method in depth through case study and found some something unexpected. We hoped that this work can help LLM RLHF training's development and give some inspiration on those who is exploring in LLM reasoning (Fig. 1).

2 Related Work

There are several key points in the ability of Large Language Model(LLM) to reason about mathematical problems, especially geometry: Chain-of-Thought, Process Reward Model (PRM), and slow thinking.

Chain-of-Thought(CoT). In 2022, Wei et al. [37] proposed CoT, which for the first time integrated the traditional human thinking process into the answers of

LLMs, and conducted experiments on multiple models on mathematical inference datasets GSM8K, SVAMP, and MAWPS. The results show that each model has been improved to some extent, which confirms the help of CoT to the reasoning task of LLMs. Based on this, Wang et al. [36] put forward self-consistency. The basic idea is that the same problem may have multiple reasoning paths, so the selection of the final answer can take into account the marginal reasoning path, and make majority voting based on the result to replace the original Greedy Decoding. The experimental results demonstrate the method in multiple models with multiple datasets has achieved an average improvement of more than 5%.

PRM Combined with Reinforcement Learning(RL). OpenAI's research [15] shows that for complex stepwise reasoning problems, rewards are given at each step, not just at the end based on the result. This intensive reward signal yielded better results. In addition, the idea of training PRM is introduced to reduce the labor cost of making process supervision datasets. In a relatively long period of time since then, people have paid more attention to how to produce higher quality datasets to fine-tune LLMs, such as the dataset MATHVISTA produced by Lu et al. [19] and the dataset MATHVERSE produced by Zhang et al. [39]. The quality of these datasets is very high, but despite fine-tuning with such high-quality datasets, the performance of LLMs has not been able to achieve a big breakthrough. In September 2024, the launch of OpenAI-o1 caused a stir again, and the model's test results on multiple inference datasets far exceeded other models at the time, so it also led to many scholars' attempts to replicate the technology used in OpenAI-o1 model. The OpenR implemented by Wang et al. [34] takes advantage of the PRM training mentioned above, uses RL technology, and uses a LLM pre-trained and fine-tuned as PRM, and trains the model by combining the classical reinforcement learning method. Three classical strategy optimization methods of RL, APPO, TPPO and GRPO, were tested. This study was tested on a small scale LLM, and the experimental results showed that although the method could not achieve the capability of OpenAI-o1, it also achieved a great improvement on small models.

Slow Thinking. The OpenR implemented by Wang et al. [34]also proposed the application of Monte Carlo Tree Search (MCTS) in the inference stage, that is, multiple new steps are generated at each step in the inference stage, and the most likely answer is finally selected as the final answer of the model. This approach uses more resources in the inference phase and the model takes longer to generate answers, which is consistent with OpenAI-o1's performance in answering inference questions. This is called slow thinking, which means that the model needs more time to think and answer questions, and more research shows that this is a new research direction for large-scale model reasoning. Similarly, Jiang et al. [12] and Min et al. [22] also tried a similar approach. Although many scholars are trying to further explore and study the reasoning problem, since OpenAI-o1 does not support the function of processing images when it is proposed, most of the research is still limited to the reasoning problem of plain text, and it is still very valuable to explore the geometric reasoning.

Before applying LLMs to geometric reasoning problems, many scholars also tried to solve geometric reasoning problems with some classical NLP combined with CV methods. For example, Chen et al. [6] tried to use pre-trained text encoder and diagram encoder to encode images and problems respectively. A combined reasoner and decoder translates all problems into a formal format and then solves them in that format. Based on this, Cao et al. [4] proposed a dual text encoder to alleviate the limitations of the original method in encoding alignment of text images. Similarly, Murphy et al. [24] also used the means of formal solution. These methods all use more or less formal solution and search methods to solve mathematical problems, but this is inconsistent with the starting point of mathematical reasoning, we often hope that the solver can find a more intelligent path to solve the corresponding geometry problems pointed out an exact development trend and direction.

Geometry Problem Solving. Prior to the adoption of large language models (LLMs) for geometric reasoning, early efforts combined classical natural language processing (NLP) with computer vision (CV) techniques. Chen et al. [40], for instance, used pre-trained encoders to extract representations from textual descriptions and diagrams, which were jointly processed by a reasoningâĂŞdecoding module to generate formal solutions. Building on this, Cao et al. [4] proposed a dual text encoder to better align linguistic and visual information. Similarly, Murphy et al. [24] explored formalization-based solution methods. While these approaches leverage formal inference and symbolic search, they often fall short of the central aim of mathematical reasoning: to uncover interpretable, intelligent solution paths tailored to the structure and semantics of geometry problems.

3 The Framework of Our Method

3.1 System Design

The process of geometry problems solving using a multimodal large model can be defined as the following sequence: $Q, I- > \{R\}- > A$. Q represents the text of the input question, I is the image input, $\{R\}$ is a set of reasoning steps, and A is the final answer. Since each state in this sequence depends only on the current state and the previous action, the process can be modeled as a Markov Decision Process (MDP). An MDP ensures logical consistency across the sequence while allowing the model to follow different $\{R\}- > A$ to reach a decision. Transformer-based large language models can maintain sequence-level coherence when generating full answers. At the same time, they can produce diverse reasoning paths using probabilistic sampling during the answering process.

In this Markov Decision Process, the LLM (Large Language Model), acting as the policy model, takes the problem input and generates a sequence of tokens.

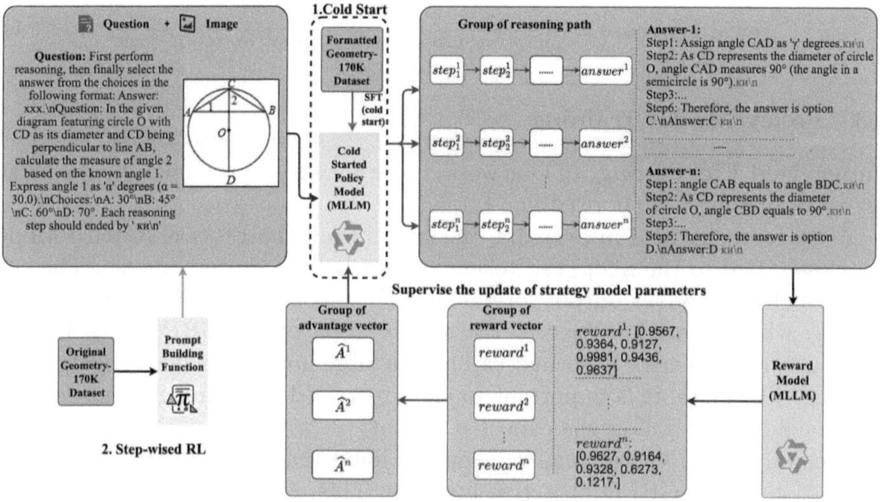

Fig. 2. The structure of the GeoGRPO training framework. First, the policy model undergoes supervised fine-tuning on the formatted Geo170K [8] dataset, which serves as the cold-start initialization. Subsequently, the model is further optimized through a step-wise GeoGRPO-RL training pipeline, where outputs are sampled from the current policy and fine-grained rewards are provided by a dedicated process-level reward model.

These tokens together form a series of reasoning steps and the final answer. Then, the PRM (Process Reward Model) provides token-level or step-level feedback based on the reasoning steps and answer. This feedback is used to guide the update of the LLM. The goal is to help the LLM optimize its policy, making it more likely to choose reasoning paths that receive higher rewards. The structure of our training methods is shown in Fig. 2.

3.2 The Design of PRM

The Process Reward Model (PRM) assigns a score to each step R_i and the action A in the $\{R\}->A$ process. Two key problems need to be addressed:

1. How to identify the end of a complete step in the sequence generated by the LLM, indicating when scoring should be applied.
2. What strategy should be used to perform the scoring.

For the design of the PRM, we follow the same approach as the openR project, where a large language model is trained as an expert scoring model. To address the first problem, the strategy model is fine-tuned to gradually learn to output a `step_tag` after completing a full step. This `step_tag` signals the PRM to perform scoring at that point. The second problem is addressed through the PRM's training method. During inference, the model takes a complete Markov Decision Process as input. At each `step_tag`, it calculates the positive softmax probability over the logits of a positive token and a negative token. This probability is used as the score for the step, and it serves as the reward:

$$\text{reward} = \text{score} = \frac{e^{\text{positive\_token\_logit}}}{e^{\text{positive\_token\_logit}} + e^{\text{negative\_token\_logit}}} \tag{1}$$

3.3 Policy Model Training

Supervised Finetuning. In Our work, there are several reasons why the strategy model requires supervised fine-tuning.

First, as mentioned in Sect. 3.2, the scores from the Process Reward Model are closely tied to the `step_tag` tokens output by the strategy model. The correctness of the model's output format directly affects the quality of the PRM's scoring, which in turn influences the computation of the GRPO objective. In this sense, a small error in format can have a large impact on the whole training pipeline. Therefore, it is important to use supervised fine-tuning to ensure that the model outputs follow the expected format.

Second, in GRPO training, we adopt an improved method that modifies the calculation of the GRPO objective function. This change requires the model's outputs not to deviate too far from those of a well-supervised model. As a result, we adopt a two-stage training strategy: supervised fine-tuning followed by GRPO training. This approach, also known as a cold start, effectively changes the model's starting point to one closer to the optimal solution. This helps speed up convergence and improves the chances of successful training.

Step level GRPO Training. After supervised fine-tuning the strategy model with the reformatted Geo170K dataset, the model has learned to effectively segment reasoning steps using the `step_tag` token in most cases. Based on this foundation, we apply the GRPO method to further encourage free exploration and enhance the model's mathematical reasoning ability.

GRPO is a variant of Proximal Policy Optimization (PPO). However, the original PPO introduces a Critic network and uses the Generalized Advantage Estimation (GAE) method to approximate the relative advantage $A(s, a)$. This Critic network often has a parameter scale comparable to the policy model, which leads to significant memory and efficiency challenges during training.

Moreover, in large language models, rewards are typically sparse—either applied only at the final token (outcome supervision) or at key steps (process supervision). As a result, training an accurate token-level advantage function estimator becomes complex and unstable.

The core idea of GRPO is to discard the GAE-based advantage estimation and instead use multiple samples for each state. The average reward of these samples is treated as a baseline. The relative advantage of each action is then computed by comparing its reward with this baseline. For outcome-supervised GRPO, the advantage is calculated as:

$$A(i, t) = \tilde{r}_i = \frac{r_i - \text{mean}(r_1, r_2, ..., r_n)}{\text{std}(r_1, r_2, ..., r_n)} \tag{2}$$

In which i is the index of the sampled response, and t represents the token position. In this setting, all tokens in the same sampled response share the same advantage, since only a single reward is given.

This approach is also adopted in Deepseek-R1 [7], known as token-level outcome-supervised GRPO. In addition to this, Deepseek-R1 introduces two innovations:

1. The reward function combines both correctness of the final answer and whether the response follows the expected answer structure. The final reward is the weighted average of the two. 2. Instead of directly computing the KL divergence, an unbiased estimate is used. A reference model (usually a supervised fine-tuned model) is introduced, and the KL term is approximated by:

$$\mathbb{D}_{KL} = \frac{\pi_{ref}}{\pi_\theta} - \log \frac{\pi_{ref}}{\pi_\theta} - 1 \tag{3}$$

This helps restrict the updated model to stay close to the reference model. Additionally, the importance weight is clipped to avoid gradient explosion.

The final GRPO objective function used in Deepseek-R1 is:

$$\begin{aligned} \mathcal{J}_{\mathrm{GRPO}}(\theta) &= \mathbb{E}[q \sim P(Q), \{o_i\}_{i=1}^{G} \sim \pi_{\theta_{\mathrm{old}}}(O|q)] \\ &= \frac{1}{G}\sum_{i=1}^{G}\frac{1}{|o_i|}\sum_{t=1}^{|o_i|}\Bigg\{ \min\left[\frac{\pi_\theta(o_{i,t}|q, o_{i,<t})}{\pi_{\theta_{\mathrm{old}}}(o_{i,t}|q, o_{i,<t})}\hat{A}_{i,t},\right. \\ &\left.\mathrm{clip}\left(\frac{\pi_\theta(o_{i,t}|q, o_{i,<t})}{\pi_{\theta_{\mathrm{old}}}(o_{i,t}|q, o_{i,<t})}, 1-\varepsilon, 1+\varepsilon\right)\hat{A}_{i,t}\right] \\ &- \beta \mathrm{D}_{\mathrm{KL}}\left[\pi_\theta || \pi_{\mathrm{ref}}\right]\Bigg\} \end{aligned} \tag{4}$$

In this work, we adopt a different approach from Deepseek-R1. Mathematical geometry reasoning is inherently divergent—like human thinking, there can be many valid reasoning paths to solve a problem. Therefore, a token-level GRPO may overly restrict the model, since every token equally contributes to the objective. This becomes problematic when minor differences (e.g., whitespace or newline tokens) carry minimal semantic weight, while critical reasoning tokens have much more impact.

To address this, we shift from token-level to step-level importance weighting, where each action corresponds to a full reasoning step, separated by `step_tag`.

Additionally, unlike general VQA tasks, correctness in intermediate steps is crucial for geometry problems. Outcome-level supervision alone is insufficient to guide the model. Process supervision, which clearly indicates which step is incorrect, provides better learning signals. This motivates the design of our Process Reward Model (PRM).

Third, unlike Deepseek-R1, which starts training from a pre-trained but unformatted model, we fine-tune our model with well-structured, formatted data.

As a result, the model already outputs responses in the expected format. This allows us to rely entirely on the process reward model, discarding format-based rewards.

We now redefine the components of Eq. (4) to form our final objective.

Given each question $q \in P(Q)$ and its associated image, we sample G responses $\{o_i\}_{i=1}^{G}$, each composed of $|o_i|$ tokens. Each response o_i can be viewed as an action sequence a_i, split by `step_tag` tokens:

$$\begin{aligned} o_i &= a_i \\ \{token_1, token_2, ..., token_{|o_i|}\}_i &= \{action_1, action_2, ..., action_{|a_i|}\}_i \end{aligned} \tag{5}$$

Each action $action_j^i$ is a sequence of tokens up to a `step_tag`:

$$\begin{aligned} action_j^i &= \{token_p, token_{p+1}, ..., \texttt{step\_tag}\} \\ p &= \sum_{q=1}^{j-1} |action_q| + 1 \end{aligned} \tag{6}$$

The probability of an action under π_θ is the product of token-level probabilities:

$$\pi_\theta(action_j^i | q, image, a_{i,<j}) = \prod_{k=p}^{p+m} \pi_\theta(token_k | q, image, a_{i,<j}, token_p, ..., token_{k-1}) \tag{7}$$

This is naturally aligned with the transformer architecture, where each token's generation depends on the previous context.

For step-level relative advantage, we move beyond outcome-level estimation. Given process-level rewards $reward_j^i$ from PRM for each action $action_j^i$, we construct a reward list:

$$\mathbf{R} = [reward_1^1, ..., reward_{|a_1|}^1, ..., reward_1^G, ..., reward_{|a_G|}^G] \tag{8}$$

Then, the relative advantage of $action_j^i$ is computed as:

$$A(i,j) = \sum_{q=j}^{|a_i|} \tilde{r}_q^i = \sum_{q=j}^{|a_i|} \frac{r_q^i - \text{mean}(\mathbf{R})}{\text{std}(\mathbf{R})} \tag{9}$$

This summation accounts for future rewards, assuming that a strong future trajectory suggests a good preceding step.

Regarding the KL divergence term: in step-level GRPO, directly applying Eq. (3) to whole actions can lead to gradient instability, since the action probability is the product of multiple token probabilities. However, since our model is already fine-tuned, we do not need to regularize it as aggressively. Thus, we drop the KL term entirely from our objective.

Our final GRPO-A (Step-level GRPO) objective is:

$$
\begin{aligned}
\mathcal{J}_{\text{GRPO-A}}(\theta) &= \mathbb{E}[(q, image) \sim P(Q), \{a_i\}_{i=1}^{G} \sim \pi_{\theta_{\text{old}}}(A|q, image)] \\
&= \frac{1}{G}\sum_{i=1}^{G}\frac{1}{|a_i|}\sum_{t=1}^{|a_i|}\Bigg\{ \min\left[\frac{\pi_\theta(a_{i,j}|q, a_{i,<j})}{\pi_{\theta_{\text{old}}}(a_{i,j}|q, a_{i,<j})}\hat{A}_{i,j},\right. \\
&\left.\text{clip}\left(\frac{\pi_\theta(a_{i,j}|q, a_{i,<j})}{\pi_{\theta_{\text{old}}}(a_{i,j}|q, a_{i,<j})}, 1-\varepsilon, 1+\varepsilon\right)\hat{A}_{i,j}\right]\Bigg\}
\end{aligned} \tag{10}
$$

In actual training, we minimize the negative of Eq. (10), which is equivalent to maximizing the original objective.

4 Experiments

The trained model was tested on GeoQA dataset [4], Geometry3K dataset [20], PGPS9K testset [48–50],and testmini geometry domain data of MathVista [19] through comparative experiments. The validity of the theoretical method deduced in Chap. 3 is explored, and some gains of GRPO method are illustrated by analyzing concrete examples.

4.1 Setting

Our process supervision model and policy model are both based on Qwen2.5-VL-7B. Since the supervision model plays a role of feedback reward, it is not appropriate to select a model with a smaller number of parameters than the policy model for training, otherwise the guidance ability is weak, and the guidance to the policy model is even worse.

Process Reward Model (PRM). The PRM model is equivalent to training the LLM as an expert model, focusing only on the logit output at each step token. Here, the PRM model is trained by lora fine-tuning [11], and the rank of lora fine-tuning is set to 8 and the scaling factor is set to 32. This makes sense because we don't care about the output logit in many places in the model, we only care about the part of the logit that is important, so we increase the scale factor.

Policy Model. The training of policy model mainly consists of two parts, one is supervision and fine-tuning, and the other is GRPO training. Since the goal of supervised fine-tuning is to adjust the model output and cold start, this part requires higher stability. For supervised fine tuning, this paper adopts the method of full parameter fine tuning, because compared with lora fine tuning, full parameter fine tuning has better stability and is suitable for cold start training.

For GRPO training method, this paper also adopts lora fine-tuning method. Here, because the mathematical geometric inference task carried out by the policy model is relatively complex and has greater explorability, the rank and scaling factor of lora fine-tuning are set to 64. In this way, the influence of lora layer on the model performance is expanded, and the improvement effect of the algorithm on the model is better reflected. In addition, for GRPO, the batch size for each set of data is set to 5, the number of answers generated for each question is also set to 5, and the number of updates for each set of data is set to 10.

4.2 Results

The three datasets used in our tests all contain a large number of length and angle problems, including a variety of elements in mathematical geometry such as circles, rectangles, and triangles. They are given in the form of multiple choice questions, and the correct rate of answers is taken as the evaluation index. In addition, using the method of program judgment and human assisted judgment, we can judge whether the answer of the policy model is correct or not, and almost make no wrong judgment. In terms of evaluation strategies, this paper adopts the top-1 judgment method for evaluation, that is, only one answer output by the policy model is taken as the final answer, rather than multiple answers at the same time, and the most answers are selected as the answer (Fig. 3).

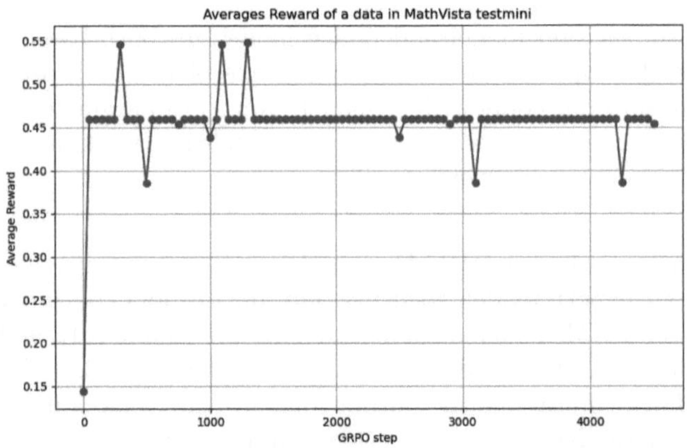

Fig. 3. The prm average reward of a data in MathVista testmini

According to the experimental results, the performance of the model was improved after using formatted Geometry170K for cold startup, indicating the effectiveness of supervised fine-tuning method for model cold start. In addition,

Table 1. Comparison with current methods in Geometry Problem Solving.

Methods	GeoQA Acc.	Geometry3K Acc.	MathVista Acc.	PGPS9K Acc.
		Neural Methods		
NGS [5]	60.5	58.8	-	46.1
Geoformer [41]	46.8	59.3	-	47.3
SCA-GPS [42]	-	76.7	-	-
PGPSNet-v2-S [43]	-	76.4	-	69.2
		General MLLMs		
Gemini-Pro [29]	-	47.0	40.4	-
GPT-4V	-	51.0	50.5	-
LLaVA-NeXT-7B [44]	17.9	19.8	32.2	20.1
Qwen2-VL-7B [45]	46.2	42.3	49.0	35.0
Qwen2.5-VL-3B [46]	56.8	41.4	42.2	31.6
Qwen2.5VL-7B [46]	72.4	51.4	66.8	46.8
InternVL2.5-8B [47]	58.1	52.6	54.3	37.4
		Geometry Centric MLLMs		
G-LLaVa-7B [8]	64.2	21.0	53.4	-
G-LLaVa-13B [8]	67.0	-	56.7	-
MAVIS-7B [51]	68.3	-	64.1	-
EAGLE-7B [52]	67.1	-	54.3	-
R-CoT-8B [53]	75.9	-	73.1	-
GenGRPO-Qw2.5-7B	**80.6**	**53.2**	**74.3**	**50.7**

GRPO method can also improve the performance of the model, especially on GeoQA and Geometry3K datasets with relatively large data volume (Table 1).

Regarding the performance degradation of our model on the MathVista testmini benchmark, we attribute the primary cause to algorithmic stability issues. It must be acknowledged that RLHF algorithms inherently exhibit poorer stability compared to SFT when fine-tuning large language models. This instability manifests more prominently with small test datasets featuring wide distributional diversity. Specifically, the Geometry Problem Solving subset in MathVista testmini contains merely 208 samples aggregated from multiple datasets with divergent distributions. We posit that the confluence of these two factors—limited sample size and heterogeneous data provenance—resulted in our model underperforming relative to its SFT-trained counterpart on this particular evaluation.

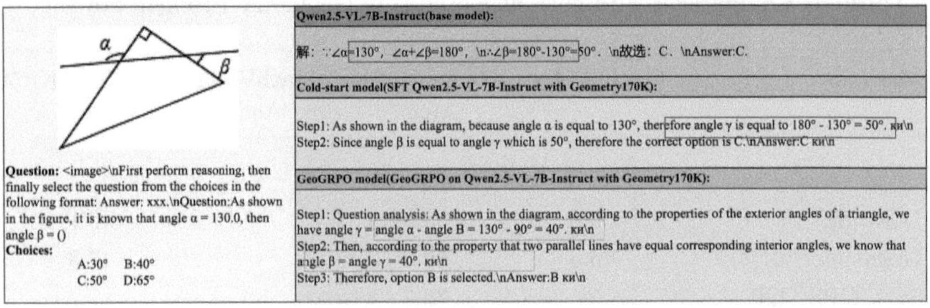

Fig. 4. Another case of 'aha moment'.

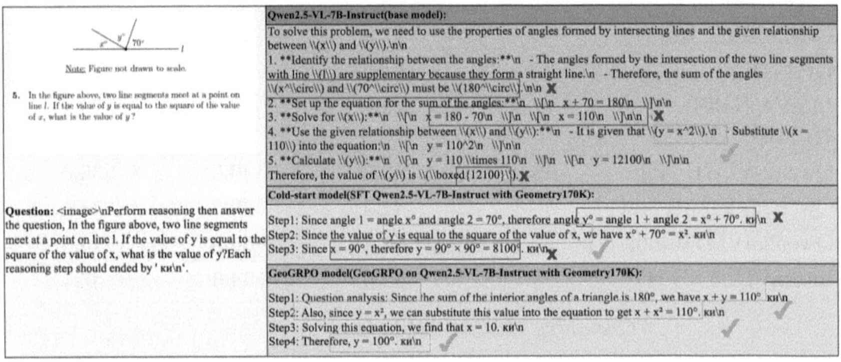

Fig. 5. A case of 'aha' moment.

4.3 Case Study

We also explored the more far-reaching effects of our GeoGRPO on the model through a few examples and found some interesting result.

Firstly, For a piece of data in MathVista, neither the original base model nor the supervised fine-tuning model could give the correct answer. But in the GRPO method with a small amount of data, the model immediately gave the correct answer. It can be seen that in the process of guiding the model to improve its reasoning ability, GRPO method can have the situation of "an aha moment", which is difficult to obtain in the supervised finetuning, as shown in Fig. 5 and Fig. 4.

Secondly, we observe that in some cases, the model learns to augment its chain-of-thought (CoT) reasoning by introducing auxiliary constructs such as auxiliary angles or helper lines, thereby producing more rigorous and structured solutions to fill-in-the-blank problems (Fig. 6).

Thirdly, there exist instances where the base model answers correctly, but the cold-start model fails; post-RL of GeoGRPO is able to recover the correct reasoning path, effectively guiding the model back on track (Fig. 7).

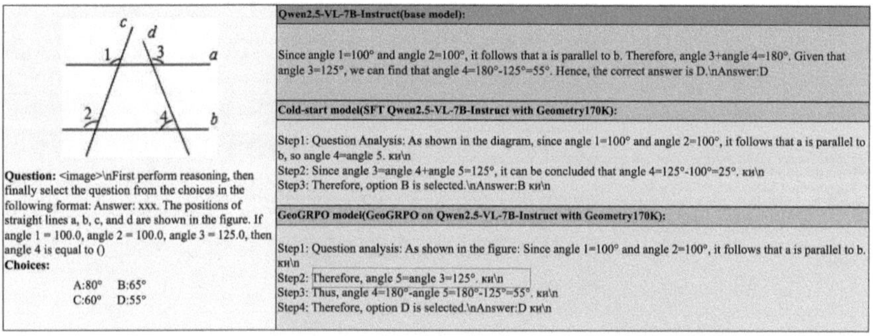

Fig. 6. A case of introducing auxiliary constructs.

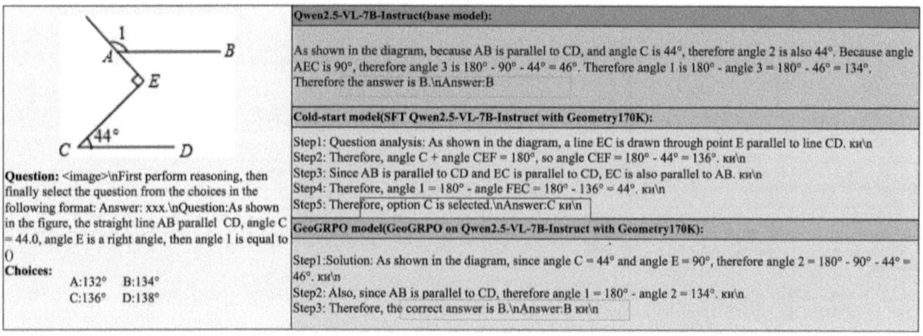

Fig. 7. A case of guiding the model back on track.

Fourthly, since we have restructured the objective function of GRPO, the influence of each token on the final outcome is no longer treated equally. As a result, models trained with our method can tolerate and ignore minor errors that do not affect the correct reasoning path, allowing them to focus more on exploring the correct path (Fig. 8).

Fifthly, the model's "aha moment" and the resulting improvements are more comprehensive. Whether the problem involves angles, lengths, or ratios in geometry, our method can achieve optimization to a certain extent. This, in turn, demonstrates the overall effectiveness and versatility of GeoGRPO (Fig. 9).

From a teaching point of view, supervised fine-tuning is more like teaching in a straight line, while the GRPO method is more leapfrog and stimulates the potential of model thinking. It is worth noting that the amount of data we trained using the GRPO method is much less than the amount of data we used using supervised fine-tuning. In the setting of this paper, GRPO updated the parameters of the model 10 times for the same set of data, which greatly improved the utilization rate of each set of data. The above experimental results are the experimental results of GRPO based on about 1000 data, which also shows that GRPO has a great advantage in data dependence compared with supervised fine-tuning.

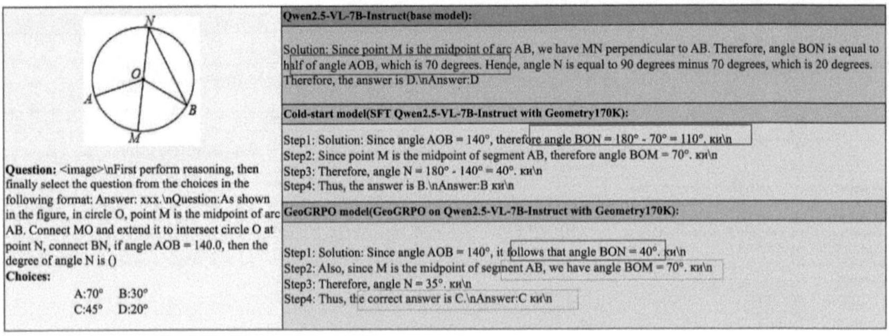

Fig. 8. A case of ignoring minor error.

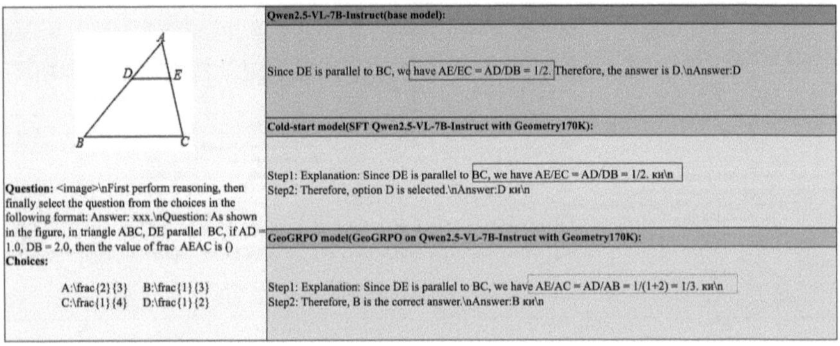

Fig. 9. A case of 'aha moment' in ratio problem.

5 Conclusion

In this work, we focuses on the domain of multimodal mathematical geometry reasoning and builds upon the GRPO method introduced in the Deepseek-R1 technical report through systematic optimization and innovative enhancements. Our most significant technical advancement lies in shifting the granularity of loss computation from the conventional token-level to a more holistic step-level. This change enables the model to better capture the semantic coherence of the entire reasoning process, rather than optimizing individual tokens in isolation. To ensure stable training dynamics and efficient convergence during the early stages, we adopt a cold-start strategy via supervised fine-tuning. This "supervised-first, reinforcement-later" hybrid paradigm effectively mitigates the randomness commonly observed in the initial phases of reinforcement learning. Using this methodology, we successfully fine-tuned the open-source Qwen2.5-VL-7B model into a specialized large language model that demonstrates strong performance on geometry problem solving task.

Acknowledgement. This work is supported by the National Natural Science Foundation of China under Grant No. 62436009 and No. 62572498.

References

1. Achiam, J., et al.: Gpt-4 technical report. arXiv preprint arXiv:2303.08774 (2023)
2. Alayrac, J.B., et al.: Flamingo: a visual language model for few-shot learning. Adv. Neural. Inf. Process. Syst. **35**, 23716–23736 (2022)
3. Bellman, R.: Dynamic programming. Science **153**(3731), 34–37 (1966)
4. Cao, J., Xiao, J.: An augmented benchmark dataset for geometric question answering through dual parallel text encoding. In: Proceedings of the 29th International Conference on Computational Linguistics, pp. 1511–1520 (2022)
5. Chen, J., et al.: Geoqa: a geometric question answering benchmark towards multimodal numerical reasoning. arXiv preprint arXiv:2105.14517 (2021)
6. Chen, L.C., Papandreou, G., Kokkinos, I., Murphy, K., Yuille, A.L.: Semantic image segmentation with deep convolutional nets and fully connected CRFs. In: ICLR (2015). http://arxiv.org/abs/1412.7062
7. DeepSeek-AI: Deepseek-r1: Incentivizing reasoning capability in LLMs via reinforcement learning (2025). https://arxiv.org/abs/2501.12948
8. Gao, J., et al.: G-llava: Solving geometric problem with multi-modal large language model. arXiv preprint arXiv:2312.11370 (2023)
9. Hariharan, B., Arbeláez, P., Girshick, R., Malik, J.: Simultaneous detection and segmentation. In: Fleet, D., Pajdla, T., Schiele, B., Tuytelaars, T. (eds.) ECCV 2014. LNCS, vol. 8695, pp. 297–312. Springer, Cham (2014). https://doi.org/10.1007/978-3-319-10584-0_20
10. Hoffmann, J., et al.: Training compute-optimal large language models. arXiv preprint arXiv:2203.15556 (2022)
11. Hu, E.J., et al.: Lora: low-rank adaptation of large language models. ICLR **1**(2), 3 (2022)
12. Jiang, J., et al.: Technical report: Enhancing LLM reasoning with reward-guided tree search. arXiv preprint arXiv:2411.11694 (2024)
13. Kaplan, J., et al.: Scaling laws for neural language models. arXiv preprint arXiv:2001.08361 (2020)
14. Lee, Y.: Qwen2-vl-finetune (2024). https://github.com/2U1/Qwen2-VL-Finetune
15. Lightman, H., et al.: Let's verify step by step. In: The Twelfth International Conference on Learning Representations (2023)
16. Liu, C., Yuen, J., Torralba, A.: Sift flow: dense correspondence across scenes and its applications. IEEE Trans. Pattern Anal. Mach. Intell. **33**(5), 978–994 (2011)
17. Liu, H., Li, C., Wu, Q., Lee, Y.J.: Visual instruction tuning. Adv. Neural. Inf. Process. Syst. **36**, 34892–34916 (2023)
18. Long, J., Shelhamer, E., Darrell, T.: Fully convolutional networks for semantic segmentation. In: Proceedings of the IEEE Conference on Computer Vision and Pattern Recognition, pp. 3431–3440 (2015)
19. Lu, P., et al.: Mathvista: Evaluating mathematical reasoning of foundation models in visual contexts. arXiv preprint arXiv:2310.02255 (2023)
20. Lu, P., et al.: Inter-gps: Interpretable geometry problem solving with formal language and symbolic reasoning. In: The Joint Conference of the 59th Annual Meeting of the Association for Computational Linguistics and the 11th International Joint Conference on Natural Language Processing (ACL-IJCNLP 2021) (2021)
21. Lu, P., et al.: Dynamic prompt learning via policy gradient for semi-structured mathematical reasoning. arXiv preprint arXiv:2209.14610 (2022)
22. Min, Y., et al.: Imitate, explore, and self-improve: A reproduction report on slow-thinking reasoning systems. arXiv preprint arXiv:2412.09413 (2024)

23. Mirzadeh, I., Alizadeh, K., Shahrokhi, H., Tuzel, O., Bengio, S., Farajtabar, M.: GSM-symbolic: understanding the limitations of mathematical reasoning in large language models. arXiv preprint arXiv:2410.05229 (2024)
24. Murphy, L., Yang, K., Sun, J., Li, Z., Anandkumar, A., Si, X.: Autoformalizing euclidean geometry. arXiv preprint arXiv:2405.17216 (2024)
25. Ouyang, L., et al.: Training language models to follow instructions with human feedback. Adv. Neural. Inf. Process. Syst. **35**, 27730–27744 (2022)
26. Wang, Q., Yang, Y., Jiang, J.: Easy RL: Reinforcement Learning Tutorial. Posts & Telecom Press, Beijing (2022). https://github.com/datawhalechina/easy-rl
27. Qiao, R., et al.: We-math: Does your large multimodal model achieve human-like mathematical reasoning? arXiv preprint arXiv:2407.01284 (2024)
28. Sutton, R.S., Barto, A.G., et al.: Reinforcement learning: An introduction, vol. 1. MIT press Cambridge (1998)
29. Team, G., et al.: Gemini: a family of highly capable multimodal models. arXiv preprint arXiv:2312.11805 (2023)
30. Team, Q.: Qwen2.5-vl (January 2025). https://qwenlm.github.io/blog/qwen2.5-vl/
31. Tighe, J., Lazebnik, S.: Finding things: image parsing with regions and per-exemplar detectors. In: Proceedings of the IEEE Conference on Computer Vision and Pattern Recognition, pp. 3001–3008 (2013)
32. Turing, A.M.: Computing machinery and intelligence. Springer (2009)
33. Vaswani, A., et al.: Attention is all you need. In: Advances in Neural Information Processing Systems, vol. 30 (2017)
34. Wang, J., et al.: Openr: An open source framework for advanced reasoning with large language models. arXiv preprint arXiv:2410.09671 (2024)
35. Wang, W., et al.: Visualprm: an effective process reward model for multimodal reasoning. arXiv preprint arXiv:2503.10291 (2025)
36. Wang, X., et al.: Self-consistency improves chain of thought reasoning in language models. arXiv preprint arXiv:2203.11171 (2022)
37. Wei, J., et al.: Chain-of-thought prompting elicits reasoning in large language models. Adv. Neural. Inf. Process. Syst. **35**, 24824–24837 (2022)
38. Yan, Y., et al.: A survey of mathematical reasoning in the era of multimodal large language model: Benchmark, method & challenges. arXiv preprint arXiv:2412.11936 (2024)
39. Zhang, R., et al.: Mathverse: Does your multi-modal LLM truly see the diagrams in visual math problems? In: European Conference on Computer Vision, pp. 169–186. Springer (2024)
40. Chen, L.C., Papandreou, G., Kokkinos, I., Murphy, K., Yuille, A.L.: "Semantic image segmentation with deep convolutional nets and fully connected CRFs," arXiv preprint arXiv:1412.7062 (2014)
41. Chen, J., et al. UniGeo: unifying geometry logical reasoning via reformulating mathematical expression. In: Proceedings of the 2022 Conference on Empirical Methods in Natural Language Processing, pp. 3313–3323 (2022)
42. Ning, M., Wang, Q.F., Huang, K., Huang, X.: A symbolic characters aware model for solving geometry problems. In: Proceedings of the 31st ACM International Conference on Multimedia, pp. 7767–7775 (2023)
43. Zhang, M.L., Li, Z.Z., Yin, F., Lin, L., Liu, C.L.: Fuse, reason and verify: Geometry problem solving with parsed clauses from diagram. CoRR 2024
44. Liu, H., et al.: LlaVA-next: Improved reasoning, OCR, and world knowledge, January 2024. https://llava-vl.github.io/blog/2024-01-30-llava-next (2024)
45. Yang, A., et al. Qwen2 Technical Report. CoRR (2024)

46. Yang, A., et al. Qwen2.5 technical report. arXiv preprint arXiv:2412.15115 (2024)
47. Chen, Z., et al. InternVL: scaling up vision foundation models and aligning for generic visual-linguistic tasks. In: Proceedings of the IEEE/CVF Conference on Computer Vision and Pattern Recognition, pages 24185–24198 (2024)
48. Zhang, M.L., Yin, F., Hao, Y.H., Liu, C.L.: Plane geometry diagram parsing. In: Proceedings of the Thirty-First International Joint Conference on Artificial Intelligence, pp. 1636–1643 (2022)
49. Zhang, M.L., Yin, F., Liu, C.L.: A multi-modal neural geometric solver with textual clauses parsed from diagram. In: Proceedings of the Thirty-Second International Joint Conference on Artificial Intelligence, pp. 3374–3382, 2023
50. Hao, Y., Zhang, M., Yin, F., Huang, L.L.: PGDP5K: a diagram parsing dataset for plane geometry problems. In: Proceedings of the 2022 26th International Conference on Pattern Recognition (ICPR), pp. 1763–1769. IEEE (2022)
51. Zhang, R., et al. Mavis: Mathematical visual instruction tuning. *arXiv e-prints*, page arXiv:2407.xxxxx, (2024)
52. Li, Z., et al.: Eagle: elevating geometric reasoning through LLM-empowered visual instruction tuning. *arXiv preprint* arXiv:2408.11397 (2024)
53. Deng, L., et al.: R-CoT: Reverse Chain-of-Thought Problem Generation for Geometric Reasoning in Large Multimodal Models. CoRR (2024)
54. Zhou, Z., et al.: Is your model really a good math reasoner? evaluating mathematical reasoning with checklist. In: International Conference on Learning Representations (ICLR) (2025)
55. Ning, M., Zhou, Z., Wang, Q.*, Huang, X., Huang, K.: GNS: Solving Plane Geometry Problems by Neural-Symbolic Reasoning with Multi-Modal LLMs. In: AAAI Conference on Artificial Intelligence (AAAI), (Oral), (2025)

Offline Handwritten Mathematical Formula Recognition Based on Primitive Representation

Yaqi Wang, Hui Wang, and Yuanping Zhu(✉)

Tianjin Normal University, Tianjin, China
zhuyuanping@tjnu.edu.cn

Abstract. The complex two-dimensional structure makes the recognition of handwritten mathematical formulas a challenging task, and offline methods cannot utilize stroke order information and trajectory information, further increasing the difficulty of accurate recognition. Traditional scene text recognition methods fail to fully exploit stable and effective feature representations of formula texts. To address this issue, this paper proposes a novel offline handwritten mathematical formula recognition method aimed at utilizing the intrinsic representation of formula images. First, elements in the feature map are modeled as nodes in an undirected graph, and pooling aggregators and weighted aggregators are used to learn the primitive representation, generating a two-dimensional feature map. Subsequently, a Graph Convolutional Network (GCN) is employed to obtain text representation, which is then input into a self-attention network along with positional encoding. By adopting the self-attention mechanism of Transformer, the model can capture long-range dependencies on the two-dimensional feature map, enabling the recognition of arbitrarily arranged and widely spaced characters. A network for learning primitive representations is constructed, utilizing visualized text representations for parallel decoding. The proposed model has been compared with other state-of-the-art (SOTA) HMER techniques on the CROHME 2014/2016/2019 datasets and achieved the best results.

Keywords: Handwritten Mathematical ExpressionÂăRecognition · Encoder-Decoder · Deep Neural Networks · Primitive Representation Learning · Graph Convolutional Network

1 Introduction

In handwritten mathematical formulas, the numerous similar characters in mathematical notation and intricate formulas, involving superscript and subscript positioning, radical ranges, and fraction structures, make the recognition highly challenging. The goal of Handwritten Mathematical Expression Recognition is to generate LaTeX expression sequences from images of handwritten mathematical expressions. Since HMER is also an image-to-text modeling task, numerous encoder-decoder models have been proposed in recent years.

L. Jin et al. (Eds.): ICDAR 2025 Workshops, LNCS 16225, pp. 362–377, 2026.
https://doi.org/10.1007/978-3-032-09368-4_22

However, the aforementioned methods still have limitations. On one hand, for CTC-based approaches, the extracted feature sequences contain redundant information, implying that certain regions of the HME image may be redundantly translated multiple times. On the other hand, attention-based encoder-decoder methods often suffer from alignment issues [1] and lack global visual information. Therefore, leveraging the intrinsic representation of images is crucial when processing formulaic text.

Unlike common CTC-based and attention-based methods, a new offline handwritten mathematical formula recognition framework is proposed. The model consists of a feature extraction module that extracts multi-scale feature maps from the input image, a primitive representation learning module that learns primitive representations and generates text representations, and a Transformer module. This model was tested on three public mathematical formula recognition datasets (CROHME 2014, 2016, 2019) and achieved the highest recognition accuracy.

2 Related Work

Mathematical expressions are indispensable for describing problems and theories in mathematics, physics, and many other fields. The recognition of handwritten mathematical expressions differs significantly from traditional text recognition, facing greater challenges such as complex geometric structures, ambiguity in handwritten input, and a strong dependence on contextual information. Depending on the input method, the recognition of mathematical formulas can be divided into two major categories: online and offline. Due to the hierarchical structure unique to mathematical formulas, offline recognition is more challenging than online recognition.

Traditional mathematical formula recognition methods were based on predefined syntax for recognition. Later, with the rapid development of deep neural networks, models based on the Encoder-Decoder framework gradually became a hot topic of research. In 2020, the Transformer model proposed by Ashish Vaswani et al. [2] achieved great success, demonstrating strong sequence modeling capabilities and significant advantages in the scope of image information extraction, effectively addressing the shortcomings of long formula recognition methods.

2.1 Syntax-Based Methods

Considering the two-dimensional structure of mathematical symbols, most methods use predefined syntax as a solution to the problem. For example, Yamamoto et al. [3] proposed a system using Stochastic context-free grammar (SCFG). Methods based on SCFG use probabilistic models to analyze the structure of expressions and resolve ambiguities in handwritten data. Alvaro and Sánchez [4] proposed a system that uses two-dimensional stochastic context-free grammar to parse expressions. MacLean and Labahn [5] proposed a Bayesian model for recognizing handwritten mathematical formulas; this model captures all recognizable input interpretations and organizes them in a parse forest. However, these syntactic rules are not data-driven but are manually designed, and they cannot benefit from large datasets.

2.2 Encoder-Decoder Based Methods

In recent years, with the innovation of computing technology, encoder-decoder networks have been specifically used to solve sequence-to-sequence learning. Encoder-decoders can handle variable-length input and output sequences [6]. Typically, RNNs are used as both encoders and decoders. The encoder encodes a sequence of variable length into a fixed-length vector. The decoder uses this vector to produce a variable-length output sequence, one word at a time. This framework has been widely applied in many fields. In the task of handwritten mathematical formula recognition, Zhang et al. [7] observed the lack of coverage and proposed the WAP model to solve the HMER task. In subsequent research, DenseWAP replaced the VGG encoder in WAP with a DenseNet [8] encoder, improving performance. Such encoder-decoder-based models have achieved excellent results in multiple CROHME competitions.

2.3 Graph Representation Learning

Representation learning has become the foundation of most deep learning methods because it makes it easier to extract useful information when building classifiers or other predictions. Feature aggregation is a common method in graph representation learning tasks. GCNs [9] aggregate adjacent vertex features by utilizing the topological structure of the graph. GraphSAGE [10] does not use all neighboring nodes but samples a few adjacent nodes using random walks [11], and feature aggregation can be completed through mean aggregators, pooling aggregators, or LSTM aggregators. Petar et al. [12] proposed the Graph Attention Network (GAT), which learns to assign different aggregation weights to different nodes by utilizing a self-attention mechanism. Inspired by the above progress, we propose learning primitive representations through global feature aggregation and using GCNs to transform primitive representations into visual text representations.

3 Methods

This section comprehensively discusses the models used and the proposed methods. First, it introduces the structural details of the proposed model, followed by a detailed description of the methods for learning the primitive representation and visualizing textual representations, as well as the Transformer module.

3.1 Network Structure

As shown in Fig. 1, the model consists of a feature extraction module, a primitive representation learning module, and a Transformer module.

Three pooling aggregators and three weighted aggregators are used to learn primitive representations from multi-scale feature mappings. The primitive representations are learned separately through pooling and weighted aggregators. Text representations are obtained from two GCNs and are summed to form a fused text representation. Spatial positional encoding is used during the Transformer's encoding and decoding phases.

This section also uses a tree structure for a detailed analysis of the formula's structure, as shown in Fig. 2. For the formula $\frac{\frac{p}{n}+v}{-H}$, "/" serves as the root node of the tree, dividing the formula's structure into upper, lower, and right parts, and so on until a complete handwritten mathematical expression tree structure is obtained.

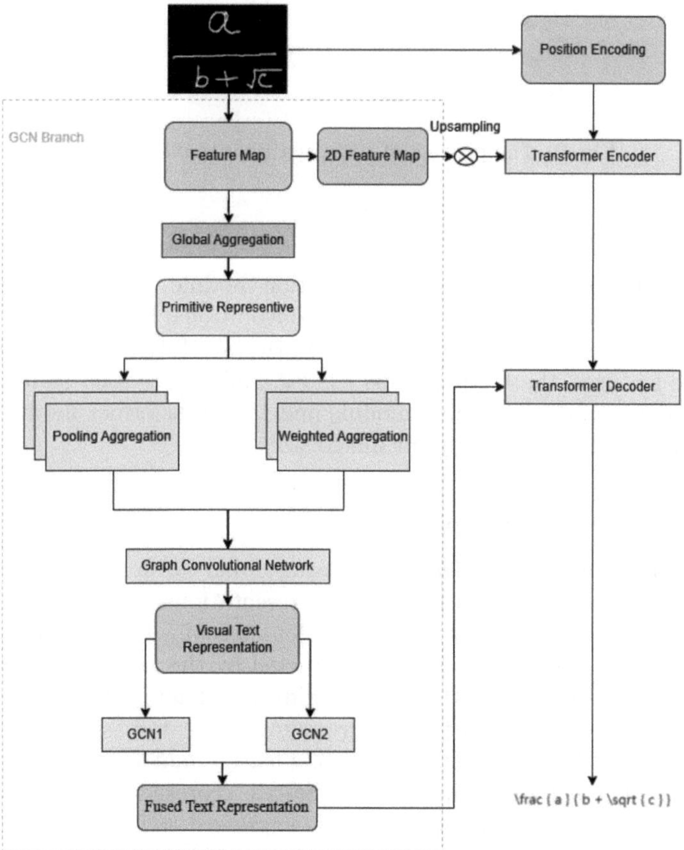

Fig. 1. The architecture of handwritten mathematical formula recognition model based on the primitive representation.

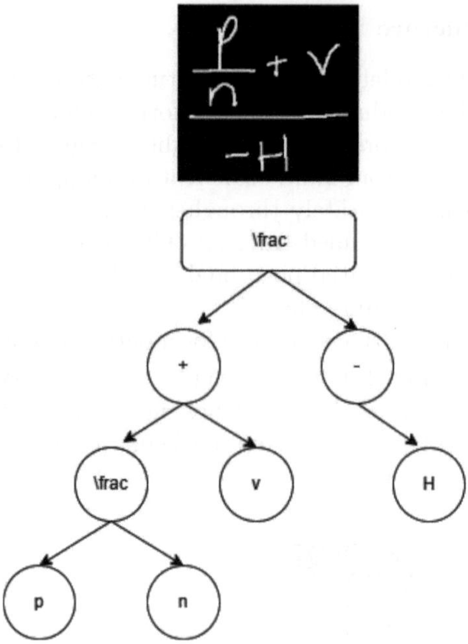

Fig. 2. The process of constructing a tree structure.

3.2 Feature Extraction Module

The Resnet50 is used as the feature extraction module, which is composed of 4 stages, each containing 3, 4, 6, and 3 small convolutional blocks, respectively. Let Fi represent the feature map output by the i-th convolutional block. To leverage multi-scale features, the feature maps F1, F2, and F4 are used as inputs to the primitive representation learning module and the Transformer module at scales of $1/8$, $1/16$, and $1/32$ of the input image size.

3.3 Primitive Representation

Primitive representations are learned by using global feature aggregation in coordinate space. In this way, the primitive representations include global information of the input image, which is beneficial for subsequent recognition processes. Let $F \in R_{d_0 \times h \times w}$ be the feature map extracted by the feature extraction module, where h, w and d_0 are the height, width, and number of channels of F, respectively. The elements in the feature map are treated as nodes of an undirected graph, that is, F is transformed into a feature matrix $X \in R_{m_0 \times d_0}$, where $m_0 = h \times w$. Let n be the number of primitive representations to be learned, and the feature aggregation process can be expressed as follows:

$$Z_i = f^{(i)}(X), i = 1, 2, \ldots, n \tag{1}$$

$$p_i = a_i \cdot Z_i, i = 1, 2, \ldots, n \tag{2}$$

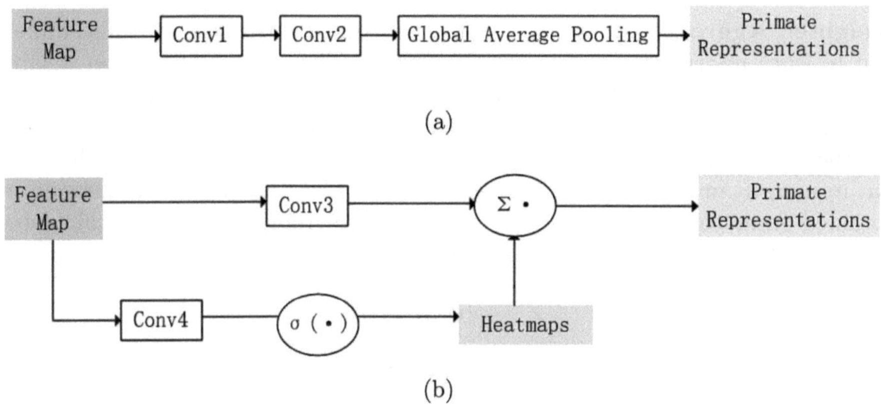

Fig. 3. Two methods for learning primitive representations. (a) Pooling aggregator. Two convolutional layers followed by a global average pooling layer to aggregate the input feature map into primitive representations. (b) Weighted aggregator. The feature map is transformed into n heat maps. Each heat map serves as an aggregation weight for a primitive representation. $\sigma(\cdot)$ is the sigmoid activation function, and $\Sigma(\cdot)$ is the scaled dot product and summation.

Here, $p_i \in R_{1\times d}$ is the i-th primitive representation. $equation f^{(i)}(\cdot)$ is the mapping function of the subnet that converts $X \in R_{m_0\times d_0}$ into a hidden representation $Z_i \in R_{m\times d}$. $a_i \in R_{1\times m}$ is the aggregation weight of the i-th primitive representation. The n primitive representations are connected to form $P = [p_1, p_2, \ldots, p_n] \in R_{n\times d}$. Let feature map, pooling aggregators, and weighted aggregators be used to learn primitive representations. Let d denote the number of channels in F4, and n be the number of primitive representations to be learned. The output dimension of each feature aggregator is $R_{n\times \frac{d}{3}}$. The outputs of the three pooling aggregators are connected and represented as P1, and the outputs of the three weighted aggregators are connected and represented as P2. Y1 and Y2 are fused to form the text representation Y.

Pooling Aggregator. As shown in Fig. 3 (a), a global average pooling layer is used for feature aggregation. Global average pooling has been proven to be an effective method for learning global information [13]. This allows all samples to share aggregation weights, utilizing the intrinsic structural information from various text instances. The function f(i) in Eq. (1) consists of two convolutions, i.e., convolutions on the original feature map f. Each convolution has a kernel size of 3 and a stride of 2. The calculation of the primitive representation can be expressed as:

$$p_i = \text{Pool}(\text{conv}_2^{(i)}(\Phi(\text{conv}_1^{(i)}(F)))) \tag{3}$$

where $\Phi(\cdot)$ is the activation function. Unlike the pooling aggregator used in GraphSAGE [11], an additional convolutional layer is added before the pooling layer to better learn the spatial information of the image.

Weighted Aggregator. Due to the diversity of text instances, it is also important to learn information specific to the sample. Therefore, aggregation weights are dynamically learned from the input features. As shown in Fig. 3 (b), a hidden representation $Z \in R_{d\times h\times w}$ is obtained through a 3×3 convolutional layer. Another 3×3 convolutional layer uses an activation function to convert the input feature map F into n heat maps $H \in R_{d\times h\times w}$. The aggregation weight can be obtained from the i-th heat map. The primitive representation is calculated through dot product and summation operations, that is:

$$Z = \phi(\mathrm{conv}_3(F)) \tag{4}$$

$$H = \sigma(\mathrm{conv}_4(F)) \tag{5}$$

$$p_i = \sum_{x=1}^{h}\sum_{y=1}^{w} H_{i,x,y} Z_{x,y} \tag{6}$$

3.4 Text Representation

Since the primitive representation contains global information of the input image, text information can be extracted from the primitive representation. A text representation is generated through a linear combination of a primitive representation and a fully connected layer, which can be expressed as:

$$Y = \phi(BPW) \tag{7}$$

Here, $P \in R_{n\times d}$ represents the primitive representation. $B \in R_{L\times n}$ is the coefficient matrix of the linear combination, L is the maximum decoding length. $W \in R_{d\times d}$ is the weight matrix, and $\phi(\cdot)$ is the activation function. The formula is implemented using GCN [9], where the coefficient matrix B plays a similar role to the adjacency matrix. Since the primitive representation does not have an explicit graph topology, B is randomly initialized and learned during the training phase. Each text representation $y_i (i = 1, 2, \ldots, L)$ represents a character to be recognized. For strings shorter than L, the excess part of Y is represented with padding symbols. The architecture of the text representation module is shown in Fig. 4.

3.5 Transformer Network

The feature extraction module is shared by the PREN and the encoder-decoder module of the improved Transformer model, upsampling the features of different levels' output and reshaping them into a 2D feature map as the input for the Transformer Encoder. In this way, the local features from the CNN branch are fed into the Transformer, enhancing the local perception capability of the Transformer branch. At this time, the self-attention mechanism performs global analysis on the feature map, as the last feature map is more effective for the recognition of larger targets, and performing self-attention on it facilitates the

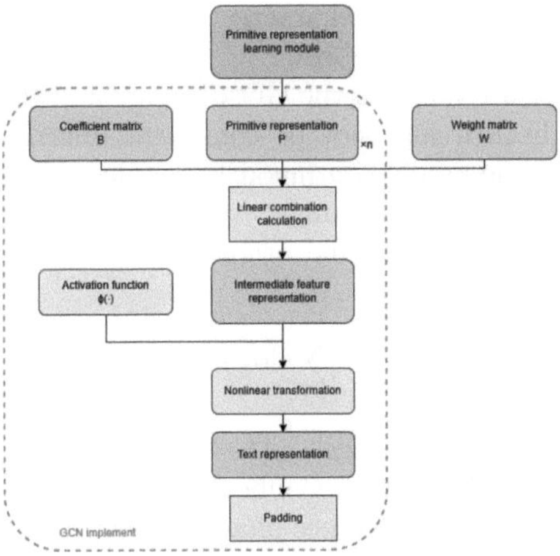

Fig. 4. The architecture of text representation module.

network to better extract the relationships between different positions and sizes of targets.

Since the Transformer itself cannot perceive the position information of the tokens, positional information needs to be injected before the tokens are passed in. To reflect the information of the image in two dimensions, Positional Encoding for the two dimensions is calculated separately and then combined with the text representation generated by the GAN. The text representation is used to enhance the character embedding of the real text during the training phase, or the previous output during the decoding phase. It can provide global guidance for the calculation of encoder-decoder attention in the Transformer model.

Finally, the global features from the Transformer branch are fed into the GCN to enhance the global perception capability of the GCN branch. The input query of the word table length M, and parallel decoding of N objects, the corresponding Transformer decoder outputs M tokens after attention and mapping. The spatial position encoding uses a two-dimensional spatial position encoding method, and the position encoding is added to the encoder's self-attention and the decoder's cross-attention, respectively.

In addition, a self-attention distillation mechanism is adopted, which wraps a Conv1D on the result of each attention layer, and then adds an additional Max-pooling layer to halve the output of each layer to highlight the main focus, effectively dealing with overly long input sequences. The architecture of the Transformer network is shown in Fig. 5.

3.6 Loss Function

End-to-end training is conducted through cross-entropy between the predicted formula and the label formula. The label formula is generated by adding an end-of-sequence symbol after the last character and extending it to the maximum decoding length with padding symbols.

Given that the feature maps in the GCN branch and the feature dimensions in the Transformer branch may not be consistent, misalignment could occur. To address this issue, the model continuously couples local features with global representations in an interactive manner.

Let l represent the length of the original text. The loss function can be expressed as follows:

$$L = -\sum_{t=1}^{l+1} \log p(g_t|I) \tag{8}$$

where I represents the input image, $g_t (t = 1, 2, \ldots, l)$ is the t-th character, and $< eos >$ is the end-of-sequence symbol g_{l+1}. Padding symbols are ignored during loss calculation. In the inference phase, the entire text is predicted in one step. During training, two cross-entropy losses are used to supervise these two classifiers separately. The importance weights of the loss functions are set to be the same. In the inference process, the outputs of these two classifiers are added together as the predicted result.

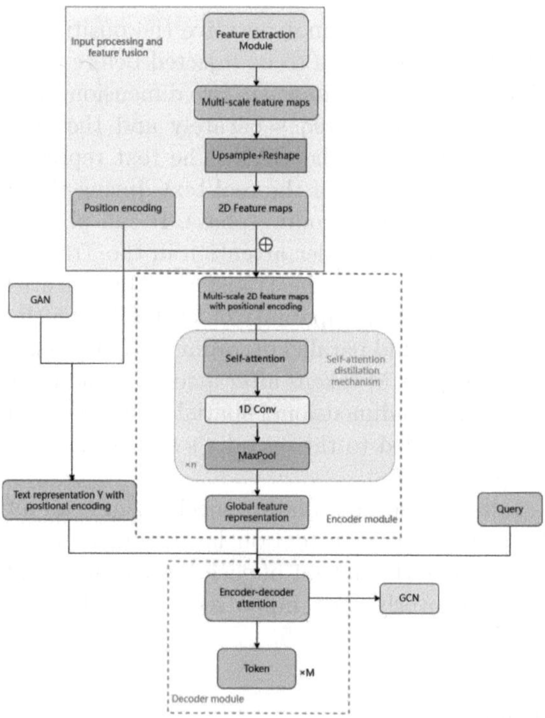

Fig. 5. The architecture of Transformer network.

4 Experiments

We conducted experiments on offline handwritten mathematical formula image recognition. Our method was compared with previous state-of-the-art methods, and a series of ablation studies were performed to explore the effects of each part of our model.

4.1 Dataset

The model is validated using the Competition on Recognition of Online Handwritten Mathematical Expressions (CROHME) dataset [14], which is currently the largest dataset for handwritten mathematical expressions. CROHME is organized by the International Conference on Frontiers in Handwriting Recognition (ICFHR). The same training dataset is used, but different test datasets are applied. The training set contains 8,836 handwritten mathematical expressions, and the test sets for CROHME 2014/2016/2019 contain 986/1,147/1,199 expressions, respectively. In the CROHME dataset, each handwritten mathematical expression is saved in an InkML file, which includes stroke trajectory information in handwriting and MathML formats. The stroke trajectory information in the InkML file is converted into bitmap format offline images for training and testing.

4.2 Training Details and Parameter Settings

In this experiment, all the images in the training set were preprocessed and normalized to 64×256 pixels. The ADADELTA optimizer was used to update the network parameters during training, with the hyperparameters $\beta 1 = 0.5$ and $\beta 2 = 0.999$. The learning rate was initialized at 0.1 and was reduced to 0.01 in the third stage and 0.001 in the fifth stage. The batch size was set to 128 during training. The number of original representations was 8. The dictionary type was 103 classes. The maximum decoding length was set to 200. The entire network model was trained for a total of 300 rounds.

The training and testing of the entire network model were conducted on the Linux Ubuntu 16.04 system. The network model was based on the deep learning framework TensorFlow and was trained on a single NVIDIA 2080Ti GPU, which has 11GB of video memory. All the codes were developed and implemented using Python 3.6. The image generation operations involved in the code were completed using the standard function library cv2 in Python 3.

4.3 Comparison with State-of-the-Art

We compared this model with the previously state-of-the-art (SOTA) methods and demonstrated its superiority. The results of some models on the CROHME 2014/2016/2019 datasets are shown in Table 1. We used the previously most advanced methods such as 'ABM' [15], 'WAP' [7], 'SAN' [16], 'CoMER' [17],

'DWAP-MSA' [8], 'DWAP-TD' [18], Ding et al. [19], 'BTTR' [20] and 'GETD' [21]. Compared with other methods, our proposed model outperformed the previous SOTA methods in almost all indicators.

In Table 1, we compare our model with various models and methods on the CROHME dataset, including several recently outstanding approaches besides the previously state-of-the-art models. For instance, the Tree-Aware Module (TAMER) [22] that enables the model to perform sequence decoding and tree structure prediction simultaneously, the multi-scale feature fusion-based co-learning model Fu et al. [23], and SATD [24], which combines the tree decoder with Transformer, similar to our model. In comparisons with these models on the CROHME 2014/2016/2019 datasets, we find that the model used in this paper outperforms Fu et al. across all metrics. On the CROHME 2014 dataset, the recognition accuracy is slightly lower than that of SATD and TAMER; on the CROHME 2016 and 2019 datasets, our model surpasses SATD and TAMER in all aspects, indicating that our method is not only effective but also competitive in the rapidly advancing field of handwritten mathematical expression recognition.

4.4 Sample Analysis

As shown in Fig. 6, we systematically list several of the more common error types in handwritten mathematical expression recognition tasks. During recognition, similar errors can easily occur for similar characters and alphanumeric symbols because the model relies only on visual information, and the formula itself lacks intrinsic logical structure.

There are two main types of errors that are prone to occur during recognition. First, the model is prone to confusion when faced with similar characters and alphanumeric symbols, resulting in similar errors. As shown in Fig. 6 (a), "9"

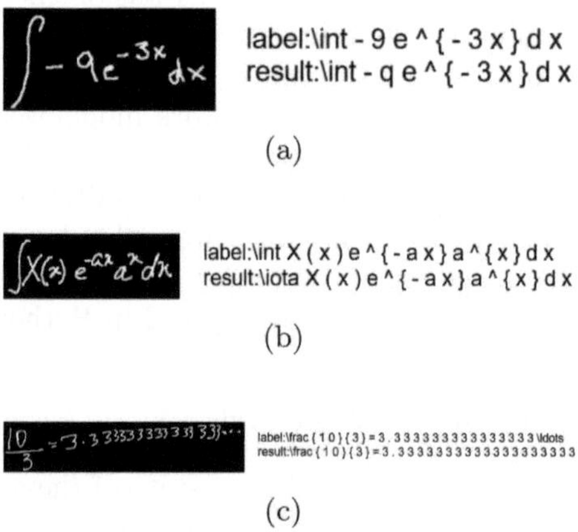

Fig. 6. The error samples of recognition results. (a) Similar character recognition error. (b) Mathematical symbol recognition error. (c) Long string recognition error.

Table 1. Performance comparison of individual models on the CROHME 2014/2016/2019 test sets(in %), where the "Accuracy", "≤1 Error" and "≤2 Error" columns represent the expression recognition rate with 0 to 2 structural or symbol errors.

Dataset	model	Accuracy	≤1 Error	≤2 Error
CROHME2014	ABM [15]	56.85	73.73	81.24
	WAP [7]	46.55	61.15	65.21
	SAN [16]	56.2	72.6	79.2
	DWAP - MSA [8]	52.8	68.1	72.0
	DWAP - TD [18]	49.1	64.2	67.8
	Ding et al. [19]	58.7	-	-
	BTTR [20]	53.96	66.02	70.28
	CoMER [17]	59.3	71.7	75.6
	Fu et al. [23]	58.88	74.92	84.14
	TAMER [22]	61.23	76.77	83.25
	SATD [24]	63.45	78.17	84.06
	Ours	59.7	75.1	79.6
CROHME2016	ABM [15]	52.92	69.66	78.73
	WAP [7]	44.55	57.1	61.55
	GETD [21]	55.27	68.43	72.62
	DWAP - MSA [8]	50.1	63.8	67.4
	DWAP - TD [18]	48.5	62.3	65.3
	SAN [16]	53.6	69.6	76.8
	Ding et al. [19]	58.7	-	-
	BTTR [20]	54.0	66.0	70.3
	CoMER [17]	59.8	66.0	70.3
	Fu et al. [23]	55.1	69.83	78.64
	TAMER [22]	60.26	76.91	84.05
	SATD [24]	60.42	75.41	81.78
	Ours	61.4	71.2	78.1
CROHME2019	ABM [15]	53.96	71.06	78.65
	DWAP - MSA [8]	47.7	59.5	63.3
	DWAP - TD [18]	51.4	66.1	69.1
	Ding et al. [19]	61.4	75.2	80.2
	BTTR [20]	52.96	65.97	69.14
	SAN [16]	62.1	74.5	81.0
	CoMER [17]	62.9	77.4	81.4
	Fu et al. [23]	57.05	73.39	79.89
	TAMER [22]	61.97	78.97	85.8
	SATD [24]	61.05	76.9	83.15
	Ours	65.5	78.3	87.2

and "q" in handwritten fonts may be very similar in visual presentation due to differences in the writing habits of the writers. When extracting features, the model finds it difficult to accurately capture the subtle differences between them, thus incorrectly recognizing "9" as "q" or vice versa. For another example, in Fig. 6 (b), the two symbols "$\int$" and "ι" in the figure only differ in the length of the symbol. It is difficult to make an accurate judgment.

Moreover, the model suffers from another problem when dealing with long formulas. Due to the mechanism and resource allocation in processing information, the model often pays insufficient attention to the second half of the long formula, which makes the model unable to fully extract the effective features of the tail characters. For example, in Fig. 6 (c), the model accurately recognizes the content of the first half, but when processing the second half, the '...' symbol is missing in the recognition result because of the decreased attention, thus affecting the recognition accuracy of the whole formula.

4.5 Parameter Selection Experiments

To investigate the impact of the number of primitive representations, we designed and conducted parameter selection experiments targeting this parameter.

In the experimental design, we treat the number of primitive representations as the core variable and conduct ablation experiments on the CROHME 2014 dataset. Under the premise of keeping other model components and the experimental environment completely consistent, we systematically adjust the value of n, the number of primitive representations, to find the value of n that achieves the highest recognition accuracy on the CROHME 2014 dataset.

Specifically, we selected a series of representative n values (integers between 6 and 11), trained complete models for each n value separately, and recorded key performance metrics of the models on the mathematical formula symbol recognition task.

From Table 2, it can be observed that in the systematic experiment on the number of primitive representations, we found that when the number of primitive representations n is set to 8, the model achieves its optimal performance in the mathematical formula symbol recognition task, with an accuracy rate of 59.7%, significantly higher than other values of n. It avoids the loss of features and inability to distinguish similar symbols due to insufficient representation capacity when n is too small ($n \leq 6$), while also preventing overfitting risks caused by increased redundant information when n is too large ($n \geq 10$).

4.6 Visualization of Attention

In the process of exploring the ability of the model to recognize two-dimensional structures, we used the method of visualization of attention. In Fig. 7, the highlighted area depicts the attention weights of the model when recognizing a picture of a handwritten mathematical expression. The attention visualization shows that the model can effectively capture the position information in the formula and can assign the attention weight corresponding to different symbols.

For example, in the attention map of Fig. 7 (a), the color of the region corresponding to the equal sign is relatively dark, because the equal sign is the key symbol for the formula to express the equality relation. In Fig. 7 (b), the region corresponding to the first digit '3' and the last digit '3' is relatively dark in the attention map, indicating that the model pays high attention to the first digit. Areas with multiple consecutive zeros in the middle are lighter in color, indicating that the model pays less attention to these zeros because consecutive zeros contribute relatively little to numeric recognition. Experiments show that the proposed model can effectively identify two-dimensional structures in mathematical formulas.

Table 2. Impact of the primitive representation quantity(in %).

Number of Primitive Representation	6	7	8	9	10	11
Accuracy	48.7	54.9	59.7	59.1	58.8	58.2

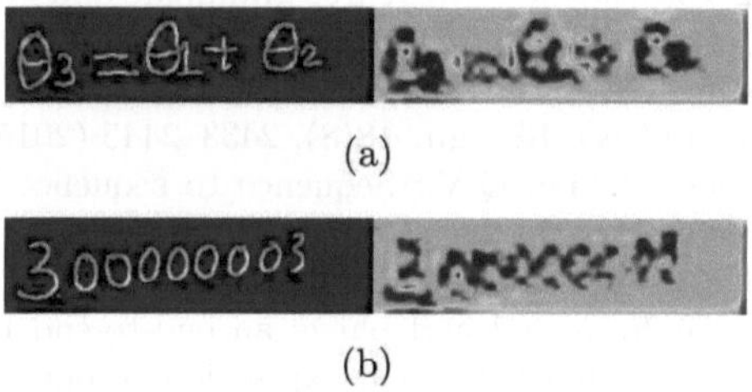

Fig. 7. Visualization of the attention map. (a) attention map sample 1. (b) attention map sample 2.

5 Conclusion

This paper presents a primitive representation learning method for offline handwritten mathematical formula recognition. The text representation generated by the primitive representation can be directly used for parallel decoding to enhance recognition performance. In addressing the complex structural characteristics of mathematical formulas, hierarchical features and spatial positional encoding are utilized to better extract spatial information.

A novel model for offline handwritten mathematical formula recognition based on primitive representation learning is proposed in this paper. Compared to previous methods, our model has the following three advantages: (1) The text representation derived from the primitive representation can be directly applied to parallel decoding, thereby improving recognition performance. (2) By leveraging hierarchical features, the model effectively addresses the issue of varying character sizes in handwritten mathematical formulas, better extracting spatial information and alleviating the problem of insufficient coverage. Experiments have demonstrated the effectiveness of the model, achieving accuracy on CROHME 2014, CROHME 2016, and CROHME 2019 of 59.7%, 61.4%, and 65.6%, respectively.

References

1. Valentini-Botinhao, C., King, S.: Detection and analysis of attention errors in sequence-to-sequence text-to-speech. In: Proceedings of the 22nd International Speech Communication Association in Interspeech 2021, pp. 2746–2750 (2021)
2. Vaswani, A., et al.: Attention is all you need. In: Proceedings of the 31st Conference on Neural Information Processing Systems, pp. 5998–6008 (2017)
3. Ota, I., Yamamoto, R., Sako, S., Sagayama, S.: Online handwritten kanji recognition based on inter-stroke grammar. In: Proceedings of the 9th International Conference on Document Analysis and Recognition, pp. 1188–1192 (2007)
4. Alvaro, F., Sánchez, J., Benedí, J.: Recognition of online handwritten mathematical expressions using 2D stochastic context-free grammars and hidden Markov models. Pattern Recogn. Lett. **35**(1), 58–67 (2014)
5. MacLean, S., Labahn, G.: A Bayesian model for recognizing handwritten mathematical expressions. Pattern Recogn. **48**(8), 2433–2445 (2015)
6. Sutskever, I., Vinyals, O., Le, Q.V.: Sequence to sequence learning with neural networks. In: Proceedings of the 28th Conference on Neural Information Processing Systems, NeurIPS 2014, pp. 3104–3112 (2014)
7. Zhang, J., et al.: Watch, attend and parse: an end-to-end neural network based approach to handwritten mathematical expression recognition. Pattern Recogn. **71**, 196–206 (2017)
8. Zhang, J., Du, J., Dai, L.: Multi-scale attention with dense encoder for handwritten mathematical expression recognition. In: Proceedings of the 24th International Conference on Pattern Recognition (ICPR), pp. 2245–2250 (2018)
9. Kipf, T.N., Welling, M.: Semi-supervised classification with graph convolutional networks. In: Proceedings of the 5th International Conference on Learning Representations (2017)
10. Hamilton, W.L., Ying, Z., Leskovec, J.: Inductive representation learning on large graphs. In: Proceedings of the 31st International Conference on Neural Information Processing Systems (NIPS), pp. 1024–1034 (2017)
11. Perozzi, B., Al-Rfou, R., Skiena, S. Deepwalk: Online learning of social representations. In: Proceedings of the 20th ACM SIGKDD International Conference on Knowledge Discovery and Data Mining, pp. 701–710 (2014)
12. Veličković, P., Cucurull, G., Casanova, A., et al. Graph attention networks. https://doi.org/10.48550/arXiv.1710.10903 (2018)
13. Lin, M., Chen, Q., Yan, S.: Network in network. In: Proceedings of the International Conference on Learning Representations (ICLR), pp. 1–13 (2013)
14. Harold Mouchère. ICDAR 2019 Competition on Recognition of Handwritten Mathematical Expressions and Typeset Formula Detection (ICDAR2019-CROHME-TDF). https://tc11.cvc.uab.es/datasets/ICDAR2019-CROHME-TDF_1 (2020)

15. Bian, X., Qin, B., Xin, X., Li, J., Su, X., Wang, Y.: Handwritten mathematical expression recognition via attention aggregation based bi-directional mutual learning. In: Proceedings of the 36th AAAI Conference on Artificial Intelligence, pp. 113–121 (2022)
16. Yuan, Y., Liu, X., Dikubab, W., Liu, H., Ji, Z., Wu, Z., Bai, X.: Syntax-aware network for handwritten mathematical expression recognition. In: Proceedings of the IEEE/CVF Conference on Computer Vision and Pattern Recognition (CVPR), pp. 4553–4562 (2022)
17. Wenqi, Z.-Liangcai, G.: CoMER: modeling coverage for transformer-based handwritten mathematical expression recognition. In: Proceedings of the 17th European Conference on Computer Vision, pp. 392–408 (2022)
18. Zhang, J., Du, J., Yang, Y., Song, Y.-Z., Wei, S., Dai, L.: A treestructured decoder for image-to-markup generation. In: Proceedings of the 37th International Conference on Machine Learning, on Proceedings of Machine Learning Research, pp. 11076–11085 (2020)
19. Ding, H., Chen, K., Huo, Q.: An encoder-decoder approach to handwritten mathematical expression recognition with multi-head attention and stacked decoder. In: Proceedings of the 16th International Conference on Document Analysis and Recognition, pp. 602–616 (2021)
20. Zhao, W., Gao, L., Yan, Z., Peng, S., Du, L., Zhang, Z.: Handwritten mathematical expression recognition with bidirectionally trained transformer. In: Lladós, J., Lopresti, D., Uchida, S. (eds.) ICDAR 2021. LNCS, vol. 12822, pp. 570–584. Springer, Cham (2021). https://doi.org/10.1007/978-3-030-86331-9_37
21. Tang, J.-M., Guo, H.-Y., Wu, J.-W., Yin, F., Huang, L.-L.: Offline handwritten mathematical expression recognition with graph encoder and transformer decoder. Pattern Recogn. **148**, 110155 (2024)
22. Zhu, J., Zhao, W., Li, Y., Hu, X., Gao, L.: TAMER: Tree-Aware Transformer for Handwritten Mathematical Expression Recognition. https://doi.org/10.48550/arXiv.2408.08578 (2024)
23. Zhao, X., et al.: Mutual learning offline handwritten mathematical expression recognition based on multi-scale feature fusion. Comput. Eng. Appl. **59**(17), 23–31 (2023)
24. Fu, P., Xiao, G., Yang, H.: SATD: syntax-aware handwritten mathematical expression recognition based on tree-structured transformer decoder. Vis. Comput. **41**(1), 883–900 (2025)

Long Math Reasoning Problem Generation

Changwei Li, Guangping Huang, Zihao Zhou, and Qiufeng Wang(✉)

Xi'an Jiaotong–Liverpool University, Suzhou, China
{Changwei.li23,Guangping.Huang22,Zihao.Zhou22}@student.xjtlu.edu.cn,
Qiufeng.Wang@xjtlu.edu.cn

Abstract. Long-text mathematical word problems (MWPs) offer a stringent probe of large-language-model (LLM) reasoning, yet reliable benchmarks scarcely exist because hand-crafting rich, distractor-laden narratives is labor-intensive. We present an automatic short-to-long pipeline that transforms abundant short MWPs (e.g., GSM8K) into context-heavy variants while rigorously preserving solution correctness. The IEC pipeline operates in three successive stages: first, an Interference step inserts answer-neutral numeric distractors; next, an Extension step uses GPT-4-turbo to expand the now-noisy prompt into a much longer narrative while preserving all original conditions; finally, an automated Check stage filters any draft that leaks reasoning steps or alters the correct answer, regenerating up to three times before rolling back to the previous round. Applying IEC pipeline to 1318 GSM8K items yields a corpus whose average length surpasses 2000 tokens—over 30 times the original—while achieving an 85% validity rate. We then benchmark three representative LLMs—DeepSeek-chat, GLM-4-plus, and Kimi—across four expansion depths. All excel on the short source set (above 0.93 accuracy) but degrade as length increases, with Kimi dropping to 0.39 after six rounds, underscoring persistent long-context weaknesses. Our results establish a scalable route to high-quality long-text MWP datasets and reveal concrete failure modes that future model architectures, training curricula, and retrieval/memory mechanisms must address.

Keywords: Large Language Models · Dataset Generation · Mathematical Reasoning

1 Introduction

Large scale language models (LLMs) now handle long documents with ease, boosting AI's capacity to process and organize information across domains [6,11,17,18,22]. Yet their strengths in summarization and retrieval do not translate to tasks that demand deeper cognition—especially mathematical word problems (MWPs), where a system must transform natural language descriptions into formal expressions and follow logical steps to a correct answer [5,10,12,18]. MWPs thus serve as an ideal testbed for probing a model's true reasoning ability beyond surface level pattern matching (Fig. 1).

L. Jin et al. (Eds.): ICDAR 2025 Workshops, LNCS 16225, pp. 378–393, 2026.
https://doi.org/10.1007/978-3-032-09368-4_23

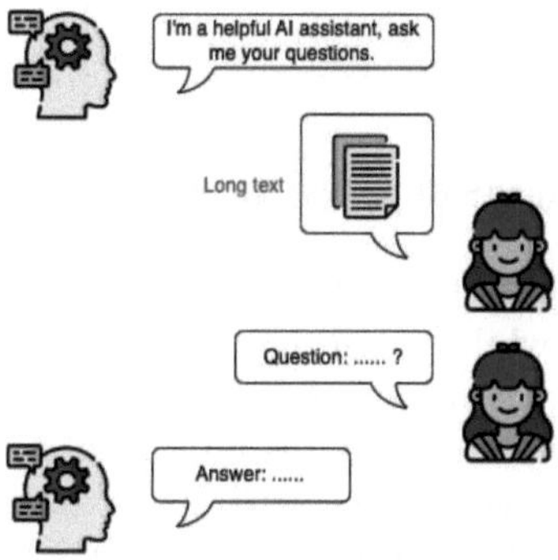

Fig. 1. Long text reasoning applications.

Current progress is largely confined to short, single-paragraph MWPs that bear little resemblance to the lengthy, context-rich problems found in real life [4]. Extending MWPs to long-text form forces models to integrate background narratives, filter extraneous details, and execute multi-step mathematical reasoning—a combination that remains challenging. Advancing research in long-text MWPs will sharpen LLMs' linguistic and mathematical understanding, opening the door to AI systems capable of tackling far more complex and realistic tasks.

Long-text mathematical word problems (MWPs) are crucial for testing deep language plus reasoning skills, yet genuine long-text datasets remain scarce because manual construction is expensive. Long-text mathematical word problems (MWPs) are crucial for testing deep language plus reasoning skills, yet genuine long-text datasets remain scarce due to the high cost of manual construction. Prior work such as Zhou et al. [27] enriches short MWPs by generating diverse question-equation pairs to improve solver generalization, but these efforts remain within short-context settings. We therefore introduce an automatic pipeline that turns abundant short-text MWPs—already paired with solutions into length and context rich versions, preserving answer correctness while adding realistic background and distractors. This scalable generator produces large-scale training and evaluation corpora that benchmark diverse models on long-text inference, offer richer contextual signals for training, and expose the points at which current systems break down. Our central goal is to perfect this short-to-long conversion so the resulting problems reflect real world complexity in both structure and content, ultimately advancing long-text processing methods and providing a solid experimental foundation for future work on robust, context aware reasoning models.

Now, in the field of natural language processing, researchers such as Xin Xu et al. first extend short MWPs with GPT-4-turbo through iterative story-style elaboration [24], but length growth soon saturates, quality varies, and dose not effectively add intrusive digital information. MathAttack [28] shows that longer and multi-step MWPs are more fragile under subtle logical perturbations.

To overcome length saturation while safeguarding logical fidelity, we introduce the IEC pipeline:

1. an Interference stage injects targeted, answer-neutral numeric distractors into each seed problem;
2. an Extension stage applies GPT-4-turbo to expand the noisy prompt into a much longer narrative while preserving all original conditions;
3. a lightweight Check stage automatically verifies that the extended item still yields the correct answer and contains no leaked reasoning. Any sample that fails this validation is regenerated, with a maximum of three retries before the round is rolled back to its pre-extension version.

Iterating IEC across multiple rounds produces multiparagraph MWPs that exceed 1500 tokens on average while maintaining high answer consistency, giving researchers a scalable source of long context problems for training and evaluating language models focused on reasoning.

We first applied the complete Interference + Extension + Check (IEC) pipeline to convert 1318 GSM8K seed problems into a large scale long text corpus. After establishing this reference set, we ran an ablation style study, first removing the Check stage (yielding I + E) and then removing both the Interference and Check stages (yielding pure E). Across six iterative rounds we tracked three metrics: average length, answer preservation, and linguistic coherence. IEC served as the upper bound baseline, while the E and I + E variants showed how each component affected length growth and validity. This experimental sequence confirms IEC's superiority for data generation and clarifies the distinct contributions of Extension and Interference in shaping long text mathematical word problems.

2 Related Work

Extended-text benchmark datasets are crucial for assessing the capabilities of Large Language Models (LLMs) in handling long-distance dependencies and complex reasoning. Current representative benchmarks include: LRA [21]: The earliest benchmark for long sequences, covering text and image tasks to stress test Transformer variants. It is valuable for architectural studies, but its small scale and vintage prior to mainstream LLMs mean most modern models have not been systematically evaluated on it. SCROLLS [20]: Aggregates government reports, television scripts and legal texts for summarization and natural-language inference. Its breadth is useful, yet limited model coverage and possible training-data leakage curb its diagnostic power. LONGBENCH/BAMBOO [3]: A bilingual (Chinese and English) suite spanning extraction, multi hop question answering, summarization and code prediction; BAMBOO adds paragraph shuffling, code completion and speaker identification tasks. Together they probe cross domain and cross language reasoning over long contexts. L-eval [2]: Mixes multiple-choice questions on video subtitles and science fiction with program tracing and financial question answering over earnings-call transcripts, filling key gaps in code and finance domains. LEGALBENCH [9]: Focuses on legal reasoning—for example, whether a ruling complies with a statute—though most tasks remain answer selection rather than deep causal chains. LooGLE [13]: Contains documents beyond 24000 tokens and 6000 fresh queries across multiple fields, explicitly crafted to test long range dependency tracking and multi step reasoning in LLMs.

These datasets and benchmarks provide essential tools and platforms for evaluating and advancing research in extended text processing and understanding. Short-text MWPs datasets offer abundant resources for research but do not typically involve the complexities of extended text processing. Some key short-text MWPs datasets include: GSM8K [7]: 8k elementary-level problems with gold solutions; the de-facto yardstick for basic arithmetic reasoning. Math23K [23]: 23k Chinese one-variable linear-equation word problems with segmented text, equation and answer, enabling large scale supervised training. DROP [8]: Paragraph-based QA requiring discrete math reasoning ("Who kicked the most field goals?"), blending reading-comprehension with computation. Others—AQuA-RAT [14], MathQA [1], ASDiv [16], MATH [10], Geometry3K [15], SVAMP [19], Ape210K [25]—further enrich the pool with program generation, multi-step algebra and geometry tasks, MathCheck [26] introduces a checklist-based benchmark to evaluate reasoning robustness and task generalization in MWPs..

3 Methodology

3.1 IEC Pipeline

The IEC pipeline is a closed loop process comprising three sequential stages—Interference (I), Extension (E), and Check (C). The control flow is as follows:

1. Interference (I). A numeric distractor sentence, carefully crafted to be topical yet neutral to the answer, is injected into the original problem stem.
2. Extension (E). The augmented stem is fed to GPT-4-turbo with a prompt for narrative expansion. The model lengthens the context while preserving the ordering and truth of all conditions.
3. Check (C). An automatic validator inspects the new item. It discards any sample that reveals the reasoning steps or the final answer, or allows the distractor to change the reference solution. Samples that fail are resubmitted for a new I + E pass; each sample is allowed at most three retries per round to avoid infinite loops (Fig. 2).

The pipeline is executed iteratively: after every successful C stage pass, the newly generated problem becomes the seed for the next round. This cyclic design sustains length growth, progressively enriches narrative complexity, and filters noise on the fly—yielding a large scale, high quality dataset suitable for training and evaluating mathematical reasoning models.

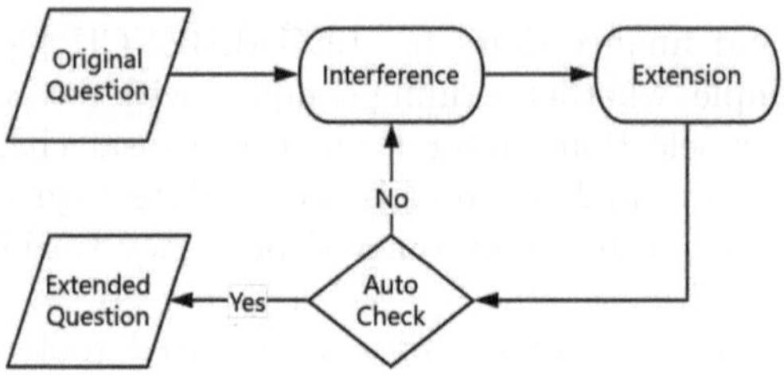

Fig. 2. The process of IEC pipeline.

3.2 Extension (E)

The Extension stage (E) of our IEC pipeline follows the strategy proposed by Xin Xu et al., who used GPT-4-turbo to enrich brief MWPs [24]. Their research methodology is founded on three core requirements:

1. the length of the problem context should be appropriately extended to enrich the narrative;
2. the answer to the original problem must remain unchanged, ensuring that the extended problem is essentially solvable in the same way as the original;
3. the conditions of the original problem and their order should not be altered during the expansion process, as illustrated in Fig. 3a.

These constraints ensure that the E stage lengthens the narrative without altering solvability, providing a clean baseline for the following I and C stages.

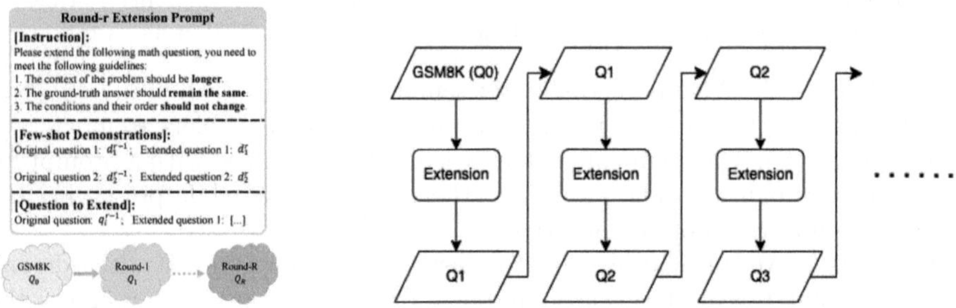

(a) Extension prompt (b) Extension of loop iteration

Fig. 3. Extension stage (E) strategy.

Initially, researchers faced a common challenge in extension: direct expansion methods struggled to achieve the length required for extended Math Word Problems (MWPs). To address this issue, they devised a continuous, iterative strategy using the GPT-4-turbo model to extend the problem text in successive rounds. Each round of expansion builds upon the outcome of the previous one, with the original problem being extended to Q1, then Q1 to Q2, and so on, as depicted in Fig. 3b. During the r-th iteration, prompts are used to extend the problem from the (r-1)-th iteration. This process employs a specific prompting template that instructs the model to maintain the conditions and answer of the original problem while generating extended problems, adding narrative text without changing the order of the conditions.

3.3 Interference (I)

Pure narrative extension increases the length of short MWPs, yet it still falls short of real world complexity: genuine problems often embed irrelevant numeric distractors that solvers must ignore before reasoning. To close this gap, we introduce the Interference (I) stage, which injects distractors into the problem stem prior to any further expansion. By doing so, the subsequent model is compelled to filter noise before performing mathematical reasoning.

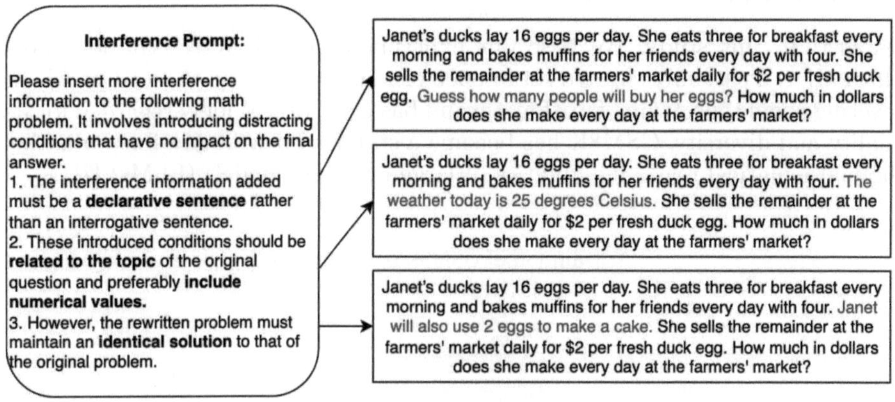

Fig. 4. Interference prompt and cases.

Figure 4 presents the prompts utilized in distractors. The intention is to incorporate numerical data as distractors, ensuring that the new problems maintain the same solution as the original. The integration of interference as an integral part of the problem could potentially obfuscate the true nature of the query. The right side of Fig. 4 demonstrates the erroneous outcomes that may arise without these restrictive conditions. Additionally, in the original extension prompt, an additional condition has been imposed: to refrain from adding reasoning steps and solutions. This measure is meant to offset the main limitation of the pure extension stage: its tendency to plateau once the text length hits a saturation point.

3.4 Check (C)

To weed out faulty distractors and stop leaked reasoning steps—yet keep the dataset growing, we introduces an automatic checking process following the extension part to assess the quality of the generated problems. The main checks include: (1) whether solution steps have been added to the problem, and (2) whether the added distractors have affected the problem's logic and final answer. If issues are found, the problem is returned to the previous round for interference and extension. To prevent deadlock in cycles, a maximum of three checks per problem per round is set. After three checks, if the problem still contains errors, the checking cycle for that round is terminated, and the result from before the extension in the current round is output, effectively abandoning the extension for that problem in the current round.

4 Experiment

4.1 Experiment Setting

Dataset: In this study, We base our experiments on GSM8K (Grade School Math 8K)—a carefully curated set of 8000 elementary level math word problems covering a wide spectrum of arithmetic and basic algebra. Widely regarded for its quality and diversity, GSM8K has become a standard benchmark for evaluating the mathematical reasoning abilities of large language models (LLMs) [7].

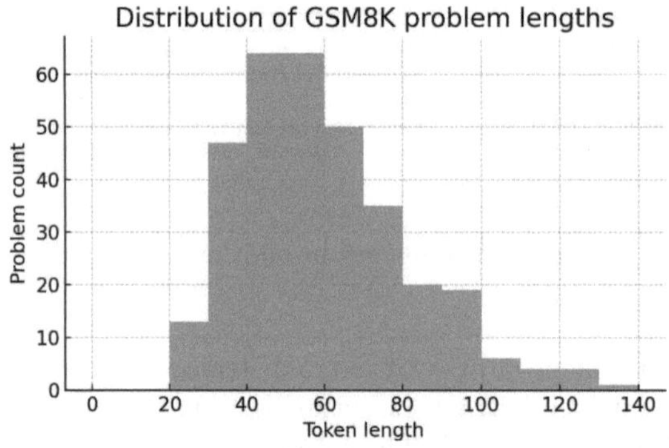

Fig. 5. Problem length of GSM8K.

To further explore the potential of the GSM8K dataset and to lay the groundwork for the generation of long-form MWPs, we conducted an in-depth analysis of the problem lengths. As depicted in Fig. 5, we observed that the majority of problems in the GSM8K dataset are contained within 200 tokens, with an average length of approximately 50 tokens. This reveals the short-text nature of the GSM8K dataset, providing significant insights for our research endeavors.

Expansion model: Each problem is passed through our three stage IEC pipeline (Interference + Extension + Check) driven by GPT-4-turbo. The model iterates up to six rounds per item, adding numeric distractors, narrative context, and automatic validation at every pass. This produces multi paragraph MWPs that exceed 2000 tokens on average while preserving the original answer.

Evaluation models: To gauge downstream utility, we ask three LLMs—DeepSeek-Chat, GLM-4-Plus, and Kimi—to solve the original GSM8K prompts, and the IEC extended versions from rounds 2, 4, and 6. Accuracy (exact answer match) is recorded to assess each model's robustness under progressively longer contexts.

4.2 Data Generation

Generation by IEC. We utilized 1,318 data entries from the GSM8K dataset as the original short-text Mathematical Word Problems (MWPs) for extension, employing IEC pipeline, which demonstrated a favorable trend in the growth of extended problem length and maintained a high accuracy rate post extension. Figure 6 illustrates the results of six rounds of extension using the DeepSeek model, with the line graph representing the average token length per round. The chart exhaustively presents the effectiveness of the IEC pipeline in extending MWPs over six consecutive iterative rounds. The horizontal axis denotes the rounds of extension, from the initial state to the sixth round, while the vertical axis corresponds to the average token length of the problem text after each round. It is observable that with the increase in iterative rounds, the average token length exhibits a stable upward trend, indicating that the I+E+C pipeline can effectively augment the length of problem text in each round without signs of saturation in growth for at least six iterations, thus validating its efficacy and sustainability in generating extended length MWPs.

Further analysis of the chart reveals a rapid growth rate in the average token length during the first two rounds, likely due to the initial brevity of the text, which is easily extended. As the iterations progress, although the growth

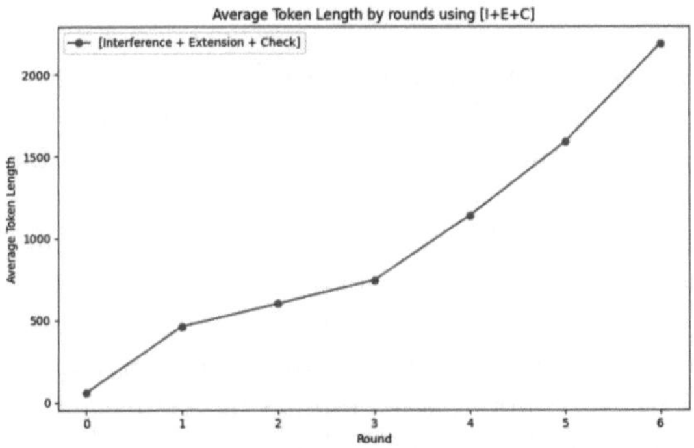

Fig. 6. Tokens length of six rounds of extension.

rate decelerates, it remains positively inclined, demonstrating the stability and robustness of the I + E + C pipeline across multiple rounds of extension. Moreover, the IEC's ability to sustain a high growth rate after six rounds is of significant practical importance for constructing large scale datasets of extended length MWPs, particularly for training and evaluating the mathematical reasoning capabilities of large language models. Consequently, the I + E + C method has not only proven its theoretical feasibility but also shown potential in generating high quality extended length data in practical applications.

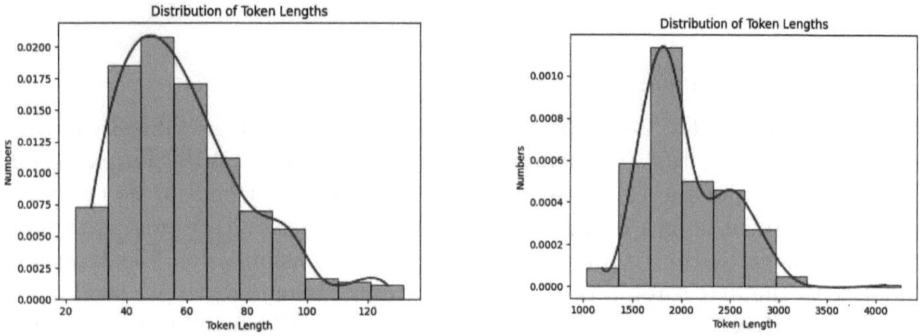

(a) Length in the GSM8K dataset

(b) MWP length after six extensions

Fig. 7. Distribution of token lengths.

Figure 7a displays the distribution of token lengths in the original GSM8K dataset. Upon examining the histogram, it is evident that most problems have a token length concentrated between 40 and 60, forming an approximate normal distribution. This indicates a moderate diversity in the length of mathematical problems within the GSM8K dataset, with the majority being of a moderate length, neither too brief nor excessively verbose. Additionally, the red curve in the figure may represent the theoretical distribution or a fitted curve of token lengths, aligning with the histogram's peak, further confirming the trend of data distribution.

The figure also shows a gradual decrease in the number of problems as the token length increases, especially after exceeding 80 tokens, where the frequency of problems drops significantly. This suggests that longer problems are relatively rare in the original GSM8K dataset.

Figure 7b illustrates the distribution of token lengths in MWPs after six rounds of extension. Compared to the distribution graph of the original GSM8K dataset, the extended dataset exhibits notable changes in token length distribution. Firstly, the range of token lengths has significantly extended, from fewer than 1,000 tokens to over 4,000 tokens, indicating a substantial increase in problem text length after multiple rounds of extension. Secondly, the peak of the distribution graph occurs around 2,000 tokens, which may imply that problem text generation is most common or effective within this length range.

In conjunction with the distribution graph of the original GSM8K dataset, several key observations can be made. Initially, most problems in the original dataset have token lengths concentrated between 40 and 60, whereas the extended dataset shows a broader distribution, indicating that the extension strategy successfully increased the length of problem text and potentially enhanced the diversity of the dataset. This also suggests that some problems may be challenging to extend to a sufficient length or may extend more slowly, while others can easily reach a length of 4,000 tokens. Secondly, in the extended dataset, problems with token lengths exceeding 1,000 have become common, which was rare in the original dataset. This shift could significantly impact model training and evaluation, as it requires models to handle longer and more complex texts. Furthermore, the distribution of token lengths in the extended dataset exhibits a multimodal characteristic, which may reflect the influence of different problem types on text length.

In summary, the distribution graph of problem lengths after six rounds of extension not only demonstrates the effectiveness of the extension strategy but also provides a new perspective for model training and evaluation.

4.3 Comparison of E, I + E, and IEC Pipelines

To demonstrate the necessity of each step of IEC, we designed three groups of methods. All three generation methods start from the same GSM8K seed set but differ in how they enrich each problem. Figure 8a shows that Extension (E) relies on narrative elaboration alone; Interference + Extension (I + E) first inserts topic related yet answer neutral numeric distractors and then performs the narrative expansion; Interference + Extension + Check (IEC) adds a lightweight automatic validator that filters outputs violating answer consistency or leaking solution steps. Consequently, E chiefly targets length, I + E adds real world noise, and IEC further enforces quality control, forming a stepwise progression from simple to robust.

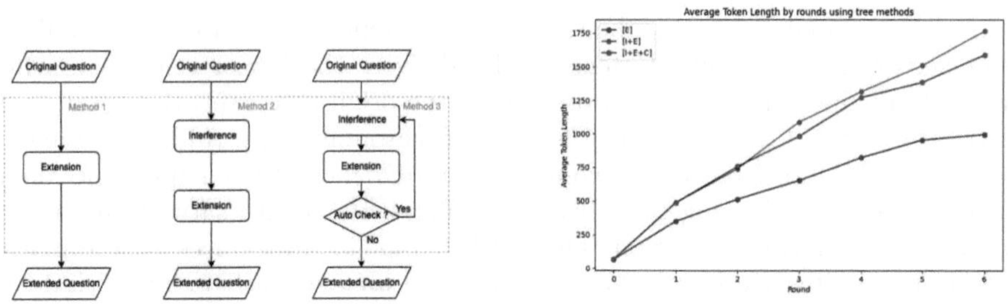

(a) Extension method E, IE, and IEC

(b) Length trend of these 3 methods

Fig. 8. Three MWP extension pipelines and their length growth trends.

Figure 8b illustrates the average length change trend of these three methods over six rounds of iteration. Method Extension passes the original question through a pure narrative elaboration block; it boosts length quickly in early rounds but soon plateaus, and because no numeric noise is added, reasoning steps often leak into the text, holding final round validity to only = 20%. Method I+E inserts an explicit "distractor" stage before expansion. The synthetic noise breaks the saturation ceiling: average length climbs past 1500 tokens in six rounds (green curve), far above E's blue curve. Quality also rises (= 60% valid items), yet faulty distractors or hidden reasoning still slip through. IEC pipeline closes that gap by looping each I+E output through an automatic validator; samples that leak reasoning or alter the answer are regenerated up to three times. This safeguard trims a little growth (red curve slightly under green) but drives validity to = 85%, giving IEC the best aggregate: it sustains long-context expansion like I+E while enforcing the reliability that E and I+E lack. In short, E is fast but shallow, I+E is long yet noisy, and IEC is the balanced solution—long, realistic, and clean.

To compare the quality of data generated by 3 methods, we examined the results of 30 data points from the sixth round of generation. The criteria for assessment were as follows:

1. If the generated problem's actual solution deviated from the original, it was deemed incorrect.
2. If the generated problem's solution remained unchanged but included additional reasoning steps and the final answer, it was also considered incorrect.

Table 1. Accuracy rates for different methods in round 6

Method	Extension	I + E	I + E + C
Accuracy rate in round 6	20%	60%	85%

As summarised in Table 1, the three pipelines exhibit a clear quality gradient after six extension rounds. Extension (E) delivers the lowest reliability, with only 20% of the generated items preserving the original answer; the absence of distractor control leads to frequent leakage of reasoning steps. Adding a numeric noise stage in Interference + Extension (I + E) more than triples the valid item rate to 60%, showing that distractors break the length saturation barrier and partly regularise generation. The full IEC pipeline introduces an automatic validator that iteratively rejects outputs whose answers drift or reveal hidden reasoning, raising accuracy to 85%.

As shown in Fig. 9, during the review process in I + E, the primary sources of errors were identified as:

1. Erroneous distractors affecting logic and solutions.

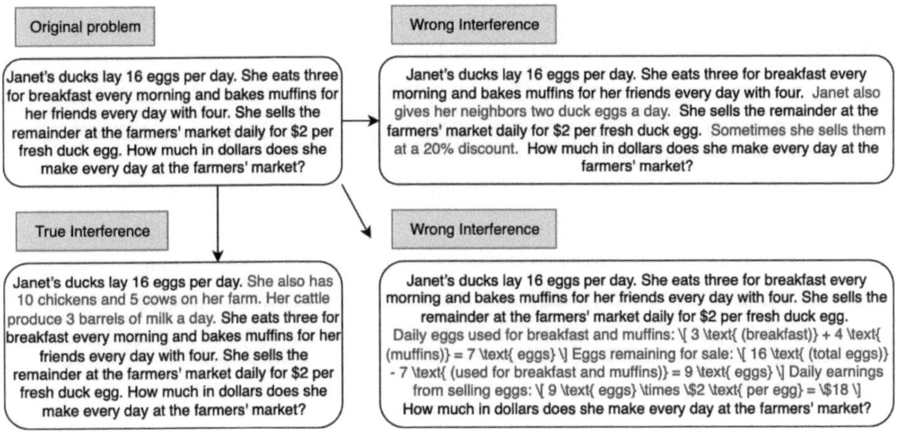

Fig. 9. Cases in generation.

2. Despite the inclusion of the restrictive condition "do not generate reasoning steps" in the extension prompts, the generation of reasoning steps was still inevitable in practice.

This impacts the quality of the problems, as the inclusion of reasoning steps and solutions in the problem statement allows the model to answer based on these steps, making it impossible to assess whether the model has actually engaged in reasoning or simply arrived at the result, thus failing to test the model's reasoning capabilities. While IEC attains the best balance between sustained text growth and data fidelity, rendering it suitable for direct large-scale training without additional filtering.

In summary, whereas pure Extension quickly encounters a length-saturation ceiling, and I+E breaks that ceiling yet remains vulnerable to faulty distractors and leaked reasoning, the additional automatic-checking loop in IEC only marginally slows growth while still producing questions exceeding 1500 tokens with markedly higher reliability. Consequently, when the goal is to balance corpus scale and data fidelity for training large language models on long-context mathematical reasoning, the IEC pipeline emerges as the most effective solution to date.

4.4 Evaluation on Long-Context LLMs

In this study, we tested three distinct large language models: DeepSeek-chat, Kimi, and GLM-4-plus, to evaluate their performance in handling extended Mathematical Word Problems (MWPs). Figures 10 and Table 2 present the results of our evaluation, where we selected data from the original state, after the second round of extension, the fourth round, and the sixth round to test the models' accuracy in answering extended-length problems. We paid particular attention to their capabilities in processing extended-length texts, a critical factor in solving complex mathematical problems.

Table 2. The accuracy of the model's answer in different rounds

Generation round	Round 0 (Original)	Round 2	Round 4	Round 6
DeepSeek	0.9419	0.7217	0.6624	0.6162
Kimi	0.9602	0.7432	0.5137	0.3940
GLM	0.9358	0.7920	0.7034	0.5615
Average	0.9459	0.7523	0.6265	0.5239

Through Table 2, it is evident that when dealing with original short-text MWPs (Round 0), all models demonstrated high accuracy rates, with Kimi leading at an accuracy rate of 0.9602, followed closely by DeepSeek-chat and GLM-4-plus at 0.9419 and 0.9358, respectively. This affirms that these models can provide high-quality solutions when confronted with unextended short-text problems.

However, as the problems underwent two rounds of extension (Round 2), a decline in the models' accuracy became apparent. DeepSeek-chat's accuracy rate dropped to 0.7217, GLM's to 0.7920, and Kimi's fell to 0.7432. This downward trend may be attributed to the increased complexity brought about by the extended problem text, posing a greater challenge to the models' comprehension and reasoning abilities. Nonetheless, the accuracy rates remained above 0.7 for all models, indicating their capability in handling extended problems.

By the fourth round of extension (Round 4), the models' accuracy further declined. DeepSeek-chat's accuracy rate fell to 0.6624, GLM's to 0.7034, and Kimi's significantly to 0.5137. This could imply that as the length of the problems increased, the models encountered more difficulties in maintaining high accuracy rates. By the sixth round of extension (Round 6), DeepSeek-chat's accuracy rate dropped to 0.6162, GLM's to 0.5615, and Kimi's further to 0.3940, the lowest among all models.

On average, Fig. 10 illustrates that the accuracy rates of the models decreased from 0.9459 in Round 0 to 0.5239 in Round 6. This overall downward trend may be correlated with the increase in problem text length, as longer texts may contain more information that could potentially interfere with the models' reasoning processes. Additionally, this trend may be related to the models' inherent limitations in processing long texts, such as difficulties in handling long-distance dependencies or extracting key information from extended texts. This suggests that in the design and training of large language models, there is a need to consider how to enhance their accuracy and robustness when dealing with long-text data. Future research may focus on developing new model architectures or training strategies to better accommodate the challenges of long-text MWPs.

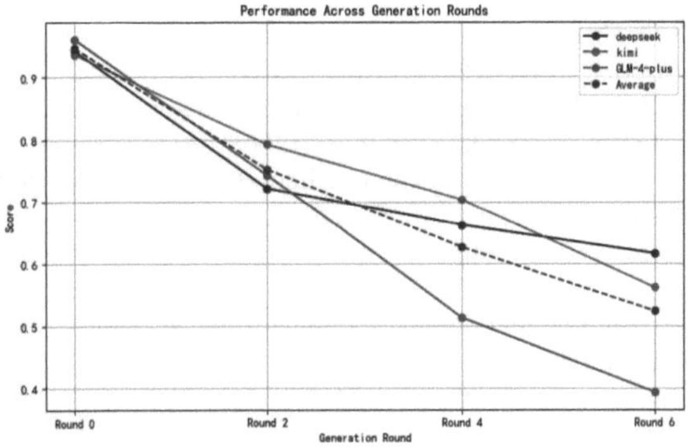

Fig. 10. The accuracy of the model's answer in different rounds.

5 Conclusions

This study introduced IEC: an Interference + Extension + Check pipeline that automatically converts short mathematical word problems (MWPs) such as GSM8K into long, distractor-rich variants while safeguarding logical fidelity. Starting from 1318 GSM8K items, IEC adds irrelevant numerical details, expands the narrative with GPT-4-turbo, and verifies answer validity through a lightweight checker. After six iterations, the average problem length reaches over 2000 tokens, which is more than 30 times the original length, while maintaining an 85% answer accuracy, significantly outperforming baselines without the Check stage. To evaluate model robustness, we tested DeepSeek-chat, GLM-4-plus, and Kimi. Although All models performed well on the original set, with accuracies exceeding 0.93, their performance dropped sharply on the extended problems—down to 0.62, 0.56, and 0.39. It reveals challenges in long-context reasoning and resistance to distraction.

Future work should therefore (i) enlarge and diversify long-text MWP corpora, (ii) benchmark a broader spectrum of models and decoding strategies, (iii) devise richer metrics that separate reading-comprehension errors from arithmetic slips, and (iv) explore training curricula and architectures that are explicitly interference-aware. Addressing these gaps is critical if the next generation of LLMs is to deliver reliable tutoring, assessment, and other real-world services that demand sustained, precise reasoning over extended texts.

Acknowledgments. The work was funded by National Natural Science Foundation of China under No. 62436009 and 62276258.

References

1. Amini, A., Gabriel, S., Lin, S., Koncel-Kedziorski, R., Choi, Y., Hajishirzi, H.: Mathqa: Towards interpretable math word problem solving with operation-based formalisms. In: Proceedings of the 2019 Conference of the North American Chapter of the Association for Computational Linguistics: Human Language Technologies, Volume 1 (Long and Short Papers), pp. 2357–2367 (2019)

2. An, C., et al.: L-eval: Instituting standardized evaluation for long context language models. arXiv preprint arXiv:2307.11088 (2023)
3. Bai, Y., et al.: Longbench: A bilingual, multitask benchmark for long context understanding. arXiv preprint arXiv:2308.14508 (2023)
4. Boaler, J.: The role of contexts in the mathematics classroom: do they make mathematics more "real"? For Learn. Math. **13**(2), 12–17 (1993)
5. Bobrow, D., et al.: Natural language input for a computer problem solving system (1964)
6. Brown, T., et al.: Language models are few-shot learners. Adv. Neural. Inf. Process. Syst. **33**, 1877–1901 (2020)
7. Cobbe, K., et al.: Training verifiers to solve math word problems. arXiv preprint arXiv:2110.14168 (2021)
8. Dua, D., Wang, Y., Dasigi, P., Stanovsky, G., Singh, S., Gardner, M.: Drop: a reading comprehension benchmark requiring discrete reasoning over paragraphs. arXiv preprint arXiv:1903.00161 (2019)
9. Guha, N., et al.: Legalbench: A collaboratively built benchmark for measuring legal reasoning in large language models. Adv. Neural Inform. Process. Syst. **36** (2024)
10. Hendrycks, D., et al.: Measuring mathematical problem solving with the math dataset. arXiv preprint arXiv:2103.03874 (2021)
11. Jiang, A.Q., et al.: Mixtral of experts. arXiv preprint arXiv:2401.04088 (2024)
12. Kahneman, D.: Thinking, Fast and Slow. Farrar, Straus and Giroux (2011)
13. Li, J., Wang, M., Zheng, Z., Zhang, M.: Loogle: Can long-context language models understand long contexts? arXiv preprint arXiv:2311.04939 (2023)
14. Ling, W., Yogatama, D., Dyer, C., Blunsom, P.: Program induction by rationale generation: Learning to solve and explain algebraic word problems. In: Proceedings of the 55th Annual Meeting of the Association for Computational Linguistics (Volume 1: Long Papers), pp. 158–167 (2017)
15. Lu, P., et al.: Inter-gps: Interpretable geometry problem solving with formal language and symbolic reasoning. In: Proceedings of the 59th Annual Meeting of the Association for Computational Linguistics and the 11th International Joint Conference on Natural Language Processing (Volume 1: Long Papers), pp. 6774–6786 (2021)
16. Miao, S.Y., Liang, C.C., Su, K.Y.: A diverse corpus for evaluating and developing English math word problem solvers. In: Proceedings of the 58th Annual Meeting of the Association for Computational Linguistics, pp. 975–984 (2020)
17. OpenAI, R.: Gpt-4 technical report. arxiv 2303.08774. View in Article **2**(5) (2023)
18. Ouyang, L., et al.: Training language models to follow instructions with human feedback. Adv. Neural. Inf. Process. Syst. **35**, 27730–27744 (2022)
19. Patel, A., Bhattamishra, S., Goyal, N.: Are NLP models really able to solve simple math word problems? In: Proceedings of the 2021 Conference of the North American Chapter of the Association for Computational Linguistics: Human Language Technologies, pp. 2080–2094 (2021)
20. Shaham, U., et al.: Scrolls: Standardized comparison over long language sequences. In: Conference on Empirical Methods in Natural Language Processing (2022)
21. Tay, Y., et al.: Long range arena: A benchmark for efficient transformers. arXiv preprint arXiv:2011.04006 (2020)
22. Touvron, H., et al.: Llama: Open and efficient foundation language models. arXiv preprint arXiv:2302.13971 (2023)
23. Wang, Y., Liu, X., Shi, S.: Deep neural solver for math word problems. In: Proceedings of the 2017 Conference on Empirical Methods in Natural Language Processing, pp. 845–854 (2017)

24. Xu, X., Xiao, T., Chao, Z., Huang, Z., Yang, C., Wang, Y.: Can LLMs solve longer math word problems better? arXiv preprint arXiv:2405.14804 (2024). https://arxiv.org/abs/2405.14804
25. Zhao, W., Shang, M., Liu, Y., Wang, L., Liu, J.: Ape210k: A large-scale and template-rich dataset of math word problems. arXiv preprint arXiv:2009.11506 (2020)
26. Zhou, Z., et al.: Is your model really a good math reasoner? evaluating mathematical reasoning with checklist. In: Proceedings of the International Conference on Learning Representations (ICLR) (2025)
27. Zhou, Z., et al.: Learning by analogy: diverse questions generation in math word problem. In: Findings of the Association for Computational Linguistics: ACL, pp. 11091–11104 (2023)
28. Zhou, Z., et al.: Mathattack: attacking large language models towards math solving ability. In: Proceedings of the AAAI Conference on Artificial Intelligence, vol. 38, pp. 19750–19758 (2024)

Author Index

L. Jin et al. (Eds.): ICDAR 2025 Workshops, LNCS 16225, pp. 395–396, 2026.
https://doi.org/10.1007/978-3-032-09368-4

MIX
Papier aus verantwortungsvollen Quellen
Paper from responsible sources
FSC® C105338

If you have any concerns about our products,
you can contact us on
ProductSafety@springernature.com

In case Publisher is established outside the EU,
the EU authorized representative is:
Springer Nature Customer Service Center GmbH
Europaplatz 3, 69115 Heidelberg, Germany

Printed by Libri Plureos GmbH
in Hamburg, Germany